Introduction to Ultrafast Phenomena

Introduction to Ultrafast Phenomena from Femtosecond Magnetism to high-harmonic Generation

Guo-ping Zhang
Georg Lefkidis
Mitsuko Murakami
Wolfgang Hübner
Thomas F. George

CRC Press
Taylor & Francis Group
Boca Raton London New York

CRC Press is an imprint of the
Taylor & Francis Group, an informa business

First edition published 2021
by CRC Press
6000 Broken Sound Parkway NW, Suite 300, Boca Raton, FL 33487-2742

and by CRC Press
2 Park Square, Milton Park, Abingdon, Oxon, OX14 4RN

© 2021 Taylor & Francis Group, LLC

CRC Press is an imprint of Taylor & Francis Group, LLC

ISBN-13: 978-1-4987-6428-5 (hbk)
ISBN-13: 978-0-367-65433-7 (pbk)

Reasonable efforts have been made to publish reliable data and information, but the author and publisher cannot assume responsibility for the validity of all materials or the consequences of their use. The authors and publishers have attempted to trace the copyright holders of all material reproduced in this publication and apologize to copyright holders if permission to publish in this form has not been obtained. If any copyright material has not been acknowledged please write and let us know so we may rectify in any future reprint.

Except as permitted under U.S. Copyright Law, no part of this book may be reprinted, reproduced, transmitted, or utilized in any form by any electronic, mechanical, or other means, now known or hereafter invented, including photocopying, microfilming, and recording, or in any information storage or retrieval system, without written permission from the publishers.

For permission to photocopy or use material electronically from this work, access www.copyright.com or contact the Copyright Clearance Center, Inc. (CCC), 222 Rosewood Drive, Danvers, MA 01923, 978-750-8400. For works that are not available on CCC please contact mpkbookspermissions@tandf.co.uk

Trademark notice: Product or corporate names may be trademarks or registered trademarks, and are used only for identification and explanation without intent to infringe.

Library of Congress Cataloging-in-Publication Data

Dedicated to our families

Contents in Brief

1	Time scales	3
2	Ultrafast phenomena: Experimental	11
3	Theoretical background	41
4	High-harmonic generation	85
5	Femtomagnetism	117
6	All-optical spin switching	157
7	Spin manipulations in magnetic nanostructures	179
8	Magnetic molecules and magnetic logic	203

Contents

Setting the Stage and Modus Operandi: "The Making of the Book" xiii

Preface xv

Authors xix

I Fundamentals 1

1 Time scales 3

1.1 Units of time and relation between energy and time 3
1.2 Time axis 5
1.3 How to describe events in space-time coordinates 6
1.4 Time scale in the hydrogen atom 7
1.5 Time scale for photoisomerization 7
1.6 Summary 9
1.7 Exercises 9

2 Ultrafast phenomena: Experimental 11

2.1 Introduction to the laser and how it works 11
2.1.1 Standing waves 11
2.1.2 Cavity 12
2.1.3 Stimulated emission and single wavelength selection 13
2.2 Short summary of nonlinear optics under cw laser excitation 15
2.3 Magneto-optics 19
2.4 Development of ultrafast lasers: Major breakthrough with the Ti-Sapphire laser 20
2.4.1 Phase alignment and mode-locking 20
2.4.2 Necessity of many modes and broad bandwidth 21
2.4.3 Emergence of ultrafast pulses from mode-locking 21
2.5 How to access the ultrafast time scale 24
2.6 Time-resolved pump-probe experiments 26
2.6.1 Basic principles 27
2.6.2 Nitty-gritties and theory behind the processes 27
2.7 Photoisomerization in bacteriorhodopsin 28
2.8 Femtochemistry 31
2.9 Metals, semiconductors and superconductors 33
2.10 Femtomagnetism 35
2.11 High-order harmonic generation and attosecond physics 37
2.12 Exercises 40

3 Theoretical background 41

3.1 Density functional theory 42
3.1.1 Hohenberg-Kohn theorem 42
3.1.2 Kohn-Sham equation 43
3.2 Time-dependent density functional theory 44
3.2.1 Solving the Kohn-Sham equation 44
3.2.2 Example: Many-electron atoms 46
3.2.3 Adiabatic approximation 50
3.2.4 Example: TDDFT of atoms in a linearly-polarized field 51
3.3 Quantum chemistry tools 58
3.3.1 Basis functions 58
3.3.2 Hartree-Fock approximation 59
3.3.3 Configuration interaction method 60
3.3.4 Coupled-cluster method 63
3.4 Solid state physics: Essentials 65
3.4.1 Crystal structure = Bravais lattice + basis 65
3.4.2 Band structure: How electronic energy disperses with crystal momentum 70
3.5 Two special features in ultrafast dynamics 75
3.5.1 Spin-orbit coupling 75
3.5.2 Interaction between laser radiation and matter 76
3.5.3 Further notes on the vector potential 78
3.6 Rotation matrices for spins 79
3.7 Software packages 80
3.8 Exercises 82

II Applications 83

4 High-harmonic generation 85

4.1 Brief history and key features of high-harmonic generation 85
4.2 Working principles of HHG 87
4.2.1 Laser electric field strength and Coulomb potential in an atom 88
4.2.2 Escaping the Coulomb potential 90
4.2.3 Ponderomotive energy 91
4.2.4 Corkum's theory: Origin of the cutoff energy of $I_p + 3.17U_p$ 92
4.3 Applications 95
4.3.1 Applications to hydrogen and neon atoms 95
4.3.2 Applications to C_{60} 100
4.3.2.1 Model 101
4.3.2.2 Time-dependent Liouville equation 103
4.3.2.3 Power spectrum 105
4.4 Experimental demonstration of high-harmonic generation in C_{60} 107
4.5 High-harmonic generation in solids 108
4.5.1 Graphene 110
4.5.2 Going to magnets 111
4.5.3 Simple picture of HHG in solids 114
4.6 Exercises 116

5 Femtomagnetism 117

5.1 History of femtomagnetism 117
5.2 Magnetic materials 118
5.2.1 General properties 118
5.2.2 Element ferromagnets and microscopic interactions 120
5.3 Time scale of laser-induced demagnetization 123
5.3.1 Time scale for electron response 123
5.3.2 Time scale for spin response 124
5.3.3 Time scale of phonon excitation 126
5.3.4 Demagnetization time 126
5.4 Sample experimental results 127
5.4.1 Fe, Ni and permalloy 127
5.4.2 Half-metallic and Heusler compounds 132
5.4.3 Short experimental summary 134
5.5 Mechanisms still under debate 137
5.5.1 Spin-orbit coupling model 138
5.5.2 Hubbard model 139
5.5.3 Heisenberg model 144
5.5.4 Time-dependent Liouville density functional theory 148
5.5.5 Time-dependent magneto-optics theory 149
5.6 Exercises 155

6 All-optical spin switching 157

6.1 Basic optics 157
6.2 Background 159
6.2.1 Ferrimagnets and magneto-optical recording 159
6.2.2 Experimental discovery 160
6.3 Key ingredients of all-optical spin switching 162
6.3.1 Composition and compensation temperature 163
6.3.2 Laser parameters 164
6.3.3 New materials 164
6.4 Theory 164
6.4.1 Birth of the first single spin switching model 166
6.4.2 Numerical solutions and MatLab codes 169
6.4.3 Reversing millions of spins 173
6.4.4 Importance of spin moments 175
6.5 Exercises 178

7 Spin manipulations in magnetic nanostructures 179

7.1 Computer memory and magnetic storage 179
7.1.1 Giant magneto-resistance 180
7.1.2 Magneto-optical recording technology 182
7.1.3 Emergence of ultrafast demagnetization 183
7.2 Experimental discovery 183
7.2.1 Experiments in permalloy 183
7.2.2 Coherent spin manipulation in NiO 184
7.3 Spin precession 187
7.4 Rabi oscillation 190

7.5 Spin-orbit coupling in an atom 192
7.6 Magnetic resonance in NiO clusters 194
7.7 Exercises 201

8 Magnetic molecules and magnetic logic 203

8.1 Λ processes in molecular systems 204
8.1.1 Degenerate case 205
8.1.2 Chirped lasers 208
8.1.3 Spectral broadening of the laser pulse 209
8.2 A closer look into electronic correlations 211
8.2.1 Correlations and interatomic distances 211
8.2.2 Correlations and ultrafast spin dynamics 212
8.3 Molecular vibrations 213
8.3.1 Electron-vibron coupling 217
8.3.2 Molecular vibrations and spin dynamics 218
8.3.3 Geometry change as a tool: Mechanical strain 221
8.4 Magnetic logic on molecules 223
8.4.1 How many magnetic centers do we need? 224
8.4.2 The Ni_3Na_2 paradigm 225
8.4.3 Elementary laser-induced processes 226
8.4.4 Spin transferability 229
8.4.5 Mapping quantum dynamics onto classical trajectories 231
8.4.6 Higher multiplicities 234
8.4.7 More complicated M and nonlinear M processes 235
8.5 First steps towards magnetic logic gates 237
8.5.1 Boolean logic on Ni_3Na_2: NAND gate 237
8.5.2 ERASE functionality 239
8.5.2.1 Laser chirp 240
8.5.2.2 Quantum interferences 242
8.5.3 Charge-spin gearbox 245
8.6 Concluding remarks 247
8.7 Exercises 248

Appendix A Appendices 253

A.1 KLI approximation 253
A.2 LDA: local density approximation 254
A.3 Self-interaction corrected LDA 256
A.4 BLYP approximation 257
A.5 Electric dipole, magnetic-dipole and other higher-order interactions 258
A.6 Code to generate ultrafast pulses 260
A.7 Code to generate figures in HHG 261
A.8 Code to compute the cutoff energy in HHG 263
A.9 Code to compute the C_{60} structure 267
A.10 Genetic algorithm example 272
A.11 Special crystal momentum points and lines 276

Bibliography 279

Index 297

Setting the Stage and Modus Operandi: "The Making of the Book"

Five years ago a group of five scientists conceived the plan to co-author a book on the exciting topic of ultrafast phenomena. While other books already existed on this overall topic, our approach is special in that it includes many realistic examples and delves into certain hot areas such as magnetism. Scientific books fall into several different categories. At one end of the spectrum is an introductory undergraduate text which, if the authors are fortunate enough, can sell thousands of copies with high profits for both the authors and the publisher. At the other end is an advanced text aimed at upper-division undergraduates, graduate students, and working scientists. The authors can review work that is in the literature, and in certain cases like ours, draw on their own research accomplishments. We have much to draw on since we have been collaborating for many years in the overall area of nanoscience/nanotechnology and ultrafast phenomena.

In carrying out our project, we are beneficiaries of the modern world of information technology, sophisticated software, and of course, the Internet. (Going back to the early 1970s, we fondly recall using the cruder forerunner of the Internet called ARPANet.) Ideas and advances can now be shared instantaneously, making collaborative research quite easy to execute. And of course, we need not all be at the same geographic location. Interestingly, the imposed working conditions due to the coronavirus (COVID-19) for many of us, especially theoreticians, are not all that different from how we have already been functioning. We have come down the home stretch of our book project in 2020 as the virus pandemic is seemingly at its height, and we sincerely hope that the world will ultimately come out stronger, with a renewed commitment to greater levels of excellence across the board in science, engineering, letters, arts, and health disciplines.

We initially intended to invite our friend and colleague Jean-Yves Bigot to write a foreword for our book, but the untimely passing of Jean-Yves Bigot and Eric Beaurepaire, two true pioneers in femtomagnetism, in the same year, made it impossible. About the initial idea of femtosecond magnetism, in Jean-Yves's email to GPZ on Friday, May 8, 2015, he recounted, "Regarding initial ideas, for me, coming from semiconductors where spin scattering near the band gap is important, or describing the SO band + the fact that I was doing surface plasmon dynamics in copper nanoparticles, it was inevitable to consider ferromagnetic systems. This is why if you refer to the invited talk at CLEO (the Conference on Lasers and Electric-Optics, Baltimore, USA (1995)) that same year, I talked both about surface plasmon dynamics in NPs and spin(s) dynamics in nickel. The interaction with you and Wolfgang after that became of course very fruitful when you came to Strasbourg and raised several other questions."

More on our personal history with the field of femtomagnetism can be found in the beginning of Chapter 5.

This book and research would not be possible without help from our colleagues, friends, and of course funding agencies. We would like to thank Dr. Mingsu Si (Lanzhou University, China), Dr. Yihua Bai (Indiana State University, USA), Dr. Peter Blaha (TU Vienna, Austria), Dr. Claudia Draxl (Humboldt University, Berlin, Germany), Dr. Stefan Mathias (University of Göttingen, Germany), Dr. Markus Münzenberg (U. Greifswald, Germany), Dr. Jean-Yves Bigot, Dr. Eric Beaurepaire and Tyler Jenkins (Indiana State University). Mitsuko and Guo-ping have been supported by the U.S. Department of Energy under Contract No. DE-FG02-06ER46304. The research used three computer clusters (Quantum, Silicon, and Obsidian) at the Indiana State University. The research also used resources of the National Energy Research Scientific Computing Center, which is supported by the Office of Science of the U.S. Department of Energy under Contract No. DE-AC02-05CH11231.

Preface

Technology driven by ultrashort and high intense lasers is revolutionizing many aspects of everyday life. It allows one to probe and detect events that were never seen before. Ultrafast phenomena represent a single most important new frontier at the intersection of many branches of sciences and engineering – from chemistry, physics and materials science to optical and electric engineering. Femtochemistry and attosecond physics are two examples of this huge development. New research developments appear on a daily basis. There have been over twenty volumes of conference proceedings on ultrafast phenomena dating back to the 1990's. These target seasoned researchers and professionals, but there have been very few books on the topic of ultrafast phenomena itself.

This is where our book comes. The book targets upper-level undergraduate and graduate students, and anyone who is interested in learning how ultrafast phenomena effects are utilized in many areas of current technology, from rapid optical communication to magnetic storage and coherent control, to quantum information processing. However, the book does not aim to cover all aspects of ultrafast phenomena; instead, it focuses on the basic principles and ideas. In contrast to many books, this book presents materials with ample examples, enabling the serious reader to see the big picture. While each chapter covers different aspects of ultrafast phenomena, the presentation of each chapter is extremely detailed without skipping steps. Because of this, some chapters contain actual research codes and webpage links. At least in physics, we find that many excellent books do not have sufficient examples and exercises. For a new researcher, this is extremely difficult. For instance, one of the authors (GPZ) had to learn from scratch how to compute laser pump-probe signals. We believe that others have the same frustrations. We hope that our book starts a new beginning, to let everybody learn how the wonderful research is done and allow others to see the research behind the scenes. We strongly support open-source research.

This book has two parts. Part I discusses fundamentals in ultrafast phenomena, and provides the basic knowledge in experiment and theory. It contains three chapters.

Chapter 1 introduces the reader time scales, with several examples in physics, chemistry and biology. We provide some useful conversions between different units that are often encountered in ultrafast phenomena. Photoisomerization is first introduced.

Chapter 2 focuses on the experimental techniques. Since a

major part of ultrafast phenomena involves the laser, we start with the basic principles of the laser, laser cavity, active lasing media, and stimulated emission. Then we review some of the basic concepts of nonlinear optics in the cw limit, where difference frequency generation, sum-frequency generation, and four-wave mixing are presented. This part can be overwhelming. The reader may skip them in the first read. We then move on to magneto-optics, which will be used in the latter part of the book. We discuss three common configurations: Faraday effect, Voigt effect, and Kerr effect. We spend a great amount of text on the development of ultrafast laser pulses. We explain why both mode-locking and broad bandwidth of lasing materials are essential to ultrafast laser pulses. Here we propose two equations to mathematically link them together. The next question is to understand how one can access these short time scale, just like you have a cake, but you can eat it too. Instead of sampling varieties of experimental configurations, we explain the time-resolved pump-probe experiment in great detail. The last four sections cover ultrafast phenomena in bacteriorhodopsin, femtochemistry, metals, semiconductors, and superconductors, concluding with femtomagnetism and high-order harmonic generation. Each topic can be branched into a subfield. The reader can select some topics from them.

Chapter 3 introduces the theoretical tool for ultrafast phenomena and contains our own understanding and research results. We begin with density functional theory and time-dependent density functional theory (TDDFT). We provide a TDDFT code through the book. In particular, we show one example of TDDFT calculations in atoms in a linearly-polarized field. We then explain in detail quantum chemistry tools, including basis functions, Hartree-Fock approximation, configuration interaction method, and coupled-cluster method. The following section reviews the essentials of solid state physics as they will be needed in later chapters. In particular, we show what the unit cell is and how the Wyckoff positions are found. This part represents a unique perspective on solids. The coordinates of special crystal momentum points and lines are given in the appendix. In connection with spin dynamics and laser-induced ultrafast demagnetization, we pay special attention to spin-orbit coupling and the vector potential. Although many textbooks involve SU(2) rotation for spins, we find a much simpler way to rotate spins, which gives the exact same results. This chapter concludes with the software packages.

Part II is on the application of ultrafast phenomena, and has five chapters. Each chapter is a subset of ultrafast phenomena. The reader can read them without a specific order. We repeat essential materials in each chapter without referring back to prior chapters too frequently. Nearly all chapters have actual research codes that we use for our own research. Chapter 4 is on high-harmonic generation. We begin with harmonic generations in atoms, with a detailed presentation on

Corkum's theory to understand the cutoff energy in harmonics. The time-dependent Schrödinger equation and TDDFT are used to compute harmonic signals from hydrogen and neon atoms, with the model application for C_{60}, graphene and magnets.

Chapter 5 introduces femtomagnetism, with the historical background and basic knowledge on magnetism given in the first two sections. We then discuss time scales of the electron, spin and phonon during laser-induced demagnetization. Next, we give both representative experimental and theoretical findings. We choose two experiments: One is on Fe, Ni, and permalloy, and the other is on half-metallic heusler compounds. Theoretical models include the spin-orbit coupling model, Hubbard and Heisenberg models. We also introduce the reader to the time-dependent Liouville density functional theory, and time-dependent magneto-optics theory. To this end, the underlying principles of laser-induced demagnetization is still under debate. The reader may explore this field herself or himself, with some of the leading references cited in the back of the book.

Chapter 6 discusses all-optical spin switching. We believe this spin switching may have an important application in magnetic recording in the future. This chapter starts with the basic experimental discovery in GdFeCo, where both helicity-dependent and helicity-independent spin switching are introduced. Then we summarize some of the key ingredients to switch spins, which include the composition of the sample and compensation temperature, and laser parameters. We provide a list of a few most frequently investigated materials, with a focus on new materials. Since most of theoretical approaches are phenomenological, we introduce our own formalism to switch spins. We provide a useful code so the reader can directly test her or his understanding.

Chapter 7 moves to magnetic nanostructures. We begin with computer memory and magnetic storage in general. We highlight three major research breakthroughs: Giant magnetoresistance, magneto-optical research, and ultrafast demagnetization. The experimental results on permalloy are first introduced, followed with more detailed discussions on NiO. The following sections present an analytic understanding of spin precession and Rabi oscillation, with a special focus on spin-orbit coupling. Finally we conclude this chapter with magnetic resonance on NiO clusters.

Chapter 8 introduces a promising research field of magnetic logic for future computing. This chapter starts with a Λ process in molecular systems for degenerate cases and chirped lasers. Then we have a closer look at electronic correlation effects and molecular vibrations, before we enter magnetic logic on molecules. A few steps to magnetic logic gates are presented in the last section, where we introduce Boolean logic and ERASE functionality.

The Appendix contains 11 sections, with details on the

KLI approximation, LDA, self-interaction correction in LDA, BLYP functional, electric dipole, magnetic-dipole and other interactions, followed by four computer codes to generate laser pulses, HHG signals, cutoff energy in atomic HHG, computing the atomic coordinate for the Buckministerfullerene, and genetic algorithm. These are research codes that are used for our own research. The last, but definitely not the least, contains two tables of special crystal momentum points and lines. These two tables are essential to any solid state calculations.

This book is appropriate for an upper-level undergraduate and beginning graduate course in science and engineering. A one-semester course may cover the first three chapters, plus one or two later chapters. A two-semester course can cover most of the book. Each chapter has exercises. The reader should find the Bibliography at the end of the book useful for further reading.

Guo-ping Zhang	Indiana State University, USA
Georgios Lefkidis	Technische Universität Kaiserslautern and Research Center OPTIMAS, Germany
Mitsuko Murakami	Indiana State University, USA
Wolfgang Hübner	Technische Universität Kaiserslautern and Research Center OPTIMAS, Germany
Thomas F. George	University of Missouri - St. Louis, USA

Authors

Guo-ping Zhang, Georgios Lefkidis, Mitsuko Murakami, Wolfgang Hübner, and Thomas F. George

Part I

Fundamentals

1 Time scales

1.1 Units of time and relation between energy and time
1.2 Time axis
1.3 How to describe events in space-time coordinates
1.4 Time scale in the hydrogen atom
1.5 Time scale for photoisomerization
1.6 Summary
1.7 Exercises

Over the course of human history, we have been fascinated by various time scales around us. Human hearts beat once every second, while mice hearts beat 10 times every second. Our eyes can only see objects at a rate of about 30 per second, which is limited by the processing speed of our nervous system. Motion pictures or television cameras take a series of still pictures at a rate of 24 (movies) or 30 (U.S. television) per second; when still pictures are projected, we perceive the appearance of motion. High-speed cameras can take more frames, so they can "slow" down the motion of a water droplet splashing on a table. Often we may think that a tree grows slowly, but do not realize that photosynthesis in leaves is extremely fast. The same thing happens to our eyes, where rhodopsin molecules absorb light energy and twist their molecular conformation on a very short time scale. This conformation change sends a stimulus to the nerve center of our brain, so that vision starts. To investigate those ephemeral events is the goal of ultrafast science.

1.1 Units of time and relation between energy and time

In sciences, the SI unit of time is the second. However, we often find that the second is not convenient. To describe the age of the universe, it is more appropriate to use years (13.82 billion years). In the ultrafast phenomena that this book is about, we need several much smaller units, such as nanosecond (1 ns = 10^{-9} second), picosecond (1 ps = 10^{-12} second), femtosecond (1 fs = 10^{-15} second), and attosecond (1 as = 10^{-18} second). The main reason why those units are important is because major activities of nuclei, electrons and spins occur on those time scales. A complication is that on these time scales, quantum effects, which cannot be described by classical physics, become important. In classical mechanics, time and energy are independent variables. In quantum mechanics, they are linked to each other. One can convert the energy unit to the time unit by

$$E = h\nu = \frac{h}{T}, \tag{1.1}$$

where E is the energy, h is the Planck constant (6.626×10^{-34}J·S), ν is the frequency in Hz, and T is the period in seconds. Then 1 eV (1.602×10^{-19}J) corresponds to (6.626×10^{-34}J $\cdot$ s)/(1.602×10^{-19}J) $= 4.13 \times 10^{-15}$ s, or 4.13 fs.[a] For this reason, we see that the energy in eV works well with time in fs. Figure 1.1 shows how a femtosecond laser pulse can switch spins within several hundred fs.

[a] A common laser pulse with photon energy of 1.6 eV has a period of 2.58 fs.

Figure 1.1
Time-resolved ultrafast demagnetization and all-optical switching under a 60-fs laser pulse at 775 nm. A spin wave propagates outward during excitation, covering hundreds of lattice sites.

Another particularly useful relation links the energy scale to the length scale. We can change Eq. 1.1 to

$$E = h\nu = h\frac{c}{\lambda}, \rightarrow E\,[\text{eV}]\times\lambda\,[\text{nm}] = hc = 1239.965 \approx 1240 \text{ nm eV},$$

where c is the speed of light. This relation allows us to convert the wavelength of light to energy in eV easily. For instance, a laser pulse with a wavelength of 800 nm has an energy of 1.55 eV. The derivation is left as a homework assignment. To describe molecular vibrations, one often uses wavenumber ($\frac{1}{\lambda}$) in the unit of cm^{-1},[b] which can be converted to eV. We start from

[b] The number of waves within 1 cm.

$$1\ (\text{eV}) = h\nu = h\frac{c}{\lambda}, \longrightarrow \frac{1}{\lambda} = 8065.544\ (\text{cm}^{-1}). \tag{1.2}$$

This means that to convert from eV to wavenumbers, one needs to multiply the energy by 8065.544. In chemistry, kcal/mol is often used as the energy unit. 1 eV is equal to 23.06035 kcal/mol.

If we want to have one photon per Å^2, the laser fluence in units of mJ/cm^2 of photon energy $h\nu$ [eV] is $1.602 \times h\nu$ [eV]. For instance, for 800-nm lasers (photon energy of 1.55 eV), the fluence is 2.48 mJ/cm^2.

Many people often use THz (10^{12} Hz) as a unit of energy. 1 THz corresponds to 4.136 meV. Some also use the temperature as a unit for energy. Its conversion is computed from k_BT, where k_B is the Boltzmann constant ($1.38064852 \times 10^{-23}$ J/K). Therefore, 1 K is equivalent to 8.161733×10^{-5} eV.

1.2 Time axis

If we arrange all the events in the universe along the time axis, it is easy to see that each discipline in science occupies a small section of time. Figure 1.2 schematically shows an example. The longest time scale is the age of the universe, which astrophysicists are interested in, while the shortest time scale is the lifetime of elementary particles, which high-energy physicists are interested in. A biologist may be interested in photosynthesis, and a climatologist may be interested in a weather pattern. However, there is a large time segment that is claimed by several disciplines. This lies between a few hundred picoseconds and a few attoseconds, where new frontiers start to emerge. For instance, a chemist may be interested in how chemical reactions occur and how reactants yield products, while physicists and engineers may be interested in how the structural changes affect material properties. Unfortunately, these regimes had been off limits for a long time, since there were no tools to detect them.

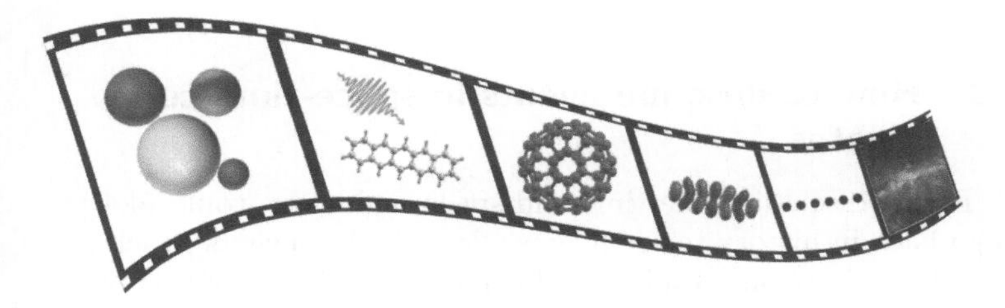

Figure 1.2
Time axis. Different disciplines in science take one portion of the time along the time axis. The lifetime of the shortest lived matter is on the order of 10^{-27} s (here we use the top quark's energy of 173 GeV). The age of the universe is on the order of 13.8 billion years (4.35×10^{17} s). The last image is taken from www.nasa.gov.

The advent of ultrafast laser technology in the 1980s has provided the necessary tools to investigate these ultrafast phenomena. The success of these investigations is glamorous. Femtochemistry is a prime example, for which Ahmed Zewail [Zewail (2000)] received the Nobel Prize in Chemistry in 1999. Investigations in atoms, molecules, nanostructures, surfaces and bulk materials have become a major focus in ultrafast phenomena. The entire field experiences an unprecedented expansion with varieties of techniques developed [Villeneuve et al. (2017)]. Figure 1.3 shows an example of velocity map imaging (VMI) in C_{60}. Here an ultrafast laser pulse first knocks out electrons from C_{60}, and those ionized electrons which carry the information of the underlying molecular orbitals are accelerated through multiple stages of electrostatic lens, so their images are magnified.

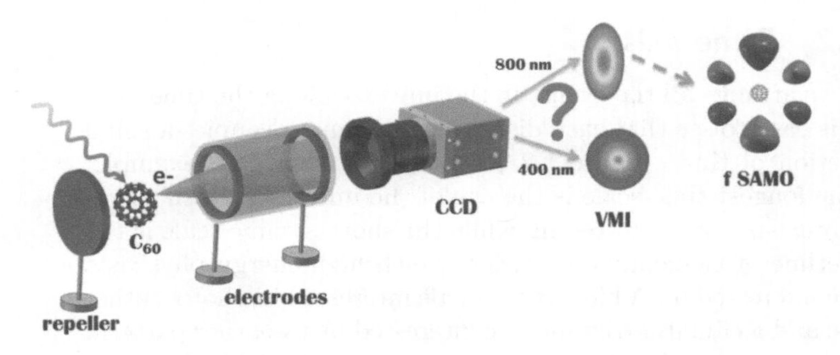

Figure 1.3
Velocity map imaging used in C_{60}, where C_{60} is subject to a laser pulse with electrons knocked out and accelerated through electrodes and then captured on a phosphor screen. The image reveals the final details of the molecular orbitals. The figure is taken from [Zhang et al. (2015d)], used with permission from World Scientific.

1.3 How to describe events in space-time coordinates

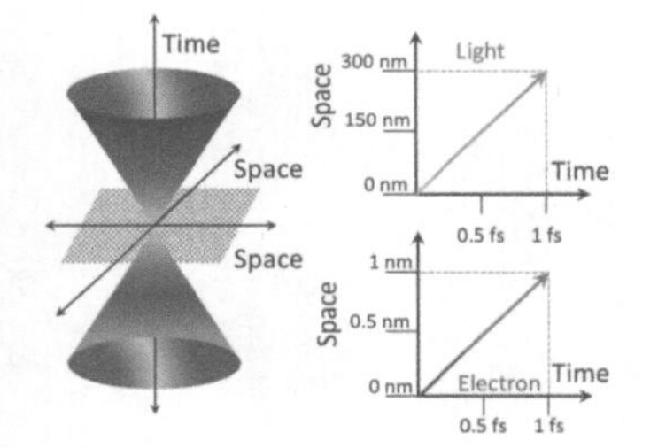

Figure 1.4
(Left) Space-time and light cone, where the vertical axis is ct and the horizontal axis is the space. (Top right) For every fs, light travels 300 nm. (Bottom right) For every fs, an electron travels 1 nm.

Before discussing the time and space scales, we would like to go back in history to see how we describe key events in science. Isaac Newton established the major foundation of classical mechanics through his three famous laws of motions, where time and space are independent of each other. A little bit more than a century ago, Albert Einstein showed in his relativity theory that time and space are interconnected. Such a relation can be viewed directly from the so-called world line plot, where the x-axis denotes space and the y-axis represents the product of the speed of light and time (ct). So, the slope is 1 (no unit) for the light signal, which forms the light cone (the left panel in Fig. 1.4). The reason why we use ct instead of t is because the inverse of the slope represents the velocity, and for light, with speed approximately 3×10^8 m/s, its slope is almost zero.

However, the above problem is not an issue for ultrafast phenomena investigations. Using femtosecond (fs) as the time unit and nanometer (nm) as the space unit, on the top right of Fig. 1.4 we plot the time axis (x-axis) in the unit of femtosecond and the space axis (y-axis) in the unit of nanometer. We see that for 1 fs, light travels 300 nm. The slope will not be extremely big or extremely small, so there is no difficulty at all. What about electrons? An electron travels 1 nm per 1 fs in metals, so it has no problem either (see the bottom right of Fig. 1.4).

1.4 Time scale in the hydrogen atom

Once we find a method for how to describe events, we can introduce concrete examples that are simple, realistic and still relevant to current research. We cannot think of a better example than the hydrogen atom, which has one proton and one electron. The proton is a thousand times heavier than the electron. It is reasonable to assume that the electron orbits around a static proton. Due to the Coulomb attraction, the electron experiences a centripetal force with magnitude (see Fig. 1.5)

$$F_c = \frac{1}{4\pi\epsilon_0}\frac{e^2}{r^2}, \tag{1.3}$$

where ϵ_0 is the permittivity in vacuum (8.85×10^{-12} C^2/(Nm^2)), $-e$ is the electron charge (1.602×10^{-19} C), and r is the distance between the electron and proton. As an estimate, we treat the electron classically and consider that the electron radius is the Bohr radius (0.529 Å), which corresponds to the 1s orbital. We then find that the electron moves at 2.18 nm/fs. From the electron speed, we find the period is 0.151 fs, or 151 as.

These time scales are typical for electrons. For instance, the Fermi velocity in metals is around 10^6 m/s, or 1 nm/s, which is plotted in Fig. 1.4. The motion of electrons is extremely fast. Therefore, it is necessary to use ultrafast laser pulses to detect the motion. For heavier atoms, the electrons are even faster, so fast that relativistic effects must be taken into account. This has some important consequences. For instance, it is part of the reason why mercury is liquid at room temperature and why gold has its golden color. Investigations of radiation in atoms have given birth to high-harmonic generation and attosecond physics [Lucchini et al. (2016)].

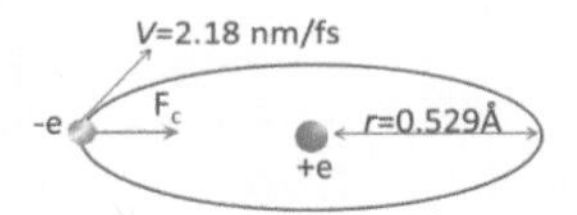

Figure 1.5
Electron in the hydrogen $1s$-orbital rotates around the proton with velocity of 2.18 nm/fs.

1.5 Time scale for photoisomerization

So far, we have discussed the electron dynamics. Nuclear dynamics is another important area for ultrafast phenomena research. Because nearly all chemical and biological reactions involve protons, the experimental detectability of the dynamical process is essential to our understanding. Nuclei are heavier, so the time scale is longer than that for electrons. There are many examples with molecules, nanostructures (nanotubes, fullerenes), films, liquids (such as water), solids (magnetic materials and superconductors), photosynthesis and myoglobin [Ferrante et al. (2016)]. Enormous research has been devoted to each topic. A particularly useful resource to explore various topics is the Web of Science or Scopus. In later chapters, we will discuss some of them in more detail. Here, we introduce rhodopsin as an example in the world of ultrafast processes.

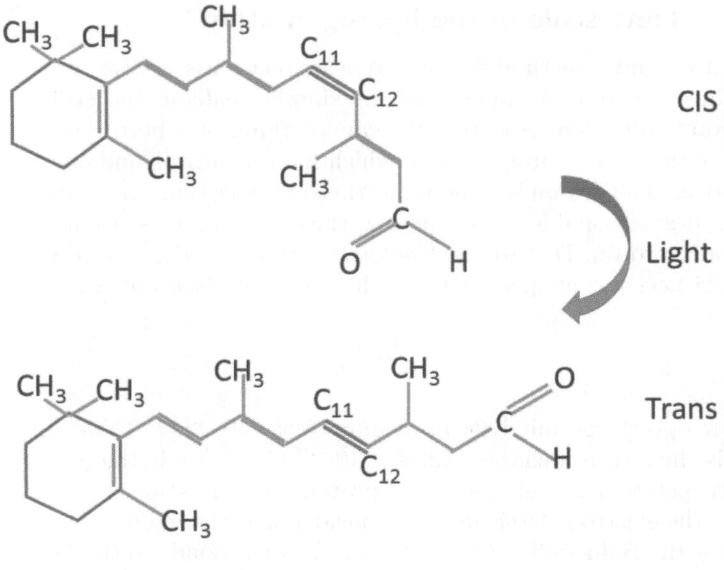

Figure 1.6
Retinal molecule in rhodopsin. Photoisomerization occurs around the C_{11}-C_{12} bond. Upon light excitation, the retinal transforms from 11-cis-retinal to all-trans-retinal within a few hundred femtoseconds.

Retinal molecules in rhodopsin are fundamental to our vision. Rhodopsin consists of seven helices.[c] Rhodopsin is a light-sensitive pigment and appears in the retinas of humans and animal eyes and bacteria. The retinal molecule is situated around the seven helices. When light strikes rhodospin, photoisomerization occurs. In mammals, light excitation loosens up the double bond around C_{11} and C_{12} (see Fig. 1.6) and allows the molecular motif to rotate around the bond, so that 11-cis-retinal is transformed to an all-trans-retinal.[d] The top figure in Fig. 1.6 shows the resting structure which has a cis structure. This is energetically unfavorable, but docking proteins create a potential that favors this configuration. Under light excitation, 11-cis retinal transforms to all-trans retinal (see the bottom figure) within a few hundred femtoseconds. Such a process is extremely efficient. In bacteriorhodopsin [Gai et al. (1998)], the situation is similar, but there is a crucial difference: Photoisomerization occurs from all-trans retinal to 13-cis retinal around the C_{13}-C_{14} bond.

The simplest potential to describe photoisomerization is

$$V(\phi) = V_0 + A\cos(\phi), \tag{1.4}$$

where ϕ is the angle between the main chain and the rest of the chain, A denotes how large the potential barrier is, and V_0 is a constant. For cis-retinal, $\phi = \pi$, and for trans-retinal, $\phi = 0$.

[c] The name refers to its pinkish color. In Greek "rhodon" ($\rho o \delta \acute{o} \nu$) means rose and "opsis" ($\acute{o}\psi\iota\varsigma$) means face or looks.

[d] The designations "cis" and "trans" describe the location of the ligands with respect to a double bond between two carbon atoms. If similar ligands are on the same side of the bond, the structure is called cis; if they are located on opposite sides of the bond, it is called trans.

1.6 Summary

Ultrafast phenomena exist in different branches of sciences. They underlie many aspects of our observation. Ultrafast laser technology finally provides us exciting opportunities to explore new frontiers from materials design, chemical reactions to photosynthesis. We can begin to appreciate the power of ultrafast phenomena by revealing the intricate working principles of nature.

1.7 Exercises

1. Show that if we use electron volt as the energy unit and nanometer as the length unit, for a photon with wavelength λ and energy E, $E[\text{eV}] \times \lambda[\text{nm}] \approx 1240$.
2. (a) Take the hydrogen atom as an example and compute the periods of the electron in each of the $2s$ and $2p$ orbitals. (b) Prove, in general, that the period of the electron in orbital n is determined by

$$T_n = \frac{4\pi^2 n^3 a_0^2 m_e}{h}, \tag{1.5}$$

   where n is the main quantum number, a_0 is the Bohr radius, m_e is the mass of the electron, and h is the Planck constant.
3. According to the light cone in relativity, draw both the electron and light in the same cone.
4. Derive the following velocity of the electron in hydrogen:

$$v_n = \frac{\hbar}{n a_0 m_e}, \tag{1.6}$$

   where n is the main quantum number of the electron, a_0 is the Bohr radius, and m_e is the electron mass.
5. Show how one can convert the electron volt to wavenumber as Eq. 1.2.

2 Ultrafast phenomena: Experimental

2.1 Introduction to the laser and how it works
2.2 Short summary of nonlinear optics under cw laser excitation
2.3 Magneto-optics
2.4 Development of ultrafast lasers: Major breakthrough with the Ti-Sapphire laser
2.5 How to access the ultrafast time scale
2.6 Time-resolved pump-probe experiments
2.7 Photoisomerization in bacteriorhodopsin
2.8 Femtochemistry
2.9 Metals, semiconductors and superconductors
2.10 Femtomagnetism
2.11 High-order harmonic generation and attosecond physics
2.12 Exercises

This chapter introduces some basic experimental techniques for ultrafast phenomena. The central theme is the laser. For this reason, we briefly review the basic features of lasers; for more detailed presentations, the reader may refer to other books. The materials we cover in this chapter are the minimum necessary for anyone to go through the rest of the chapters without any major difficulties. It is important to point out that the laser itself is quantum mechanical. We recall that in quantum mechanics, energies are discrete and consist of many levels. Some levels have higher and some lower energies. We will focus on how ultrafast lasers are constructed in the simplest possible terms. In some sense, this chapter is a direct continuation of the preceding chapter, but samples a larger scope of materials and systems.

2.1 Introduction to the laser and how it works

The term laser stands for Light Amplification of Stimulated Emission of Radiation. On May 16, 1960, Theodore H. Maiman from the Hughes Aircraft Company in California, USA, demonstrated the first functioning laser in the world. He once said, "A laser is a solution seeking a problem." To understand how a laser works, we should go back to the history of how things started.

2.1.1 Standing waves

The laser is based on a similar concept that was used to develop the maser (microwave amplification of stimulated emission of radiation), due to Charles H. Townes and his associates James P. Gordon and Herbert J. Zeiger at Columbia University in 1953. This complicated name has two crucial components: amplification and stimulated emission. Just as diodes and triodes amplify electric signals, the laser amplifies light signals. The amplification of light signals is entirely different from that of electric signals. Consider two light sources with electric fields $\mathbf{E}_1(t)$ and $\mathbf{E}_2(t)$, where t is time. The total field is

$$\mathbf{E}_{\rm total} = \mathbf{E}_1(t) + \mathbf{E}_2(t). \quad (2.1)$$

[a] Light intensity is defined as power per unit area. Fluence is defined as energy per unit area.

The laser transient intensity (ι)[a] in vacuum is

$$\iota(t) = \epsilon_0 c |\mathbf{E}_{\rm total}(t)|^2, \tag{2.2}$$

where ϵ_0 is the permittivity in vacuum and c is the speed of light. If we use SI units for ϵ_0, c and the electric field, $\iota(t)$ has units of W/m^2 [Butcher and Cotter (1990)]. The laser intensity I is the time average of $\iota(t)$. In order to amplify the signal, the resultant laser intensity I should be larger than their algebraic sum, i.e., $I > I_1 + I_2$, where $I_1(I_2)$ is the light intensity of $\mathbf{E}_{1(2)}(t)$. If $\mathbf{E}_1(t)$ and $\mathbf{E}_2(t)$ oscillate with arbitrarily different frequencies, amplification is not possible. To see this clearly, we temporally assume that $\mathbf{E}_1(t)$ and $\mathbf{E}_2(t)$ are along the same direction, so that we can use their scalar forms and the intensity is

$$\begin{aligned}\iota(t) &= \epsilon_0 c[E_1(t) + E_2(t)]^2 \\ &= \epsilon_0 c[E_1^2(t) + E_2^2(t) + 2E_1(t)E_2(t)].\end{aligned} \tag{2.3}$$

Assume that $E_1(t) = A_1\cos(\omega_1 t + \phi_1)$ and $E_2(t) = A_2\cos(\omega_2 t + \phi_2)$, where $\omega_{1(2)}$ is the angular frequency and $\phi_{1(2)}$ is the initial phase. If ω_1/ω_2 is not a rational number or $\phi_{1(2)}$ is all different, then the last term on the right side is time-averaged to zero. So, we only have $I = I_1 + I_2$, and there is no amplification. For this reason, amplification in a laser only works if a single frequency or a multiple of the same frequency, such as 1ω, 2ω, ..., is used, so that the second term is time-averaged to a nonzero number. The solution is the standing wave.

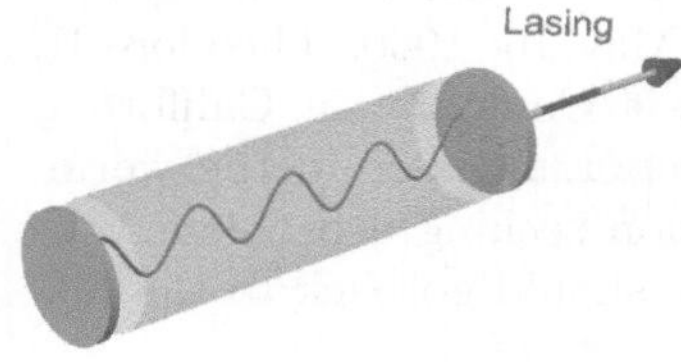

Figure 2.1
A laser cavity has two mirrors at the ends. One mirror has nearly perfect reflection, while the other is a partially transmitting mirror, which allows the laser light to come out.

2.1.2 Cavity

Experimentally, the standing wave is achieved in a cavity with two mirrors at the ends (Fig. 2.1). Suppose the distance between the two mirrors is L. If an external field generates various waves inside the cavity, only waves whose wavelengths λ_n satisfy the equation

$$n\lambda_n/2 = L \tag{2.4}$$

can survive. They are standing waves, and the harmonic order n is an integer, 1, 2, 3, ...

A wavelength can be converted to a frequency ν via

$$\nu = c/\lambda, \tag{2.5}$$

or an angular frequency ω via

$$\omega = 2\pi\nu = 2\pi c/\lambda. \tag{2.6}$$

The lowest frequency ω_0, also called the fundamental frequency, is

$$\omega_0 = 2\pi c/2L = \pi c/L. \tag{2.7}$$

For all the other waves, once they bounce from these two parallel mirrors, they cancel themselves out. The mirrors must be parallel to each other so that many rounds of reflection are possible. Any misalignment between these mirrors may deteriorate the performance of the cavity. By adjusting the distance between the mirrors, we can change the wavelength.

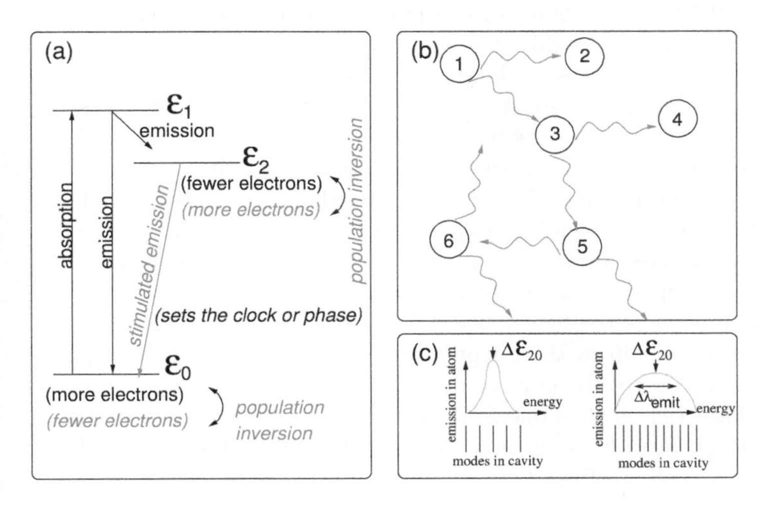

Figure 2.2
(a) Three-level system for lasing. The ground state is the lowest energy state of the system, with energy $\mathcal{E}_0$. The first-excited state energy is $\mathcal{E}_1$ and the second $\mathcal{E}_2$. The system absorbs one photon or other forms of energy from an external field, so one electron is promoted to the excited state. This process is called absorption. Emission is just the opposite, where the electron transits back to the ground state and emits a photon with energy $\hbar\omega$. Therefore, for absorption, $\hbar\omega = \mathcal{E}_1 - \mathcal{E}_0$. We use italic fonts to highlight the stimulated emission. (b) Stimulated emission starts with atom 1, progresses to atom 2, and so on. (c) The spectral linewidth $\Delta\lambda_{\mathrm{emit}}$ of the atom absorption determines which and how many modes are amplified. The vertical lines denote the energies of the standing wave modes. Only those standing waves whose energy falls in the absorption spectrum of the atom can survive. Therefore, the absorption spectrum of the medium (in this case, atoms) is a frequency selector. Left: The medium only selects three modes. Right: The medium selects ten modes. A large number of modes is crucial to the generation of ultrafast laser pulses.

2.1.3 Stimulated emission and single wavelength selection

If two fields have the same frequency and are in phase, i.e., $E_1(t) = E_2(t)$, the total field is $E_{\mathrm{tot}}(t) = 2E_1(t)$. Since the intensity is proportional to the square of the field, $I_{\mathrm{tot}} = 2^2 I_1$. For N fields, $I_{\mathrm{tot}} = N^2 I_1 = N I_{\mathrm{uncorrelated}}$, where $I_{\mathrm{uncorrelated}}$ refers to the intensity for the uncorrelated light sources. If N is large, then the enhancement is huge. How are we able to get so many atoms emitting light with the same frequency simultaneously?

To understand this, we need to discuss the absorption and emission of atoms, molecules and solids (see Figs. 2.2(a) and (b)). We take atoms as an example. If the photon energy $\hbar\omega$ matches the energy difference between two energy levels, $\mathcal{E}_1 - \mathcal{E}_0$, the atom absorbs the light energy, with one electron excited from the ground state 0 to excited state 1. The emission is opposite, where the electronic transition back to the ground state emits a photon with the same energy as the difference between levels 1 and 0. If there is another level 2 between 1 and 0, then emission is also possible from 1 to 2, as far as the selection rules allow. The key feature for normal emission is that the emission is spontaneous and has no control over the initial phase of the fields. Because their phases are all random, there is no gain in the output signal as seen above. How can we then control the emission phase?

The trick is that one finds a system where a special middle level, level 2, is situated between the ground state 0 and excited state 1 (Fig. 2.2(a)). One pumps lots of electrons in many atoms from 0 to 1. The electrons transit from 1 to 2 and can pile up on level 2 of the atoms, leaving very few electrons on level 0. In other words, we have more electrons in the higher energy level (level 2 in this case)[b] than in the lower energy level (ground state in this case). This process is called population inversion, because it is opposite to thermal equilibrium where more electrons are in the lower level. Population inversion is a crucial step to lasing. Figure 2.2(b) schematically shows the entire process. The emission starts with the spontaneous emission from an arbitrary atom first. In Fig. 2.2(b), we assume that this is atom 1. The emitted light from atom 1 stimulates the neighboring atoms 2 and 3 to emit light. The emitted light from atoms 2 and 3 continues the chain reaction. Because two ending mirrors reflect light back and forth and because the speed of light is extremely high, the remaining unstimulated atoms get stimulated quickly; almost immediately, the emission is in phase, and we have stimulated emission or lasing. The strength of the emitted light is proportional to the number of atoms in the cavity and is amplified enormously. To allow laser light to come out of the cavity, one mirror is partially transmitting. In Fig. 2.1(a), the laser light comes out of the right mirror.

[b]Level 2 is a metastable state and stores the electrons transitioned from level 1 temporarily. Since we want electrons ultimately to transition back from level 2 to level 0, level 2 must be metastable until a stimulus acts upon it. When we say the electrons pile up at level 2, what we really mean is that electrons pile up at level 2 of many atoms.

In the cavity, there are lots of possible standing waves or modes as far as their wavelengths satisfy Eq. 2.4, but not all modes are amplified. This is different from the electric amplification, where signals with various frequencies can be amplified since usual electric amplification is not frequency-dependent. In a laser, we only amplify those frequencies selected by the atomic absorption spectrum (see Fig. 2.2(c)). The emitted light from level 2 to 0 has wavelength $\lambda_{\rm emit}$:

$$\lambda_{\rm emit} = \frac{hc}{\mathcal{E}_2 - \mathcal{E}_0}. \quad (2.8)$$

Suppose that the emission line width is $\Delta\lambda_{\rm emit}$. Then the emission covers all the wavelengths within $\lambda_{\rm emit} \pm \Delta\lambda_{\rm emit}$. If $\Delta\lambda_{\rm emit} \to 0$ and if $\lambda_{\rm emit}$ happens to be one of the wavelengths given in Eq. 2.4, this is the only wavelength that is amplified, and the rest decay to zero. In other words, to get amplified, the modes must satisfy two independent conditions simultaneously:

1. Cavity requirement (Eq. 2.4)

   Their wavelengths must match one of those standing wavelengths that are permitted by the cavity.

2. Medium requirement (Eq. 2.8)

   Their wavelengths must be within the range defined by the emission spectrum of the laser medium, i.e., $\lambda_n \in (\lambda_{\rm emit} \pm \Delta\lambda_{\rm emit})$.

If $\lambda_{\rm emit}$ does not match any wavelengths given in Eq. 2.4, no lasing occurs. This means that the cavity length cannot be arbitrary, but must be fine-tuned such that at least one wavelength λ_n of Eq. 2.4 falls in $(\lambda_{\rm emit} \pm \Delta\lambda_{\rm emit})$ of the medium.

It is important to stress that if the laser medium has nearly zero linewidth, then the emitted laser light has only a single wavelength. This leads to a continuous wave (cw), not a pulse. The left part of Fig. 2.2(c) includes these three modes; only these three modes are amplified, and the rest are unenhanced and decay. The right part of Fig. 2.2(c) covers ten modes, which are enhanced. The emitted laser contains ten frequencies. The broad linewidth of the emission spectrum of the medium is essential to the generation of ultrafast laser pulses. We will come back to this below.

Key message Although the cavity allows a multiple of the fundamental wavelength, the laser medium decides which wavelengths are amplified. Changing lasing media changes the laser wavelength. That is why we have a variety of lasers available on the market, such as CO_2, He-Ne and titanium sapphire lasers

2.2 Short summary of nonlinear optics under cw laser excitation

Microscopically, under laser excitation, the polarization $\mathbf{P}$ is changed. It can be written as a summation over the linear response, second-order response and other higher-order terms (in SI units) as[c]

$$\mathbf{P} = \epsilon_0\Big(\chi^{(1)} \cdot \mathbf{E}_1 + \chi^{(2)} :: \mathbf{E}_1\mathbf{E}_2 + \chi^{(3)} ::: \mathbf{E}_1\mathbf{E}_2\mathbf{E}_3 + \ldots\Big), \quad (2.9)$$

[c] We assume that the sample has zero polarization without the laser field.

where ϵ_0 is the permittivity in vacuum, $\chi^{(1)}$, $\chi^{(2)}$ and $\chi^{(3)}$ are the first-, second- and third-order susceptibilities, respectively, and $\mathbf{E}_1$, $\mathbf{E}_2$ and $\mathbf{E}_3$ are the electric fields of light. Here $\mathbf{E}_1$, $\mathbf{E}_2$ and $\mathbf{E}_3$ may represent the same field of the same incident light or different fields of different incident light. These susceptibilities (in SI units)[d] are tensors with the units of $(\mathrm{m/V})^{n-1}$ [Butcher and Cotter (1990)], where n is the order. For a linear medium, only $\chi^{(1)}$ (no units in SI) is nonzero. The number of indices of susceptibility is always $n+1$. For $n=1$, the linear case, there are two indices, so in a three-dimensional system, this is a 3×3 matrix with 3^2 elements, where "3" stands for the three coordinates, x, y and z. We normally write $\chi^{(1)}$ as $\chi^{(1)}_{xx}$ and $\chi^{(1)}_{xy}$, which means that the incident light electric field is along the x-axis (the first subscript of $\chi^{(1)}$), and the polarization is measured along the x- and y-axes (the second subscript of $\chi^{(1)}$). Other terms have a similar meaning. For $n=3$, there are four indices, and the susceptibility has 81 (3^4) elements. Since $\mathbf{P}$ is a three-dimensional vector and is real, regardless of orders, those indices must be summed over so that the final results have only three components. If we are interested in a particular component of the susceptibilities, we can purposely designate special directions for $\mathbf{E}_1$, $\mathbf{E}_2$ and $\mathbf{E}_3$.

[d] Paul Butcher and David Cotter's book is among the first books that use SI units for nonlinear optics. This makes the conversion much easier for magneto-optical susceptibility (dimensionless) and conductivity ($1/(\Omega\mathrm{m})$ to 1/s).

We should emphasize that Eq. 2.9 is not a most generic equation that covers all different kinds of nonlinear optical interactions, because the only terms that are related to the material properties are the susceptibility. χ may depend on motions of nuclei or magnetic field. Next, we show some examples of the most popular processes in nonlinear optics (Fig. 2.3). Suppose that two laser fields with carrier frequencies ω_1 and ω_2 propagate along the $\mathbf{k}_1$ and $\mathbf{k}_2$ directions and interact with a nonlinear medium. We use Ω to denote the generated signal frequency.

Sum-frequency generation and harmonic generation When two waves are incident on a nonlinear medium, a new frequency $\Omega=\omega_1+\omega_2$ is generated along the $\mathbf{k}_3=\mathbf{k}_1+\mathbf{k}_2$ direction (Fig. 2.3(a)). Because the new frequency is a sum of two incident frequencies, this is called *sum-frequency generation.*

If ω_1 is equal to ω_2, so that $\Omega=2\omega_1$, this process is further called second-harmonic generation. If there are three incident waves with all frequencies the same, we have third-order harmonic generation (Fig. 2.3(b)).

Difference-frequency generation Opposite to sum-frequency generation is difference-frequency generation, where a new frequency Ω is the difference between two incident frequencies ($\Omega=\omega_1-\omega_2$). The propagation direction is along $\mathbf{k}_3=\mathbf{k}_1-\mathbf{k}_2$ (Fig. 2.3(c)). If ω_1 is equal to ω_2, this process ($\Omega=0$) is further called optical rectification.

Raman scattering Raman scattering concerns the frequency of generated light that is not a simple combination of

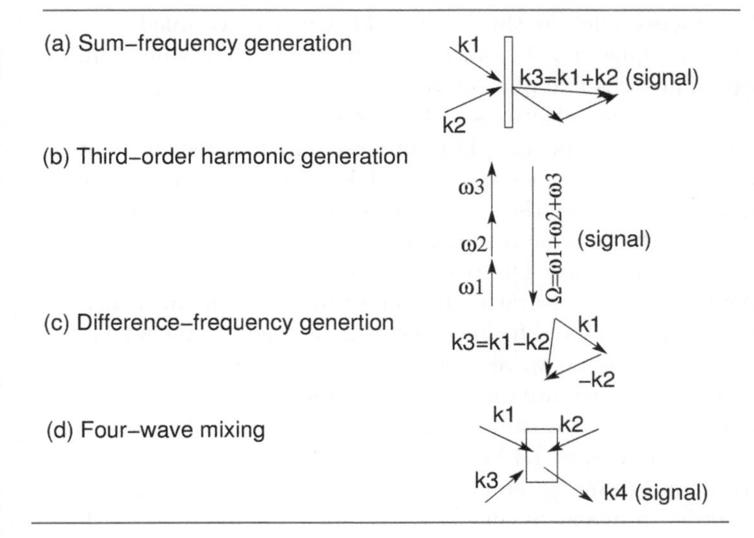

Figure 2.3
Some of the most important nonlinear optical processes. They normally involve different waves represented by the wavevectors $\mathbf{k}_1$, $\mathbf{k}_2$, $\mathbf{k}_3$ and $\mathbf{k}_4$ (signal). (a) Sum-frequency generation. The signal frequency is the sum of two incident frequencies, $\Omega = \omega_1 + \omega_2$. The signal propagates along the $\mathbf{k}_1 + \mathbf{k}_2$ direction. (b) Third-order harmonic generation. This is a special case of sum-frequency generation, where all the incident frequencies, ω_1, ω_2 and ω_3 are the same. (c) Difference-frequency generation. The signal frequency is the difference of the incident frequencies. (d) Four-wave mixing. Two incident waves act as a spatial grating, and a third incident wave is diffracted from the grating to generate the signal. This is a background free process, where along the signal wavevector direction there is no incident wave.

the incident frequencies. The medium's fundamental frequencies are mixed into the frequency of the generated light. For instance, if the medium's vibrations participate in the excitation process, then besides these frequencies that originate from the laser frequencies themselves, new frequencies appear. These new frequencies normally appear around the laser incident frequencies or a combination of the laser frequencies. When we probe with a spectrometer, the frequency is shifted away from the laser fundamental frequency combination. This is called stimulated Raman scattering, which becomes a very powerful tool to probe phonon vibrations, nuclear rotation and other elementary excitations. This is an example, where Eq. 2.9 does not cover the new frequency generated by phonons or other elementary excitations.

Two-photon absorption The system absorbs two photons $\hbar\omega_1 + \hbar\omega_2$ and is excited to a high-level state. Two-photon absorption (TPA) is often an intermediate step in a higher-order process, such as third-order. TPA can be coupled with other techniques. For instance, in two-photon photoemission, a system absorbs two photons and electrons are ionized, so we detect the ionized electrons instead of the emitted photons.

Four-wave mixing This proceeds with three incident waves with wavevectors $\mathbf{k}_1$, $\mathbf{k}_2$ and $\mathbf{k}_3$ and one signal wave (Fig. 2.3(d)). If these three incident waves are all independent, there are eight possible combinations for the signal wavevector $\mathbf{k}_4 = \pm\mathbf{k}_1 \pm \mathbf{k}_2 \pm \mathbf{k}_3$. The basic principle is that two incident waves form a spatial grating on the sample, and a third incident wave is diffracted from the grating to generate the signal. Because there is no wave along the signal direction to start with, the four-wave mixing is a background free process. If incident waves have the same frequency, this is called degenerate four-wave mixing, or DFWM.

It is important to emphasize that four-wave mixing is a third-order process. Specifically, the medium interacts with the incident laser fields three times (the fourth wave does not count because it is a signal coming out of the sample).

To summarize this section, we point out that even a fixed-order[e] nonlinear optical process has many variations and can be very complicated. We take two incident fields ($\mathbf{E}_1$ with carrier frequency centered at ω_1 and $\mathbf{E}_2$ with carrier frequency centered at ω_2) as an example, and suppose that we are interested in a third-order process. There are four major kinds. One may view this as an application of the above discussion.

[e] The order of nonlinear optics is determined by the number of times that the system interacts with the laser fields. One should not overly emphasize the order because besides the dipole interaction with the electric field, there are many other interactions present. For instance, if a superconductor enters the superconducting state, Cooper pairs are formed, so the electrons behave very differently even before the laser interacts with the superconductor. These interactions are not characterized by the order itself.

Case 1 The system can interact with $\mathbf{E}_1$ alone three times. If the medium has an inversion symmetry, the signal frequency could be ω_1, ω_1^*, $3\omega_1$, ... The first ω_1 is from the linear response, but the second ω_1^*, which is numerically equal to ω_1, is from a third-order process. The system first absorbs a photon $\hbar\omega_1$, emits a photon $\hbar\omega_1$, and finally absorbs a photon $\hbar\omega_1$, so that the resultant frequency is $\Omega = \omega_1 - \omega_1 + \omega_1 \equiv \omega_1^*$. In other words, even for a single frequency, the first-order and third-order can be mixed together. The $3\omega_1$ process means that the system absorbs a photon $\hbar\omega_1$ three times for third-order harmonic generation.

Case 2 Same as the above, but for $\mathbf{E}_2$.

Case 3 The system can interact with $\mathbf{E}_1$ twice and $\mathbf{E}_2$ once.[f] If the system absorbs two photons from $\mathbf{E}_1$ and one photon from $\mathbf{E}_2$, the signal has a frequency of $\Omega = 2\omega_1 + \omega_2$ along the $2\mathbf{k}_1 + \mathbf{k}_2$ direction. Then we have sum-frequency generation. If the system absorbs one photon from $\mathbf{E}_1$ ($\hbar\omega_1$), emits one photon ($-\hbar\omega_1$), and absorbs one photon from $\mathbf{E}_2$ ($\hbar\omega_2$), the signal has a frequency of $\Omega = \omega_1 + \omega_2 - \omega_1 = \omega_2$ along the $\mathbf{k}_1 + \mathbf{k}_2 - \mathbf{k}_1 = \mathbf{k}_2$ direction. This is often used in the pump-probe experiment (see Section 2.7 below).

[f] The orders in the expression are crucial. $\mathbf{k}_1 + \mathbf{k}_2 - \mathbf{k}_1$ means that the system interacts first with the pump and then the probe, and finally with the pump again. $\mathbf{k}_1 - \mathbf{k}_1 + \mathbf{k}_2$ means that the system interacts with the pump twice through a difference frequency process and then interacts with the probe.

Case 4 Same as 3 but $\mathbf{E}_1$ and $\mathbf{E}_2$ swap their roles.

2.3 Magneto-optics

All the above nonlinear optical processes discussed so far have no magnetic field. With the presence of magnetism, we have magneto-optics, so Eq. 2.9 is not adequate; we must include magnetization. Both experimental and theoretical efforts on magneto-optics already date back a century ago. Its widespread application is much more recent. Magneto-optics has three types of experimental effects: Faraday, Voigt and Kerr (see Fig. 2.4). Common to all of them is that one measures the rotation angle and ellipticity change of the light polarization.

Transmission

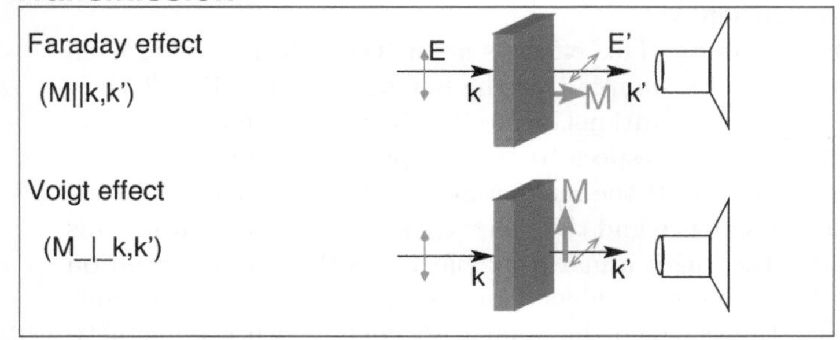

Reflection

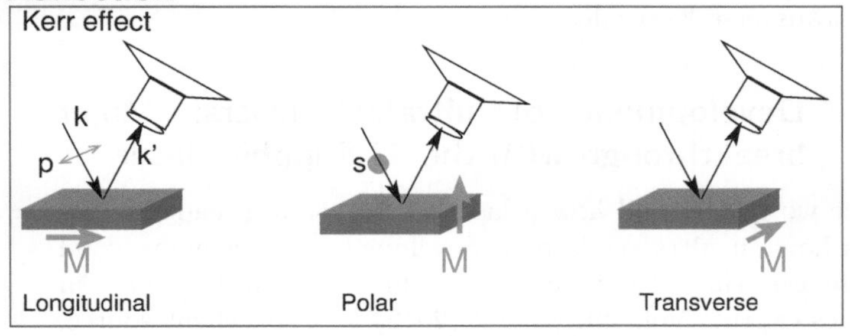

Figure 2.4
Magneto-optics is categorized according to how the signal is probed. If the transmitted signal is probed, these are called a Faraday ($\mathbf{M} \parallel \mathbf{k}(\mathbf{k}')$) or Voigt effect ($\mathbf{M} \perp \mathbf{k}(\mathbf{k}')$), where $\mathbf{M}$ and $\mathbf{k}$ are the magnetization of the sample and the wavevector of the light. If the reflected signal is detected, this is called a Kerr effect. Based on how $\mathbf{M}$ is oriented with respect to the plane of incidence which is formed by the incident wavevector $\mathbf{k}$ and reflected one $\mathbf{k}'$, the Kerr effect is further categorized according to three configurations. In the longitudinal and polar Kerr effects, $\mathbf{M}$ is in the incident plane, while in the transverse effect, $\mathbf{M}$ is perpendicular to the incident plane. Longitudinal and polar Kerr effects differ as to whether $\mathbf{M}$ is out of the sample plane. In the Kerr effect, the light polarization can lie in the plane of incidence (parallel, p) or perpendicular to the plane of incidence (s) (senkrecht in German, meaning perpendicular or vertical). Regardless of geometries, one probes the polarization ($\mathbf{E}$) change.

We briefly explain each effect below.

Faraday effect This concerns the transmission of light through a sample. The sample can be a liquid or solid. The magnetization of the sample is in the same or the opposite direction of the light propagation (see the top figure of Fig. 2.4). One measures the light polarization change with respect to the incident polarization. The change of polarization is proportional to the magnetization of the sample.

Voigt effect Similar to the Faraday effect, the Voigt effect also concerns the transmitted light through a sample. In the Voigt geometry, the sample magnetization is perpendicular to the light propagation direction (see the top figure of Fig. 2.4). The polarization change is an even function of the magnetization M^2.

Kerr effect This effect is about the light reflection from the surface of a sample (see the bottom figure of Fig. 2.4). It has three configurations, according to how the magnetization is oriented with respect to the sample surface and the plane of incidence. (a) If the magnetization of the sample lies in the plane of incidence and the reflective surface of the sample, this is called the longitudinal Kerr effect. (b) If the magnetization is in the plane of incidence and perpendicular to the sample surface, this is called the polar Kerr effect. (c) If the magnetization is perpendicular to the plane of incidence, this is called the transverse Kerr effect.

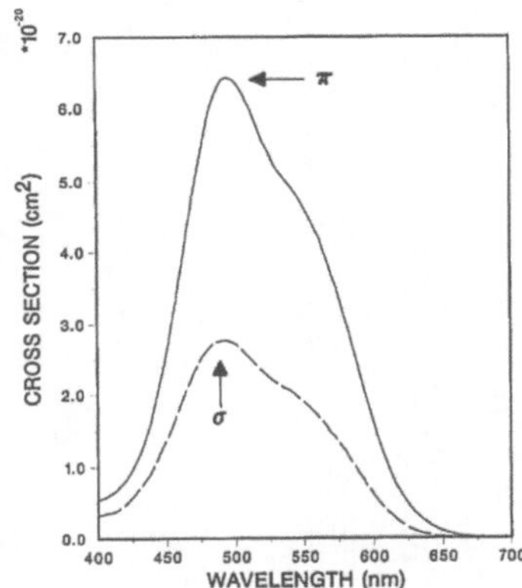

Figure 2.5
Polarized absorption cross sections for the $^2T_2 \rightarrow {}^2E$ transition in Ti:Al_2O_3. The baseline is arbitrarily set to zero for both polarizations at 700 nm [Moulton (1986)]. Used with permission from the Optical Society of America.

2.4 Development of ultrafast lasers: Major breakthrough with the Ti-Sapphire laser

Once we understand how a laser works, we are ready to discuss how an ultrafast laser is developed. Among many good references, the focus issue on "Modular Ultrafast Lasers" in Optics express Vol. **20**, issue 7 (2012) is an excellent source. Generating an ultrafast pulse involves three crucial elements: phase alignment (mode-locking), many modes, and a broad bandwidth medium.

2.4.1 Phase alignment and mode-locking

In the above discussion, we ignored the phase information of the electric fields. This is not a major issue for a CW laser, but it is critical to ultrashort laser pulses. Suppose that we have two electric fields, in their scalar forms, $E_1(t) = A_1 \cos(\omega t + \phi_1)$ and $E_2(t) = A_1 \cos(\omega t + \phi_2)$, where ϕ_1 and ϕ_2 are the phases of the fields. If these two phases are arbitrary or random, there is no way to generate a pulse. Therefore, we need to lock their phases, i.e., mode-locking [Sibbett et al. (2012)].

There are a variety of ways to achieve mode-locking [Haken (1984); Silfvast (2003); Kärtner (2004); Ye and Cundiff (2005)], but the purpose is the same in that all the fields have the same phase. A common mode-locking is to use a shutter inside the laser cavity [Silfvast (2003)] before a pulse hits one mirror. This shutter opens when the pulse arrives, and once the pulse passes through the shutter, the shutter closes until the next arrival of the same pulse. This potentially cuts off all other pulses with a different phase. The shutter can be an acousto-optic device controlled externally by a radio frequency power supply, in which case this is called an active shutter. A passive shutter uses the principle of nonlinearity. If the pulse intensity is high, the shutter opens, but otherwise, it closes. This diminishes the weak pulses. For detailed accounts of mode-locking, see [Kärtner (2004)].

2.4.2 Necessity of many modes and broad bandwidth

According to the basic principle of a Fourier transform, a narrower time domain must come with a broader frequency domain. This means that to build a shorter pulse, we want to have a broader frequency or energy spectrum, or a larger bandwidth. To construct an ultrafast pulse, we need many modes so "different" frequencies appear.[g] As stated above, the cavity only allows those eigenmodes with wavelengths given in Eq. 2.4. If the medium spectral line width is too narrow like the one on the left of Fig. 2.2(c), then it is not possible to have multiple modes or frequencies. This happens with the helium-neon laser that has a bandwidth of 0.002 nm at 633 nm [Donnelly and Grossman (1998)]. What we need is a medium that has a broad bandwidth like the one on the right of Fig. 2.2(c), where many modes are covered and boosted by stimulated emission.

[g] This is not an arbitrarily different frequency. These frequencies must be multiples of one fundamental frequency.

Sapphire (Al_2O_3) crystal doped with titanium (Ti:sapphire) is a popular lasing medium that covers a broad range [Moulton (1986)]. Figure 2.5 shows the absorption spectrum of Ti:sapphire for two polarizations, circular (σ) and linear (π). Note that the absorption is centered around 500 nm. Figure 2.6 shows the fluorescence spectra for how the system emits light, of central importance to our discussion. One can see how broad the spectrum is. It covers from 600 nm to 1050 nm. The gain is centered around 800 nm. This makes Ti:sapphire so special that it has become the main lasing medium for many femtosecond laser pulses.

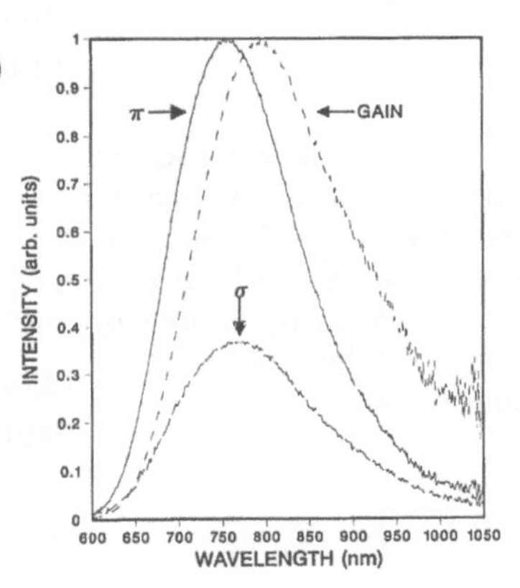

Figure 2.6
Polarized fluorescence spectra and calculated gain lineshape for $Ti:Al_2O_3$ [Moulton (1986)]. The gain is mainly around 800 nm. Used with permission from the Optical Society of America.

2.4.3 Emergence of ultrafast pulses from mode-locking

To appreciate how ultrafast pulses are generated, we suppose that a broad band of the emission spectrum covers modes with angular frequencies from $M\omega_0$ to $N\omega_0$ (such as the right inset in Fig. 2.2(c)). Here ω_0 is the fundamental angular frequency (Eq. 2.7) and is the smallest frequency that the cavity allows. N and M are the lower and upper limits of harmonic order that are covered by the emission spectrum. For mode j, its angular frequency is $\omega_j = j\omega_0$, where j runs from M through

N, with the complex electric field,

$$\mathbf{E}_j(t) = \mathbf{E}_0 \exp[i(\mathbf{k}_j \cdot \mathbf{r}_j - \omega_j t + \phi_j)], \tag{2.10}$$

where $\mathbf{k}_j$ is the wavevector, $\mathbf{r}_j$ is the position, and ϕ_j is the phase. We purposely adopt a complex form for the field as the reason below makes clear. Here we assume that all the amplitudes are the same; for different amplitudes, the key conclusion is the same, and we will show some numerical examples below. Since all the modes are phase-locked, all the phases are the same; in the following discussion, we can ignore ϕ_j. The reader may be surprised as to why we do not use the cosine form for the electric field as before. The complex field has a beauty in that it allows us to know in which direction the wave propagates. If we want them to propagate in the same direction, we can choose all the $\mathbf{k}_i$ along the same direction. Mathematically, the calculation is much simpler.

To ease our derivation, we assume all the fields oscillate along the same direction, so we can replace all the vectors by scalars. The total field, after we set $\phi_j = 0$, is

$$E_{\text{total}} = \sum_{j=M}^{N} E_0 \exp[i(\mathbf{k}_j \cdot \mathbf{r}_j - j\omega_0 t)], \tag{2.11}$$

where we have replaced ω_j by $j\omega_0$. This expression is very similar to a Fourier transform, so even if E_0 is not assumed constant among different modes, the final form of the laser pulse is still a Fourier-transformed mode amplitude. What is different from the ordinary Fourier transform is that here the upper and lower limits do not run from negative infinity to positive infinity, but instead from $M\omega_0$ to $N\omega_0$.

Setting $\mathbf{k}_j \cdot \mathbf{r}_j$ to zero,[h] we have

[h] We do not have to set $\mathbf{k}_j \cdot \mathbf{r}_j$ to zero. Because the magnitude of $\mathbf{k}_j$ is $2\pi/\lambda_j = \omega_j/c$, where λ_j is the wavelength, $-t$ in the equation is replaced by $-t + x/c$. The final result remains the same.

$$E_{\text{total}} = \sum_{j=M}^{N} E_0 \exp[-i(j\omega_0 t)]. \tag{2.12}$$

After a straightforward calculation, we can reduce the above equation to

$$E_{\text{total}}(t) = E_0 \exp(-i((M+N+1)\omega_0 t/2)) \frac{\sin((N-M)\omega_0 t/2)}{\sin(\omega_0 t/2)}. \tag{2.13}$$

Its detailed derivation is left as a homework assignment. Because the instantaneous intensity is $\iota(t) = \epsilon_0 c |E_{total}(t)|^2$, we have

$$\iota(t) = \epsilon_0 c E_0^2 \left(\frac{\sin((N-M)\omega_0 t/2)}{\sin(\omega_0 t/2)} \right)^2. \tag{2.14}$$

This expression is exactly the same as Haken's (see page 214 in [Haken (1984)]. The maxima appear at

$$t = j2\pi/\omega_0, \tag{2.15}$$

where j is an integer and ω_0 is the fundamental frequency of the cavity. From the equation, we see that the smallest time interval Δt between two adjacent peaks is $2\pi/\omega_0 = 2L/c$. If the cavity length is 1 meter, Δt is $(2/3)\times10^{-8}$ s. The inverse of Δt is the laser repetition rate. So, in this case, the repetition rate reaches the MHz region (150 MHz). One can of course reduce it if a chopper is used.

The peak intensity is obtained by taking $t \to 0$ in Eq. 2.14:

$$I_{\text{peak}} = \epsilon_0 c E_0^2 (N-M)^2. \tag{2.16}$$

If we do not have phase-locking, $I_{\text{peak,unlock}} = E_0^2(N-M)$. This $(N-M)$ times boost comes from the number of modes participating in the lasing process under the broad band.

The pulse duration is defined as twice the time from the peak to the first zero. Consider the first minimum

$$(N-M)\omega_0 t_1/2 = \pi. \tag{2.17}$$

Then our duration is

$$\tau = 2 \times \frac{2\pi}{(N-M)\omega_0} = \frac{4L}{c(M-N)}, \tag{2.18}$$

where we have used Eq. 2.7. If we have more modes, the duration becomes shorter. Suppose $L = 1$ m and $M - N = 1$. We have $\tau \propto 10^{-8}$ s or 10 ns. To have a 100 fs pulse, the number of modes is around 10^5. This explains why pulse compression is more important for obtaining shorter laser pulses instead of seeking more modes.

We take $\lambda = 800$ nm as an example to show how the number of modes affects pulse duration and peak intensity. We use Eq. 1.1 to find its corresponding energy $E = h\nu = hc/\lambda = 1.55$ eV and period $T = \lambda/c = 2.667$ fs. The angular frequency is $\omega = 2\pi/T = (3/4)\pi \times 10^{15}$ Hz, where we approximate the speed of light by 3×10^8 m/s. Figure 2.7(a) shows how mode-locking is crucial to the generation of ultrashort pulses. Here we use 20 modes, or $N - M = 20$. If we choose the phase of each mode as $j^2 \times 1.34$ (see the solid line in Fig. 2.7(a)), where j identifies the mode itself (see Eq. 2.13), the envelope of the total electric field of the laser (ignoring the amplitude of each mode) is very small (see the solid line), and there is no enhancement. If all the modes have the same phase of 0 (see the dotted line), we see a huge sharp peak, with the maximum at 20^2.[i] The peak occurs every few femtoseconds. Its inverse is the repetition rate of the laser. The phase requirement is not that stringent as one may believe. For instance, if the phase between each mode is 1.34 or the phase changes as $j \times 1.34$, a sharp peak still emerges, but the peak is now shifted from 0 fs. Figure 2.7(b) shows that as the number of modes increases from 2 to 4, the peak grows sharply, with the peak height being exactly at 2^2, 3^2 and 4^2, while the pulse becomes much sharper, and the pulse duration is reduced. This is just one

[i] Note that $N - M = 20$.

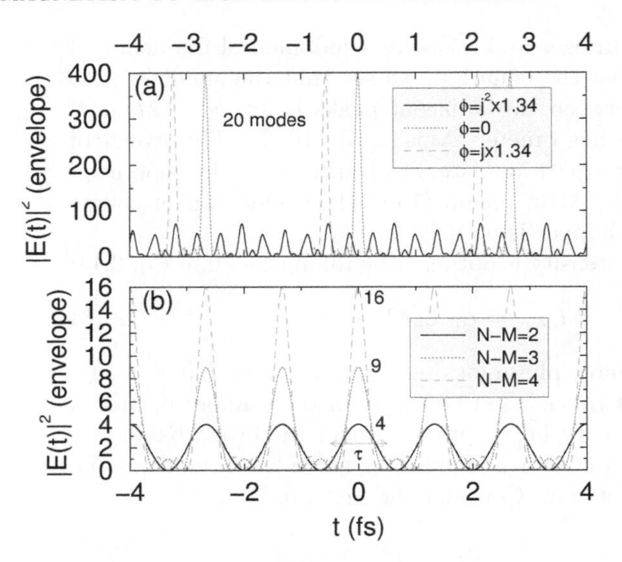

Figure 2.7
(a) Phase-locking is critical to the generation of the ultrafast pulses. We choose 20 modes with three different phases. Solid line: $\phi_j = j^2 \times 1.34$, simulating the random phase. Dotted line: zero phase. Dashed line: $\phi_j = j \times 1.34$. (b) Dependence of the pulse peak intensity and pulse duration on the number of modes covered in the broad band spectrum of a medium.

simple example. There are many sophisticated ways to manipulate the laser pulse duration, phase, shape and intensity. A complete description is not provided here, but it is worth pointing out that there is a limit as to how much one can compress a laser pulse. For a single laser pulse, the Heisenberg uncertainty principle[j] must be obeyed since a single pulse is considered as a single event, so

[j] If one has two pulses or events, the uncertainty principle only applies to each pulse. The time delay between them is not subject to the principle.

$$\Delta E(\text{same event})\Delta t(\text{same event}) \geq \hbar/2. \tag{2.19}$$

When the pulse becomes shorter in the time domain, the energy uncertainty becomes larger. This is the same principle as the Fourier transform, where the frequency domain is broader and the time domain narrower.

2.5 How to access the ultrafast time scale

Ultrafast phenomena cover many different fields of research, each of which forms its own field. The true power of the laser comes from the fact that it can generate other ultrafast pulses. For instance, it can generate ultrafast electron and ultrafast nuclear pulses, each of which has its own important applications. This opens the door to several unexpected frontiers. Figure 2.8 summarizes the various techniques that are based on either photon-in or electron-in or ion-in processes. After the electron, photon or ion interacts with the system, one probes their responses to learn the material properties. For instance, in x-ray scattering experiments, one sends in the x-ray photon and collects scattered photons. In time-resolved photoemission, an ultrafast laser pulse excites an electron out of a sample; the ejected electron is collected and detected by an energy analyzer. These techniques are instrumental to modern scientific research. Also, depending on the detection scheme, one can have variants of different kinds.

Photon in	Electron in	Ion in
Atoms	Superconductors	Compounds
Molecules	Photosynthesis	Alloys
Clusters	Ferroelectrics	Vacuum
Nanomaterials	DNA	Cells, blood, neurons
Magnets	Protein	many more systems
Photon out	Electron out	Ion out

Figure 2.8
Ultrafast laser pulses can generate other ultrafast pulses, like ultrafast electron and ultrafast ion pulses. This greatly expands various research frontiers.

An ultrafast laser provides a powerful tool to probe ultrafast dynamics. However, generation of the ultrafast pulses is only the first step. In order to use it, one must find a way to detect the signal after interaction with a sample. But nearly all the electronic devices are intrinsically slow, typically on a nanosecond time scale, though new instruments can access several hundred picoseconds. Using them would mean that we only get a time-averaged signal, defeating the purpose of ultrafast probing. How do we solve this problem? The answer is in the optical delay stage (see Fig. 2.9).

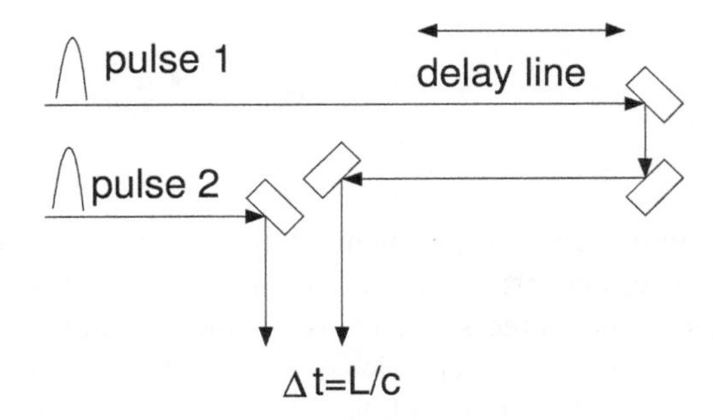

Figure 2.9
Delay lines between two pulses. The time delay between two pulses is determined by the path length difference L, which can be changed mechanically.

Suppose that the path length difference between two pulses is L. Then the time delay between these two pulses is $\Delta t = L/c$, where c is the speed of light. If $L = 1$ m, $\Delta t = 3.3$ ns. For a time step of 1 fs, $L = 0.3$ μm. Such a path length difference is achievable experimentally using piezoelectric transducer [Rosker et al. (1986)]. Conversely, the delay time can be used as a pulse position detector [Balistreri et al. (2001)].

Almost all the times detected in experiments are not absolute time; instead, they are the time delay between two pulses. This delay time can be arbitrarily large or small, or zero or negative, and is not subject to the quantum mechanical uncertainty principle for the time and energy relation, because two pulses are independent events. For the same pulse, it must follow the uncertainty principle, and if the pulse duration becomes shorter, the energy spectrum has to be broader.

2.6 Time-resolved pump-probe experiments

There are numerous techniques that we wish to cover, but it is impossible to do so in the space constraints of this chapter. Instead of giving for each a brief overview, here we focus on just the pump-probe technique. Once one understands how this works, it is easier to understand other related techniques.

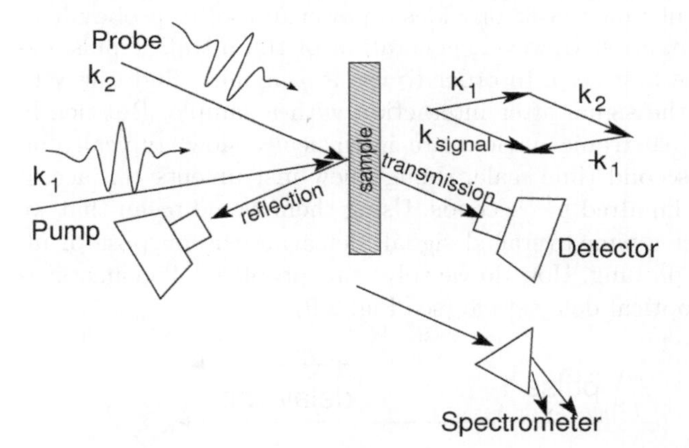

Figure 2.10
Time-resolved pump-probe experiment. A pump pulse $(\mathbf{k}_1)$ first excites the sample, and after a time delay (Δt), one sends in a probe pulse $(\mathbf{k}_2)$ and detects the transmittance change in the probe pulse along $\mathbf{k}_{\rm signal} = \mathbf{k}_1 + \mathbf{k}_2 - \mathbf{k}_1$ as a function of Δt between the pump and probe pulses. The detector can detect the reflectivity change as well (see the left). If the signal passes through a spectrometer, it can be dispersed into a spectrum.

2.6.1 Basic principles

In a typical time-resolved pump-probe experiment, one employs two laser fields (Fig. 2.10), a strong pump field $\mathbf{E}_1(t)$ propagating along the $\mathbf{k}_1$ direction and a weak probe field $\mathbf{E}_2(t)$ along the $\mathbf{k}_2$ direction. $\mathbf{E}_1(t)$ impinges on a sample first, and after a time delay Δt, $\mathbf{E}_2(t)$ arrives at the sample. The detector, in the $\mathbf{k}_2$ direction, detects the signal difference with and without the pump as a function of Δt. Without the pump, the detector probes the transmittance due to the probe alone, or T(probe only), while with the pump, T(pump + probe).

It is clear that the transmittance T depends on the delay time Δt and angular frequencies of the pump (ω_1) and probe (ω_2), or $T(\omega_1, \omega_2, \Delta t)$. Experimentally, the relative change is used:

$$\frac{\Delta T}{T(\text{probe only})} = \frac{T(\text{pump} + \text{probe}) - T(\text{probe only})}{T(\text{probe only})}. \tag{2.20}$$

Another option is that one replaces the single detector by a spectrometer, so ΔT is now dispersed into the frequency domain, but is still a function of the time delay between the pump and probe pulses. This is useful to disentangle complex dynamics. One can also use multiple delays to construct a two-dimensional spectrum, just like 2D NMR [Fayer (2013)].

2.6.2 Nitty-gritties and theory behind the processes

Before we conclude this section, we deem it necessary to dig into a bit of detail as to what theoretically happens in these pump-probe experiments in the simplest possible terms, so that anyone, with a moderate quantum mechanical background, can follow it.

In a pump-probe experiment, one probes the signal along the same direction as the original probe field. For this reason, we say the pump-probe experiment has a background contribution. Theoretically, this is strange. The signal's wave vector is also $\mathbf{k}_{\text{signal}} = \mathbf{k}_2$, but how is it possible to separate it from the original one? In fact, one ought to write $\mathbf{k}_{\text{signal}} = \mathbf{k}_1 + \mathbf{k}_2 - \mathbf{k}_1$, meaning that the system absorbs a photon from the pump, then absorbs another photon from the probe field (see Fig. 2.11), and finally emits a photon with the same energy as the pump's photon energy. It is these interactions that introduce the high-order optical nonlinearity. This problem also leads to several theoretical difficulties in computing the signals along different directions. First of all, the signal wave propagates along the $\mathbf{k}_{\text{signal}} = \mathbf{k}_2$ direction with $\exp(i\mathbf{k}_{\text{signal}}\cdot\mathbf{r})$. One might think that the Fourier transform can be used to compute the direction, but this is not possible because the system is too small, in comparison to the light wavelengths (except x-rays). The phase factor $(\mathbf{k}_{\text{signal}}\cdot\mathbf{r})$ is close to zero. Therefore, one has to separate the responses according to their frequencies before a signal can be computed.

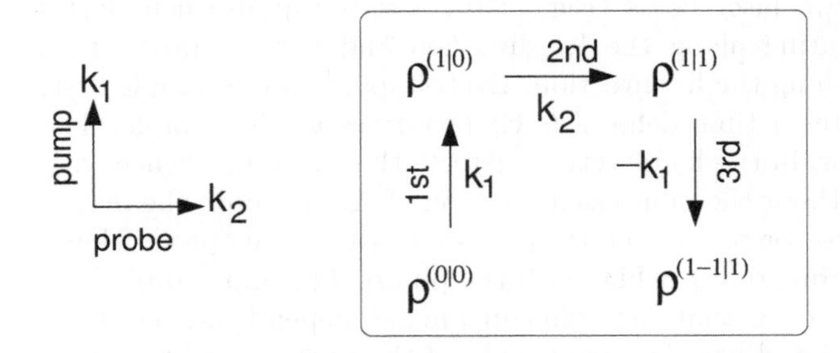

Figure 2.11
Pump-probe excitation diagram shows how the optical signal is calculated. The vertical arrows denote the wavevector $\mathbf{k}_1$ for the pump pulse, and the horizontal ones the wavevector $\mathbf{k}_2$ for the probe pulse. Initially the system is in the ground state, represented by the ground state density matrix $\rho^{(0|0)}$. The entire process starts with the first interaction with the pump field to generate a first-order density matrix $\rho^{(1|0)}$, where the number "1" before | in the superscript refers to the first-order interaction with the pump, and the number "0" behind | denotes the interaction with the probe. Then the system interacts with the probe field to generate the second-order density matrix, $\rho^{(1|1)}$. In the end, the system interacts with the pump again in the $-\mathbf{k}_1$ direction, so thus the third-order density matrix $\rho^{(1-1|1)}$.

Suppose that a density matrix [k] $\rho(t)$ represents the electron density at time t. One has to rewrite it as $\rho(t)^{(n|m)}$, where n and m signify the number of times that the system interacts with the pump (ω_1) and probe (ω_2), respectively. $\rho(t)^{(n|m)}$ oscillates with frequency $n\omega_1 + m\omega_2$.

The first order of interaction only leads to two density matrices, $\rho(t)^{(1|0)}$, and $\rho(t)^{(0|1)}$. In the second order, one has more terms (the details are left as exercises at the end of this chapter), among which $\rho(t)^{(1|1)}$ appears. This term interacts with the pump and probe fields once each. Finally the system emits a photon with the same energy as the pump photon (see Fig. 2.11), so the density matrix is written as $\rho(t)^{(1-1|1)}$. Equation 5.39 in Chapter 5 shows an example.

[k] The density matrix ρ is formed by two state vectors of length I and J. The dimension of ρ is determined by the length of these two vectors, so it is a $I \times J$ matrix. In the time domain, each element of ρ is time-dependent, i.e., $\rho(t)$.

2.7 Photoisomerization in bacteriorhodopsin

As mentioned in Chapter 1, rhodopsin is the key molecule for our vision. Rhodopsin is not only found in mammals, but also in bacteria. Bacteriorhodopsin is an example, with 15 carbons forming the backbone of the molecule, in contrast to rhodopsin which has 13. Photoisomerization refers to the photon-induced molecular structural change without change in constituents.

The molecule may have multiple structures (called isomers), which have different total energies and are on different points on potential energy surfaces.

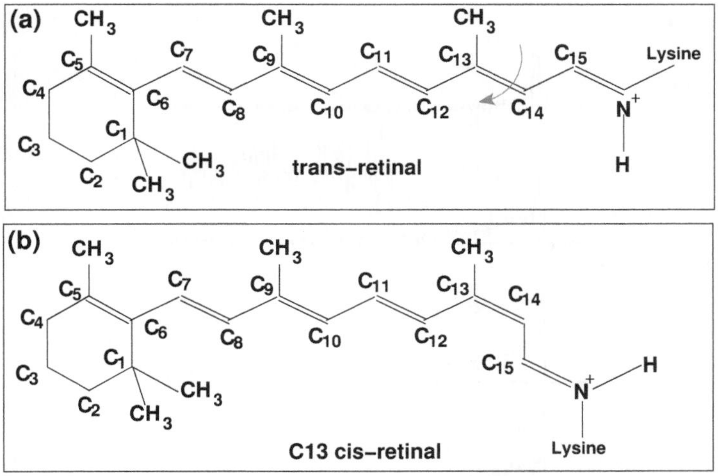

Figure 2.12
Structure of (a) trans-retinal Schiff base and (b) C_{13} cis-retinal Schiff base. During the photoisomerization, there are many steps. The first and also highly efficient step is the trans to cis transition at C_{13}. The arrow in (a) shows the place where the isomerization occurs.

To convert one structure to another, external energy must be provided. If one uses light energy, this is called photoisomerization. Figure 2.12 shows two conformations or isomers of the same retinal. (a) is trans-retinal, where C_{15} and C_{12} are on opposite sides of the double bond C_{13}-C_{14}. The nitrogen ion is linked to lysine. The arrow shows the rotation around C_{13}-C_{14}. (b) is cis-retinal, where C_{15} and C_{12} are on the same side of the double bond C_{13}-C_{14}. The interest in this molecule is that when light illuminates it, it undergoes a very rapid isomerization change from trans-retinal to cis-retinal. Caution: This is opposite to rhodopsin in bovine or human eyes, where the isomerization occurs from cis- to trans-isomers. Since the lysine is connected to the nitrogen ion at the end, the twisting of the retinal molecule sends in a stimulus to the nerve center, where vision begins. Such a rapid isomerization is difficult to catch without an ultrafast laser pulse.

Figure 2.13 shows the transmittance change as a function of time delay between the pump and probe pulses [Kobayashi et al. (2001)]. The laser duration is sub-5 fs, and its wavelength is tunable from 520 to 750 nm. $\Delta T/T$ is strongly modulated by molecular vibrations (Raman active modes). Thus, oscillations in the transmittance change are the signature of rapid vibrations and conformation twisting around the bond which connects C_{13} and C_{14}. With different probe wavelengths, one probes different paths of isomerization.

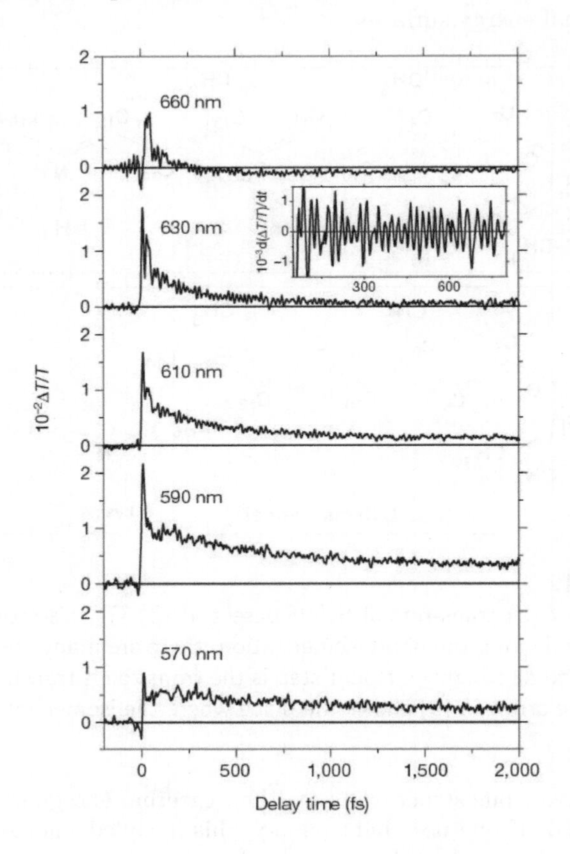

Figure 2.13
Time dependence of the transmittance change ($\Delta T/T$) following sub-5-fs pulse excitation of a suspension of the light-adapted purple membrane of halobacterium salinarum containing bR568 at room temperature [Kobayashi et al. (2001)]. From the top to bottom, five wavelengths are used for the probe pulse. The inset shows an enlarged trace monitored at 630 nm. The figure and caption are used with permission from the Nature Publishing Group.

If we Fourier-transform the time evolution of the transmittance change, we can detect the frequency change due to the photoisomerization. These frequency changes can be correlated to the molecular motion. For the symmetric double bond $C = C$ and $C = N$ stretching modes, their frequency is in the wavenumber range 1500-1550 cm^{-1}. We can convert wavenumbers to eV using $E = hc/\lambda$, where h is the Planck constant, c is the speed of light, and λ is the wavelength. Recall from Chapter 1 that the wavenumber is simply the inverse of the wavelength, $1/\lambda$, but the wavelength must be expressed in the unit of length (cm). 1 eV is equal to 8065.6 cm^{-1}, so the above wavenumber is about 0.19 eV, which is very high for a molecular vibration. For the in-plane $C = C - H$ bending, the wavenumber is 1150-1250 cm^{-1}, lower than the above. The hydrogen out-of-plane mode has a wavenumber at 900-1000 cm^{-1}. This mode is associated with the C_{13}-C_{14} bond, so it can signal whether the isomerization really occurs or not. Therefore, its appearance is of central importance to photoisomerization. Ultrafast lasers are capable of detecting when this occurs. This frequency only appears within 300 fs.

2.8 Femtochemistry

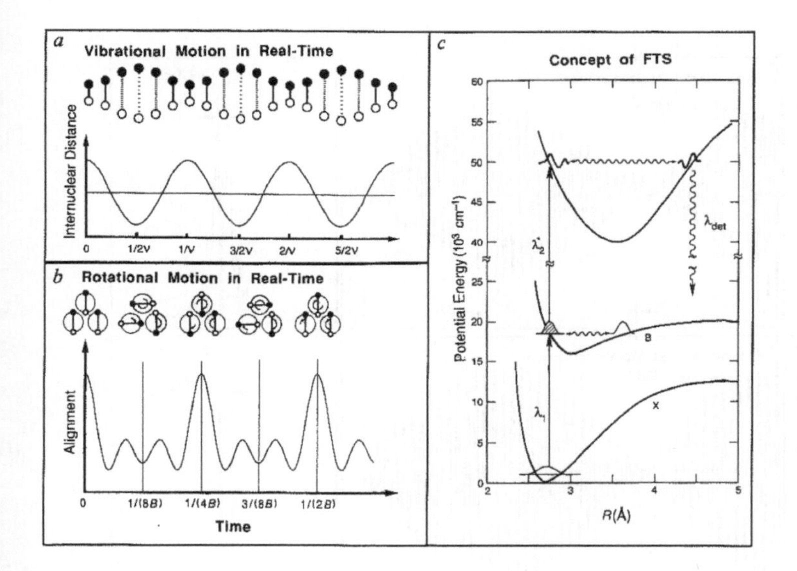

Figure 2.14
(a) Schematic picture of single vibrational motion, where the internuclear distance changes as a function of time (in the unit of inverse frequency, $1/\nu$). (b) Rotational motion in real time. B is the rotational constant. (c) Working principle of the experiment [Dantus et al. (1990)]. Used with permission from the Nature Publishing Group.

Femtochemistry deals with molecular vibrations, rotations, ionization, dissociation and chemical reactions on a femtosecond time scale. This field is so broad that it needs several volumes to cover the entire topic. We choose one of the simplest systems, molecular iodine, or I_2. This molecule can vibrate or rotate. Suppose the vibrational frequency is ν and its period is $T = 1/\nu$. Figure 2.14(a) schematically shows the internuclear separation R of I_2 changes in real time (in the unit of $1/\nu$). This is purely harmonic with a single frequency. Figure 2.14(b) depicts the rotational motion, where B is the rotational constant. Figure 2.14(c) explains how the experiment is carried out. Before laser excitation, I_2 vibrates on the potential energy surface X. The x-axis denotes R in the unit of Å, and the y-axis is the potential energy measured in the unit of 1000 cm^{-1}, or 1000 wavenumbers.

One first pumps I_2 with wavelength λ_1 from X to the bound state B (see Fig. 2.14(c)). This prepares the system in one or more vibrational states. These vibrational states refer to the vibronic states which include the vibrational states and the electronic states. In physics, such a notation is rarely necessary since the nuclear vibration is minor, but in chemistry it is extremely important because the molecule may move into a highly-nonharmonic region (see Fig. 2.14(c)) and dissociate.

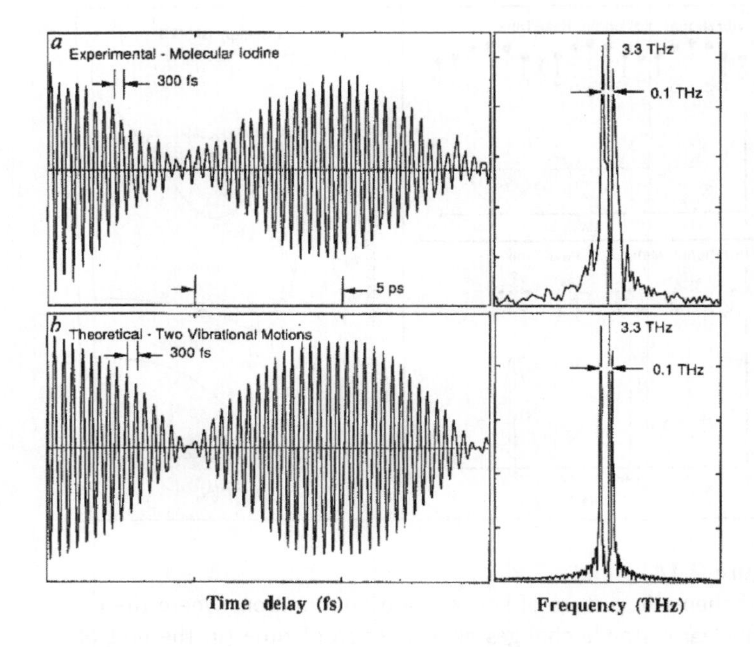

Figure 2.15
(a) Transient fluorescence change as a function of time delay between the pump and probe pulses. Right: Fourier-transformed spectrum. (b) Theoretical simulation by [Dantus et al. (1990)]. Used with permission from the Nature Publishing Group.

After pumping the nuclei vibrate. Then we send in a probe pulse with wavelength λ_2. These wavelengths are not chosen arbitrarily. They must be able to reach those desired states, so the probe light can be absorbed again (in this experiment), and the molecule is excited to a vibrational level in a higher electronic state. Here, $\lambda_1 = 620$ nm and $\lambda_2 = 310$ nm. Normally, the laser has a fixed wavelength. To get λ_2 from $\lambda_1 = 620$ nm, one can use sum-frequency generation (as discussed before), where a nonlinear optical crystal (such as β barium borate) is placed in the direction of the laser beam, and after passing through the crystal, the frequency is doubled or the wavelength is shortened by half. Once the electron reaches an upper level state, it starts to decay by fluorescence. The experiment measures the laser-induced fluorescence, instead of the transmission as sketched in Fig. 2.10, as a function of the time delay between the pump and probe pulses.

As the atoms vibrate away from and back to their equilibrium positions, the fluorescence signal varies. This produces the beautiful oscillation in Fig. 2.15(a) with a period of 300 fs. This is direct evidence of the molecular vibration in real time. Besides the rapid oscillation, there is a modulation over the background, which is due to the anharmonicity of the vibration. If we Fourier-transform to the frequency domain, we see that there are two main peaks around $\mu = 3.3$ THz, with a frequency difference of 0.1 THz. This result is consistent with the high-resolution spectroscopy of iodine. The key significance of this experiment is that it allows us to observe the vibration in real time.

2.9 Metals, semiconductors and superconductors

Metals, semiconductors and superconductors are among earlier intensively investigated materials through ultrafast laser technology. To this end, insulators, strongly correlated materials, topological insulators, graphene and other crystals have been examined. However, theory lags behind experimental research significantly. Most theories are not designed for highly excited states in the time domain. The most successful theory is probably the semiconductor theory [Haug and Koch (2009)] noticeably for GaAs. The reason is clear. In semiconductors, one only deals with a few energy levels and a few k points with very simple s and p states.

Among the first few physical quantities measured are the electron-phonon coupling constants λ (not to be confused with the wavelength above) in Cu, Au, Cr, Ti, W, Nb, V, Pb, NbN and V_3Ga [Brorson et al. (1990)]. Phonons are the quanta of vibration, so λ is an important parameter to measure as to how strongly electrons and phonons are coupled with each other. The importance of λ should not be underestimated. In conventional low-transition-temperature superconductors, it is this coupling that leads to electron pairing (Cooper pairs), through the narrow and effective attractive potential around the Fermi sphere, to overcome the strong Coulombic repulsion between electrons.

Experimentally, the measured quantity is the reflectivity of the probe beam, which is different from the above experiment.

In order to measure λ, one has to assume that the electron temperature change ΔT_e and the reflectivity change ΔR are proportional to each other,

$$\Delta R(t) = a\Delta T_e(t) + b\Delta T_l(t), \tag{2.21}$$

where a and b are constant coefficients describing how electron heating and lattice heating affect R. ΔT_e and ΔT_l are the electronic and lattice temperature changes, respectively. The entire treatment is similar to the two-temperature model. The experiment employs a balanced colliding-pulse mode-locked dye laser with 60-fs duration and at repetition rate of 100 MHz.[1] The average output power is 10 mW, and the wavelength is 630 nm.

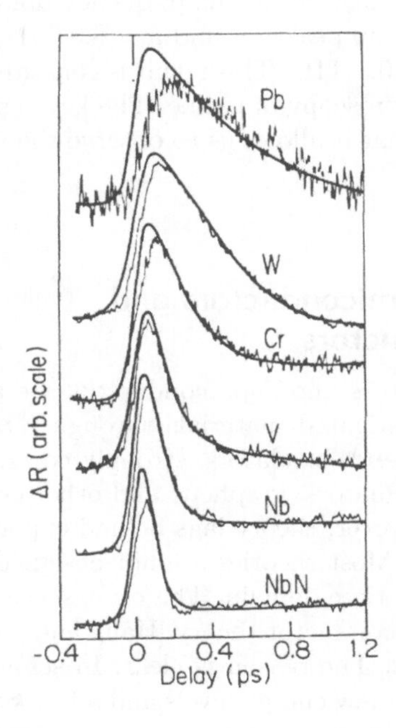

Figure 2.16
Change in the reflectivity of the probe beam as a function of the time delay between the pump and probe pulses [Brorson et al. (1990)]. The pump and probe polarizations are orthogonal to each other. The laser duration is 60 fs, the average output power is 10 mW, and the wavelength is 630 nm or 1.98 eV. Pb has a longer decay rate than NbN. Used with permission from the American Physical Society.

The results are shown in Fig. 2.16. t is the time delay. For $t = 0$ the delay between between pump and probe pulses is zero. In addition, here the electron temperature concept is assumed to be valid over a time scale of several hundred femtoseconds, which is difficult to justify rigorously because tem-

[1] This repetition rate is comparable to that discussed in Section 2.4.3.

Table 2.1
Experimental values for the electron-phonon coupling $\lambda_{\rm exp}$ [Brorson et al. (1990)]. $T_e(0)$ is the electronic temperature, with 20% error (see the discussion in the text). $\lambda_{\rm lit}$ is the literature value obtained through other means. The data are used with permission from the American Physical Society.

Material	$T_e(0)$(K)	$\lambda_{\rm exp}$	$\lambda_{\rm lit}$
Cu	590	0.08 ± 0.01	0.10
Au	650	0.13 ± 0.02	0.15
Cr	716	0.13 ± 0.02	...
W	1200	0.26 ± 0.04	0.26
V	700	0.80 ± 0.06	0.82
Nb	790	1.16 ± 0.11	1.04
Ti	820	0.58 ± 0.05	0.54
Pb	570	1.45 ± 0.16	1.55
NbN	1070	0.95 ± 0.06	1.46
V_3Ga	1110	0.83 ± 0.13	1.12

perature is a statistical concept. Without an advanced theory, these assumptions are necessary to extract some useful information from the experiments and should be thoroughly investigated in the future. What is clear from Fig. 2.16 is that different materials respond to the laser beam differently. Pb shows a slow decay, while NbN's response almost overlaps with the laser pulse shape. One extremely important feature for the success of this experiment is that the samples are coated with a thin layer (4 nm) of copper, so that each material can absorb the same amount of light; otherwise, materials with very different absorption coefficients would be difficult to compare quantitatively.

Table 2.1 shows the experimental results for the electron-phonon coupling $\lambda_{\rm exp}$. One sees that the agreement with other literature values is very good. This shows the power of the ultrafast laser pulse to determine such a microscopic constant.

2.10 Femtomagnetism

Magnetism is at the center of the information technology revolution. Magnetic media have the capability to store bits on a hard drive in terms of magnetic domains. These domains are the minimum storage unit, also called a bit. If we align them along one direction called "1," then the bits along the opposite direction are "0." That is how the storage is achieved. It is well known that these domains are the origin of the hysteresis loop, as shown in Fig. 2.17. The positive and negative maxima of the magnetization correspond to "1' and "0", respectively. If the area of the hysteresis loop is reduced, the magnetization must be reduced. To store and process these bits takes time. Therefore, speed is a key concern, in particular with 10 or larger terabyte hard drive disks and logic devices in production. Femtomagnetism may present an opportunity.

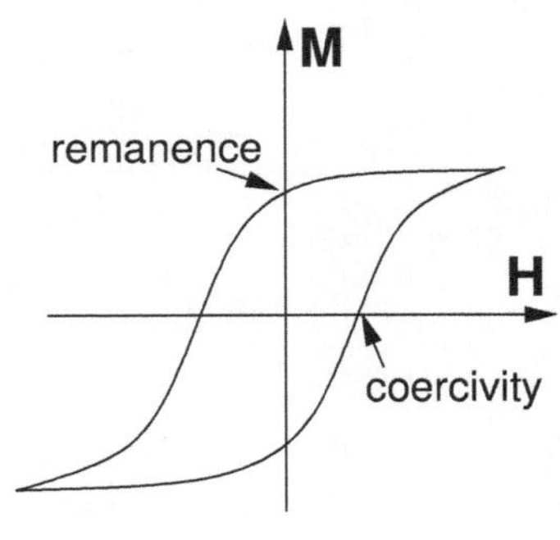

Figure 2.17
Typical hysteresis loop of magnetization M versus an external magnetic field H for a ferromagnet. The magnetic remanence is the value of the magnetization at the zero field limit. The coercivity refers to the needed external field in order to diminish the magnetization.

Investigations on the interaction between light and magnetism have a long history. It started with the Faraday effect (see Section 2.3), where a light beam changes its polarization direction once it passes through a medium under the influence of a magnetic field along the same direction as the beam. The reflected version is called the Kerr effect. For over a century, the main interest has been in the static interaction between the light and magnetic properties of a medium. With the advent of ultrafast laser technology in 1990, experimentalists have started to increasingly focus on the dynamic response on a time scale of several hundred picoseconds.

In 1996, a team from Strasbourg, France [Beaurepaire et al. (1996)] reported that when a 60-fs laser pulse impinges on a 22-nm ferromagnetic nickel thin film, the film loses the magnetization within 1 ps. Their experiment was based on the time-resolved magneto-optical Kerr effect, or TRMOKE, where they measured the longitudinal MOKE signal as a function of the time delay between the pump and probe pulses. Their experimental setup is shown in Fig. 2.18(a). Without the pump, a typical hysteresis loop appears with a nearly square shape (see Fig. 2.18(b)). But with the pump, the hysteresis loop is reduced sharply (see the hysteresis loop at $\Delta t = 2.3$ ps in Fig. 2.18(b)). This shows that the magnetization is reduced.

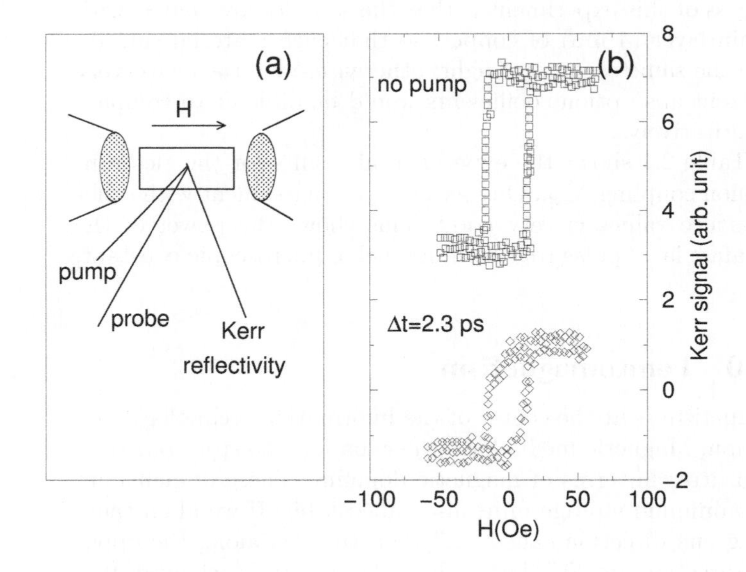

Figure 2.18
(a) Time-resolved magneto-optical pump-probe experiment. (b) Hysteresis loops without and with pump. Figures are replotted. See the original figure in [Beaurepaire et al. (1996)].

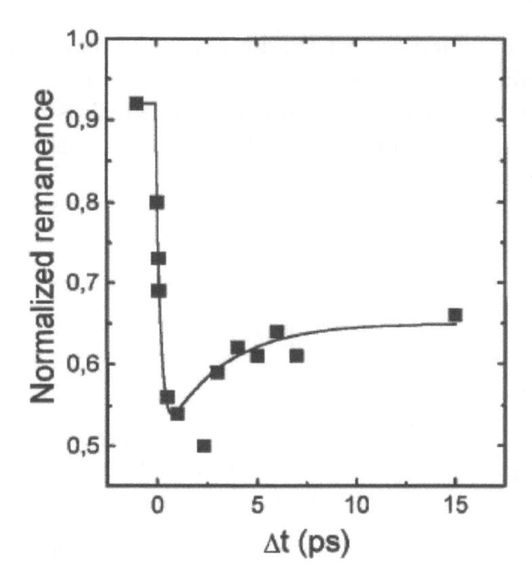

Figure 2.19
Pioneering experimental results for transient normalized remanence as a function of time delay between the pump and probe pulses. The normalization is done with respect to the unpumped signal [Beaurepaire et al. (1996)]. Used with permission from the American Physical Society.

Figure 2.19 shows the time-resolved remnant longitudinal MOKE signal as a function of the time delay between the pump and probe pulses. Under a pulse of 7 mJ/cm^2, the magnetization is reduced by up to 50% within an unprecedented 1 ps. Such an unusually fast demagnetization caught the entire physics community by surprise. Growing experimental evidence supported the original finding. A new frontier, femtomagnetism, was born, opening a door to all-optical spin switching [Stanciu et al. (2007)]. These topics will be discussed in later chapters.

2.11 High-order harmonic generation and attosecond physics

The initial development of nonlinear optics was mainly in the perturbative regime, where the higher the order is, the weaker the nonlinearity gets. This misses the nonperturbative regime, which is directly responsible for high-order harmonic generation (HHG).

This started in 1987, when McPherson and colleagues [McPherson et al. (1987)] radiated four rare gases – Ne, Ar, Kr and Xe – with intense ultraviolet light (wavelength of 248 nm and intensity of $10^{15} - 10^{16}$ W/cm^2). Figure 2.20 shows their experimental setup. A laser beam from the KrF system directly aims at the pulsed gas jet, which is mounted 11.5

cm before the entrance slit of the grating. The light emitted from the jet hits the spherical gold coated grating (600 lines per mm). A single-stage microchannel plate, which is mounted tangentially to the Rowland circle of the spectrometer,[l] detects the dispersed light from the grating.

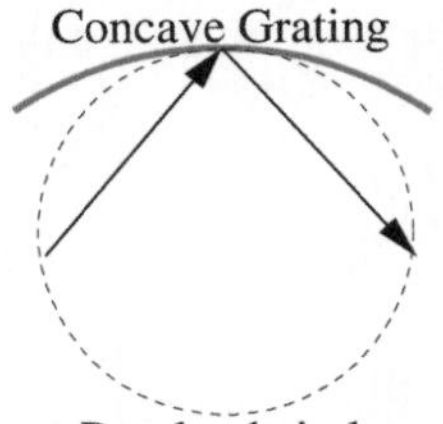

[l] Rowland circle
A Rowland circle is a circle that is tangential to the diffraction grating at its middle point. An incident ray hits the circle along any direction, and its reflected or dispersed ray can be probed around the circle.

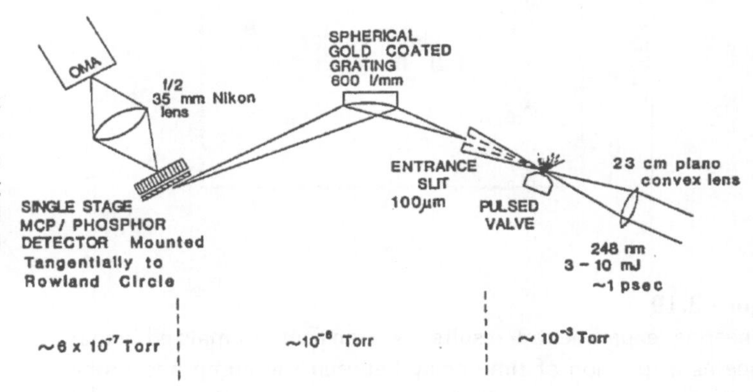

Figure 2.20
Experimental setup for high-harmonic generation. The laser beam is incident on the pulse gas jet [McPherson et al. (1987)]. The emitted light is detected by the spectrometer. Used with permission from the Optical Society of America.

Upon laser excitation, the target gas undergoes the following reaction:

$$N\gamma(\omega) + X \rightarrow X^{q+} + qe^- + \gamma'(N'\omega), \tag{2.22}$$

where $\gamma(\omega)$ refers to a photon with frequency ω and $\gamma'(N'\omega)$ refers to a photon with frequency $N'\omega$. Here, a number N of incident photons with frequency ω interact with the target X to create q-charged ions and a number q of electrons, and to emit new photons with frequency at $N'\omega$. X^{q+} refers to X missing $+q$ electrons, so it has a net $+q$ charge. With laser intensities from 10^{15} to 10^{16} W/cm^2, harmonic generation is observed at much higher orders in He and Ne and falls rather slowly as the harmonic order increases (see Fig. 2.21), in contrast to the perturbation theory prediction.

We can make this semiquantitative. Figure 2.21 shows that for the first three harmonics (5th, 7th and 9th), their efficiencies decrease with the harmonic order, as predicted by perturbation theory, and they fall on the same straight line. But from the 11th to 17th order, they do not fall on the same line as above. Instead, there is a plateau. This is quite different from the perturbative prediction. Later experiments [Ferray et al. (1988)] confirmed this earlier observation.

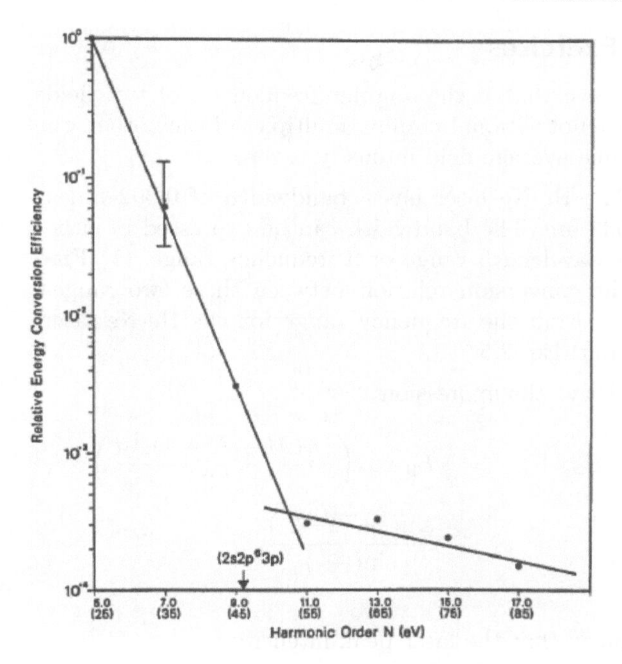

Figure 2.21
Logarithmic plot of the relative energy-conversion efficiency in Ne versus harmonic order N (or energy below in eV) [McPherson et al. (1987)]. The efficiency has a slope change between the 9th and 11th harmonics. The absolute efficiency of the 13th harmonic is about 2×10^{-11}. The arrow denotes the energy at which a $2s$ electron is excited. Used with permission from the Optical Society of America.

The research on HHG has broadened its scope on the attosecond time scale. Energetically, they can reach up to several hundred eV, or attosecond time scales. These time scales capture electron dynamics, with a potential attosecond metrology envisioned [Hentschel et al. (2001)]. While most discoveries are made for gas targets and small molecules, recent focus has shifted toward solids.

2.12 Exercises

1. Prove that if the angular frequencies of two fields are not rational number multiples of each other, the time-average field intensity is zero.

2. The He-Ne laser has a bandwidth of 0.002 nm at 633 nm. The bandwidth can be expressed in either a wavelength range or a frequency range. (1) Find the conversion relation between these two ranges. (2) Find the frequency range for the He-Ne laser. Hint: Eq. 2.5.

3. Derive the expression

$$E_{\rm total}(t) = E_0 \exp\left(-i\frac{(M+N+1)\omega_0 t}{2}\right) \times \frac{\sin(\frac{(N-M)\omega_0 t}{2})}{\sin(\frac{\omega_0 t}{2})}.$$

4. Show that the laser peak intensity is

$$I_{\rm peak} = E_0^2 (M-N)^2.$$

5. Use Eq. 2.13 and choose several different phases for each mode to compute the total field. You can use an Excel sheet to solve it numerically.

6. A nonlinear medium interacts with two fields with wavevectors $\mathbf{k}_1$ and $\mathbf{k}_2$. If the medium interacts with the fields twice, write down all the possible wavevectors generated. Hint: Follow the diagram in Fig. 2.11.

7. If we have a pump and a probe pulse, write down all the possible combinations of the density matrices up to second order. Hint: Follow the example in Section 2.6.2.

8. (a) Show that 1 eV is equivalent to a wavenumber of 8065.6 cm^{-1}. (b) If we convert the y-axis of Fig. 2.14(c) to eV, plot a new figure.

3 Theoretical background

3.1 Density functional theory
3.2 Time-dependent density functional theory
3.3 Quantum chemistry tools
3.4 Solid state physics: Essentials
3.5 Two special features in ultrafast dynamics
3.6 Rotation matrices for spins
3.7 Software packages
3.8 Exercises

When a theorist ponders how to tackle a specific problem, the first thing that comes to mind is which theoretical method to use. Intuition and experience matter. The Hodgkin-Huxley model replaces the lipid bilayer by a capacitor, voltage-gated and leak ion channels by linear and nonlinear conductance, electrochemical potential gradients by a power supply, and ion pumps by current sources, thus converting each element of biology into physics and converting biological complex processes into a set of differential equations, solvable mathematically. For their work, Alan Hodgkin and Andrew Huxley received the Nobel Prize in Physiology or Medicine in 1963. When dealing with quantum effects, or a combination of classical and quantum effects, two possibilities exist. The first is to use a model Hamiltonian such as the Hückel model, Hubbard model, tight-binding model, and so on. This approach is appealing if the model really reflects reality, but it heavily depends on one's understanding and experience of a particular system. The second approach is to use an *ab initio* method to describe the system in a possibly exact manner, and to extract the studied effects from the system's Hamiltonian. The word "exact," however, should never be understood literally, as the exactness depends on the quality of the Hamiltonian used.

The main conceptual difference between the two approaches is not so much the accuracy of the results, but rather the need (or not) for **empirical parameters**.[a] Although model Hamiltonians, like the Heisenberg or the Dirac-van-Fleck Hamiltonians, try to minimize the number of empirical parameters to a minimum, they still require experimental input in order to determine those parameters. *Ab initio* (also called first-principles) methods do not require experimental input. Of course, it is the ultimate comparison with the experiment that assesses the validity and quality of the theoretical method, provided that the experimental results are accurate.

[a] The **Heisenberg** model Hamiltonian is $\hat{H} = -\sum_{i,j} J_{i,j}\hat{\mathbf{S}}_i \cdot \hat{\mathbf{S}}_j$, where $J_{i,j}$ is the empirical exchange parameter and must be found out beforehand, and $\hat{\mathbf{S}}_i$ is the electron spin operator at lattice site i.

Roughly speaking, one can further subdivide the *ab initio* methods in two categories: the density functional theory (DFT) methods (and alike), including the time-dependent version (time-dependent density functional theory, TDDFT) and quantum chemical calculations. Because all the resultant equations are differential equations, numerically one either uses discrete mesh grids or basis functions. This chapter covers some of the basic techniques and tools that are needed to investigate ultrafast phenomena.

3.1 Density functional theory

In general, an N-electron Hamiltonian consists of kinetic energy, potential energy and electron-electron interaction terms,

$$\hat{H} = \sum_i^N \left(-\frac{\hbar^2}{2m}\nabla_i^2 + V(\mathbf{r}_i) + \sum_{j>i} U(\mathbf{r}_i, \mathbf{r}_j) \right), \quad (3.1)$$

where the first term is the usual kinetic energy of the electron i at position $\mathbf{r}_i$, the second is the electrostatic potential due to the ion cores, and $U(\mathbf{r}_i, \mathbf{r}_j)$ is the electron-electron interaction energy. The summation in Eq. 3.1 is over electron i. The many-body Schrödinger equation is

$$\hat{H}\Psi(\mathbf{r}_1, \mathbf{r}_2, \cdots, \mathbf{r}_N) = E\Psi(\mathbf{r}_1, \mathbf{r}_2, \cdots, \mathbf{r}_N), \quad (3.2)$$

where $\Psi(\mathbf{r}_1, \mathbf{r}_2, \cdots, \mathbf{r}_N)$ is the many-body wavefunction.[b] Although the Hamiltonian has a simple structure, it is difficult to solve it exactly because the complexity in the many-body wavefunction $\Psi(\mathbf{r}_1, \mathbf{r}_2, \cdots, \mathbf{r}_N)$ is proportional to 3^N. When the number of electrons increases, solving the Schrödinger equation represents a major impediment to modern computations.

[b] A crucial property of a many-body electron wavefunction is that if one permutes two electrons, say 1 versus 2, the wavefunction changes a sign like $\Psi(\mathbf{r}_1, \mathbf{r}_2)$=-$\Psi(\mathbf{r}_2, \mathbf{r}_1)$. This is the consequence of Fermions.

3.1.1 Hohenberg-Kohn theorem

Because of the challenges in the many-body wavefunction, Hohenberg and Kohn [Hohenberg and Kohn (1964)] proposed to use the single particle density as the basic variable for the total energy. The true single electron density $n(\mathbf{r}_1)$ is

$$N\int d\mathbf{r}_2 \cdots \int d\mathbf{r}_N \Psi^*(\mathbf{r}_1, \mathbf{r}_2, \cdots, \mathbf{r}_N)\Psi(\mathbf{r}_1, \mathbf{r}_2, \cdots, \mathbf{r}_N), \quad (3.3)$$

where N is the number of electrons, Ψ is the many-body wavefunction and the integration is over all the electron coordinates except electron 1. If the integration is integrated over all the electron coordinates, the result is N because Ψ is normalized.

Hohenberg and Kohn first noticed that $n(\mathbf{r}_1)$ is a functional[c] of $V(\mathbf{r}_1)$. The functional derivative is $\delta n(\mathbf{r}_1)/\delta V(\mathbf{r}_1)$. This can be seen from the product $\Psi^*\Psi$ in Eq. 3.3,

$$\frac{\delta\Psi^*}{\delta V}\Psi + \frac{\delta\Psi}{\delta V}\Psi^* = ((E-\hat{H})^{-1}\Psi^*)\Psi + \Psi^*((E-\hat{H})^{-1}\Psi, \quad (3.4)$$

where the second equation uses the functional derivative of Eq. 3.2. It is interesting to note that the functional derivative of the density with respect to the potential has a singularity at the exact wavefunction.

[c] Functional is defined as a function's function. For instance, if $f(g) = \sin(g)$ and $g = \sin(x)$, we say function f is a functional of function g, and g is a function of x. Functional derivatives and integrations follow the same rule as regular derivatives and integrations, but for derivatives, we denote it as $\delta f/\delta g$, instead of dg/dx.

Then they set to prove that conversely $V(\mathbf{r})$ is a unique functional of $\delta n(\mathbf{r}_1)$ by *reductio ad absurdum*, i.e., one potential gives one density. In the following, we will drop the subscript 1 since only the single electron enters the proof. The proof starts with an assumption that another potential $V'(\mathbf{r})$, with ground state eigenfunction Ψ' and eigenvalue E', gives the *same* density $n(\mathbf{r})$. To simplify our notation in Eq. 3.1, we denote three terms in our Hamiltonian as $H = T + U + W$,[d] where T is the kinetic energy operator and W represents the potential energy. For potential W', we have $H' = T + U + W'$, so we have $H - H' = W - W'$. We use the Dirac notation for our many body wavefunction as $|\Psi\rangle$. Since E' is the ground state energy of H', we have

[d] The external potential energy W only depends on the single-electron density through $W = \int V(\mathbf{r})n(\mathbf{r})d\mathbf{r}$.

$$E' = \langle\Psi'|H'|\Psi'\rangle. \tag{3.5}$$

If we replace Ψ' by Ψ, the energy that we obtain is higher than E' since E' is the ground state energy,

$$\begin{aligned} E' < \langle\Psi|H'|\Psi\rangle &= \langle\Psi|H' - H + H|\Psi\rangle & (3.6)\\ &= \langle\Psi|W' - W|\Psi\rangle + E & (3.7)\\ &= E + \int [V'(\mathbf{r}) - V(\mathbf{r})]n(\mathbf{r})d\mathbf{r}, & (3.8)\end{aligned}$$

where $V(\mathbf{r})$ is the potential. Now, we exchange primed and unprimed potential and energy, while keeping the density $n(\mathbf{r})$ unchanged, and we find

$$E < E' + \int [V(\mathbf{r}) - V'(\mathbf{r})]n(\mathbf{r})d\mathbf{r}. \tag{3.9}$$

If we add both sides of Eqs. 3.8 and 3.9, we have

$$E + E' < E + E'. \tag{3.10}$$

This is clearly untrue. Therefore, $v(\mathbf{r})$ is a unique functional of $n(\mathbf{r})$. The density functional theory (DFT) is built upon the single electron density $n(\mathbf{r})$, which only has one variable $\mathbf{r}$, thus greatly simplifying the problem. The idea is to find an alternative equation to find $n(\mathbf{r})$, instead of Ψ, which matches the true single electron density.

3.1.2 Kohn-Sham equation

Minimizing the total energy with respect to this density leads to the noninteracting Kohn-Sham equation [Kohn and Sham (1965)],

$$\left[-\frac{\hbar^2\nabla^2}{2m_e} + V(\mathbf{r}) + V_{\rm H} + V_{xc}[n](\mathbf{r})\right]\psi_{k\sigma}(\mathbf{r}) = E_{k\sigma}\psi_{k\sigma}(\mathbf{r}), \tag{3.11}$$

where $\psi_{k\sigma}(\mathbf{r})$ is called the Kohn-Sham wavefunction which may or may not be related to actual wavefunctions, although they are often used for other calculations. σ is the spin index. Here $V_{\rm H}(\mathbf{r})$ is the Hartree term in atomic units

$$V_{\rm H}(\mathbf{r}) = \int \frac{n(\mathbf{r}')}{|\mathbf{r} - \mathbf{r}'|}d\mathbf{r}'. \tag{3.12}$$

To convert it to SI unit, we need to multiply it by $1/(4\pi\epsilon_0)$.

We note that the electron index i is removed in the Kohn-Sham equation (Eq. 3.11), since the entire calculation is a single electron calculation. This greatly eases numerical calculations. However, DFT introduces an unknown quantity, the exchange-correlation functional $V_{xc}[n](\mathbf{r})$, which replaces the electron-electron interaction term in Eq. 3.1. There are lots of approximate functionals available, with different qualities. For some authors, DFT using approximate functionals is not considered to be strictly *ab initio*. The great popularity of DFT methods lies in the fact that it can handle both extended (such as solids) and localized systems (such as atoms and molecules), with high reproducibility [Lejaeghere et al. (2016)] and reasonable computing time and complexity.

To solve the Kohn-Sham equation (Eq. 3.11), one can use a real grid mesh or basis function. For atoms and molecules, Gaussian functions are often used (see below). In solids a planewave basis is used, together with a pseudopotential for ion cores.[e] The hybrid basis functions are also used. Augmented planewave basis functions are an example. One uses dual basis functions, one for the core electrons in atomic sphere, called Muffin-tin sphere, and the other the planewave basis for the interstitial regions between atoms. Both basis functions are matched at the sphere boundary in both the value and the slope of the basis. Matching allows the spherical basis to get the crystal momentum index. For time-dependent problems, it is customary to employ the time-dependent density functional theory. The detailed accounts of time-dependent DFT will be given in the next section.

[e] Core wavefunctions are local and have a rapid oscillation, so one needs lots of planewaves to describe them. The pseudopotentials replace the true potential with a smooth one, so one can use a fewer number of planewaves.

3.2 Time-dependent density functional theory

3.2.1 Solving the Kohn-Sham equation

We rewrite the Kohn-Sham equation as [Kohn and Sham (1965)]

$$\hat{h}_0^{\mathrm{KS}}\psi_{i\sigma}(\mathbf{r}) = \varepsilon_{i\sigma}\psi_{i\sigma}(\mathbf{r}), \tag{3.13}$$

where $\sigma \in \{\uparrow, \downarrow\}$ specifies the electron spin, $i = 1, 2, ..., N_\sigma$ for each σ, and $\hat{h}_0^{\mathrm{KS}}$ is the single-electron Hamiltonian of the form

$$\hat{h}_0^{\mathrm{KS}} = -\frac{\hbar^2}{2m}\nabla^2 + V(\mathbf{r}) + v_\sigma^{\mathrm{KS}}[n_\uparrow, n_\downarrow](\mathbf{r}). \tag{3.14}$$

Here, we use a lower-case $\hat{h}_0$ to emphasize the fact that it is a single-electron Hamiltonian, rather than the many-electron Hamiltonian.

The Kohn-Sham wavefunctions $\psi_{i\sigma}(\mathbf{r})$ in Eq. 3.13 are subject to an electron-electron interaction potential $v_\sigma^{\mathrm{KS}}[n_\uparrow, n_\downarrow](\mathbf{r})$

which is a *functional*[f] of spin electron densities given by

$$n_\sigma(\mathbf{r}) = \sum_{i=1}^{N_\sigma} n_{i\sigma}(\mathbf{r}) = \sum_{i=1}^{N_\sigma} |\psi_{i\sigma}(\mathbf{r})|^2 = \sum_{i=1}^{N_\sigma} \psi_{i\sigma}^*(\mathbf{r})\psi_{i\sigma}(\mathbf{r}). \tag{3.15}$$

[f] As mentioned in sidemargin c, mathematically while a function f maps a number x onto another number $f(x)$, a functional F maps a function f to a number $F[f]$. For instance, the expectation value $\langle\Psi A|\Psi\rangle$ of an operator A is a functional of a wavefunction Ψ.

In general, $v_\sigma^{\text{KS}}[n_\uparrow, n_\downarrow](\mathbf{r})$ consists of two terms,

$$v_\sigma^{\text{KS}}[n_\uparrow, n_\downarrow](\mathbf{r}) = V_{\text{H}}[n](\mathbf{r}) + V_{\text{xc}}[n_{i\sigma}](\mathbf{r}). \tag{3.16}$$

The first term in Eq. 3.16 is called the Hartree potential and is given by

$$V_{\text{H}}[n](\mathbf{r}) = \int \frac{n(\mathbf{r}')}{|\mathbf{r}-\mathbf{r}'|}\,d\mathbf{r}', \tag{3.17}$$

which is a functional of the total electron density $n(\mathbf{r}) = \sum_\sigma n_\sigma(\mathbf{r})$. When expressed in differential forms, the above definition becomes a Poisson's equation:

$$\nabla^2 V_{\text{H}}[n](\mathbf{r}) = -4\pi\, n(\mathbf{r}). \tag{3.18}$$

On the other hand, the second term $V_{\text{xc}}[n_{i\sigma}](\mathbf{r})$ in Eq. 3.16 is called the exchange-correlation potential, and it is a functional of the orbital spin electron density $n_{i\sigma}(\mathbf{r}) = |\psi_{i\sigma}(\mathbf{r})|^2$.

Equation 3.13 is a set of 3-dimensional equations, which is much easier to handle than the 3^N-dimensional (even without spin) Schrödinger equation. However, DFT introduces an unknown quantity, the exchange-correlation functional $V_{\text{xc}}[n_{i\sigma}](\mathbf{r})$. At the heart of DFT is the assumption that there exists a noninteracting system described by a set of Kohn-Sham equations (3.13) that possesses the same total electron density (Eq. 3.3) as the interacting system described by the many-electron Schrödinger equation [Marques and Gross (2005)]. Then, DFT is exact only if an exact exchange-correlation functional $V_{\text{xc}}[n_{i\sigma}]$ is used, but its exact form is unknown and needs to be approximated in practice. Some of the most commonly used approximations for $V_{\text{xc}}[n_{i\sigma}]$ are discussed in Appendices A.1-A.4, namely,

1. KLI (Krieger-Li-Iafrate) approximation,
2. LDA (local density approximation),
3. LDA with self interaction correction (LDA-SIC),
4. BLYP (Becke-Lee-Yang-Parr) approximation.

Since the single-electron Hamiltonian $\hat{h}_0^{\text{KS}}$ of the Kohn-Sham equations depends on the orbital spin electron density $n_{i\sigma} = |\psi_{i\sigma}|^2$ and therefore on the solutions $\psi_{i\sigma}$ themselves, the system of N Kohn-Sham equations (3.13) needs to be solved iteratively, by minimizing the total energy E of the system. The total energy E of N electrons in terms of the Kohn-Sham wavefunctions and their spin electron densities is given by

$$E[n_\uparrow, n_\downarrow] = \langle T_s\rangle + U[n] + J[n] + E_{\text{xc}}[n_\uparrow, n_\downarrow], \tag{3.19}$$

where $\langle T_s \rangle$ is the non-interacting kinetic energy

$$\langle T_s \rangle = \sum_{\sigma} \sum_{i=1}^{N_\sigma} \int d^3\mathbf{r}\, \psi_{i\sigma}^*(\mathbf{r}) \left(\frac{-\hbar^2}{2m} \nabla^2 \right) \psi_{i\sigma}(\mathbf{r}), \tag{3.20}$$

$U[n]$ is the Coulomb energy

$$U[n] = \int d^3\mathbf{r}\, V(\mathbf{r}) n(\mathbf{r}), \tag{3.21}$$

$V(\mathbf{r})$ is a single-electron Coulomb potential (the second term in Eq. 3.14), and

$$J[n] = \frac{1}{2} \int d^3\mathbf{r} \int d^3\mathbf{r}' \, \frac{n(\mathbf{r})n(\mathbf{r}')}{|\mathbf{r} - \mathbf{r}'|}. \tag{3.22}$$

The exchange-correlation energy functional $E_{\text{xc}}[n_\uparrow, n_\downarrow]$ in Eq. 3.19 is unknown and needs to be approximated in practice. Minimization of Eq. 3.19 subject to the constraint

$$\int d^3\mathbf{r}\, n_\sigma(\mathbf{r}) = N_\sigma \tag{3.23}$$

gives rise to the definition of the exchange-correlation potential as the *functional derivative*[1] of the exchange-correlation energy $E_{\text{xc}}[n_\uparrow, n_\downarrow]$, i.e.,

$$v_\sigma^{\text{xc}}[n_\uparrow, n_\downarrow](\mathbf{r}) = \frac{\delta E_{\text{xc}}[n_\uparrow, n_\downarrow]}{\delta n_\sigma(\mathbf{r})}. \tag{3.24}$$

3.2.2 Example: Many-electron atoms

As an example, consider the ground states of many-electron atoms. Unlike molecules or crystalline solids, the atom has only one binding nucleus, so that the nuclear potential in Eq. 3.14 is simply $V(\mathbf{r}) = -Ze/r$, where the nucleus is placed at the origin of a coordinate system ($\mathbf{R} = 0$). Moreover, since $V(\mathbf{r})$ in this case is a central function, we may assume a separable solution of the form

$$\psi_{i\sigma}(\mathbf{r}) = \frac{R_{i\sigma}^\ell(r)}{r} Y_\ell^m(\theta, \phi), \tag{3.25}$$

where $Y_\ell^m(\theta, \phi)$ are the spherical harmonics

$$Y_\ell^m(\theta, \phi) = \sqrt{\frac{(2\ell+1)}{4\pi} \frac{(\ell-m)!}{(\ell+m)!}}\, P_\ell^m(\cos\theta)\, e^{im\phi}, \tag{3.26}$$

and $P_\ell^m(\cos\theta)$ are the associated Legendre functions. Each spin orbital $\psi_{i\sigma}(\mathbf{r})$ is $(2\ell + 1)$-fold degenerate for different m's

[1] For the formal definition of a functional derivative, see Appendix A in [Parr and Yang (1989)].

unless $\ell = 0$. If we take an average over these degenerate states, then the orbital spin electron density is spherically symmetric and is given by using the addition theorem of spherical harmonics as

$$\bar{n}_{i\sigma}(r) = \left(\frac{1}{2\ell+1}\right)\frac{|R^{\ell}_{i\sigma}(r)|^2}{r^2}\sum_{m=-\ell}^{\ell}|Y_{\ell m}(\theta,\phi)|^2 = \frac{|R^{\ell_i}_{i\sigma}(r)|^2}{4\pi r^2}, \tag{3.27}$$

where ℓ_i is the angular momentum of $\psi_{i\sigma}(\mathbf{r})$. The corresponding spin electron density $\bar{n}_\sigma(r) = \sum_{i=1}^{N_\sigma}\bar{n}_{i\sigma}(r)$ and the total electron density $\bar{n}(\mathbf{r}) = \sum_\sigma \bar{n}_\sigma(\mathbf{r})$ are also spherically symmetric. Note that ℓ_i in Eq. 3.27 is not a summation index but specific to each i-th orbital. In the limit of the central field approximation (Eq. 3.27), the Kohn-Sham effective potential (Eq. 3.16) reduces to a function of the radial coordinate only. Accordingly, the eigenvalue problem for the Kohn-Sham equation (Eq. 3.13) becomes diagonal in each ℓ, such that

$$\hat{h}_0^{\mathrm{KS}}(\ell)R^{\ell}_{i\sigma}(r) = \varepsilon_{i\sigma}R^{\ell}_{i\sigma}(r), \tag{3.28}$$

where

$$\hat{h}_0^{\mathrm{KS}}(\ell) = \frac{-\hbar^2}{2m}\frac{\partial^2}{\partial r^2} + \frac{\ell(\ell+1)}{2r^2} - \frac{Ze}{r} + V_{\mathrm{H}}[\bar{n}](r) + v_\sigma^{\mathrm{xc}}[\bar{n}_{i\sigma}](r). \tag{3.29}$$

Eq. 3.28 can be discretized on a one-dimensional grid using the finite difference method [Schafer (2008)] or the pseudo-spectral method [Tong and Chu (1997b)]. Then, it can be solved by using the LAPACK subroutine `DSYEV`.

To begin the iteration, we could start with the hydrogen-atom solution as a reasonable guess:

```
1    ! Solve for the hydrogen groundstates
2    do l=0,lmax
3      do j=1,Nb
4        V(j,l,1) = Vatom(r(xi(j)),l,1)
5        V(j,l,2) = Vatom(r(xi(j)),l,1)
6      end do
7    end do
8    call eigen(U0,V(:,:,1),xi,E0,Nk0)
```

Here, `V(1:Nb,0:lmax,1:2)` is an array for the net potential for spin-up (`beta=1`) and spin-down (`beta=2`) electrons, and `Vatom(r,l,Z)` is a function which returns the nuclear potential for the atomic number `Z`. `xi(0:Np)` is an array of Legendre-Lobatto collocation points which map a set of `Nb` radial coordinate points onto an interval $[0:1]$ in the pseudospectral grid, and $r(x)$ is a function which defines the mapping: $x \in [0:1] \to r \in [0:\mathrm{rmax}]$. These collocation points are weighed by a factor `wi(0:Np)` depending on the resolution of the grid which is denser at the origin. The subroutine **`eigen`** returns the eigenstates `U0` and eigenvalues `E0` and the number of eigenstates `Nk0` whose energies are less than 5 a.u. The free-electron states whose energy is greater than 5 a.u. are too noisy to be used in the energy representation of the stationary Hamiltonian. If using the LDA-SIC approximation, in particular, each iteration proceeds as follows:

```
1  ! copy U0 to U(1:Nr,1:Nk,0:lmax,1:2)
2        U(:,:,:,1) = U0(:,:,:)
3        U(:,:,:,2) = U0(:,:,:) !
4        E(:,:,1) = E0(:,:) ! copy E0 to E(1:Nk,0:lmax,1:2)
5        E(:,:,2) = E0(:,:) !
6
7        Nitr=0
8        Enet=E0(1,0) ! hydrogen GSE
9        Enet0=0d0
10       do ! start LDA-SIC iterative loop -----
11
12       ! Exit-condition of the loop
13       if (abs(Enet-Enet0).lt.1e-6 .or. Nitr.gt.100) EXIT
14
15          Enet0=Enet ! Update Enet0 with a previous Enet
16
17          Vold(:,:,:)=V(:,:,:) ! Copy V to Vold
18
19          ! Initialize psi(x)
20          psi(:,:,:)=0d0
21          do beta=1,2 ! spins
22          do j=1,Nb
23            do s=1,Nsigma(beta) ! orbitals
24              if (s.lt.3) then
25                psi(j,s,beta) = U(j,s,0,beta) ! 1s/2s-states
26              else if (s.lt.6) then
27                psi(j,s,beta) = U(j,1,1,beta) ! 2p-states
28              else if (s.eq.6) then
29                psi(j,s,beta) = U(j,3,0,beta) ! 3s-states
30              else if (s.lt.10) then
31                psi(j,s,beta) = U(j,2,1,beta) ! 3p-states
32              else
33                STOP
34              end if
35            end do
36          end do
37          end do
38
39          ! Calculate normalized Rpsi(r) in terms of
40          ! normalized Legendre polynomial (LPN)
41          do beta=1,2
42          Rpsi(:,:,beta)=0d0
43          do s=1,Nsigma(beta)
44            do j=1,Nb
45              Rpsi(j,s,beta) = psi(j,s,beta) * LPN(j)
46     &                          *dsqrt(dfloat(Np*(Np+1))/2d0)
47            end do
48          end do
49          end do
50
51          ! find orbital-dependent potentials
52          call hartree(VH,Rpsi,xi,wi)
```

```
53          call LDA(Vxc,Rpsi,xi)
54
55          ! find Vsic
56          call SIC(Vsic,VH,Vxc,Rpsi,xi,wi)
57
58          ! find orbital-independent potentials
59          do j=1,Nb
60            VH0(j) = sum(VH(j,1:Nup,1)) + sum(VH(j,1:Ndn,2))
61          end do
62          call LDA0(Vxc0,Rpsi,xi)
63
64          ! Set up KS potential
65          do beta=1,2
66          do l=0,lmax
67          do j=1,Nb
68            V(j,l,beta) = Vatom(r(xi(j)),l,Z)
69      &                 + VH0(j) + Vxc0(j,beta) - Vsic(j,beta)
70            ! damping
71            V(j,l,beta) = V(j,l,beta)*0.618d0
72      &                 + Vold(j,l,beta)*0.382d0
73          end do
74          end do
75          end do
76
77          ! Find all bound/continuous eigenvalues and states
78          call eigen(U(:,:,:,1),V(:,:,1),xi,E(:,:,1),Nk(:,1))
79          call eigen(U(:,:,:,2),V(:,:,2),xi,E(:,:,2),Nk(:,2))
80
81         ! Copy E to Eocc
82          do beta=1,2
83          do s=1,Nsigma(beta)
84            if (s.lt.3) then
85              Eocc(s,beta) = E(s,0,beta) ! 1s/2s-states
86            else if (s.lt.6) then
87              Eocc(s,beta) = E(1,1,beta) ! 2p-states
88            else if (s.eq.6) then
89              Eocc(s,beta) = E(3,0,beta) ! 3s-states
90            else if (s.lt.10) then
91              Eocc(s,beta) = E(2,1,beta) ! 3p-states
92            else
93              STOP
94            end if
95          end do
96          end do !beta
97
98          ! Sum Eocc
99          Enet = sum(Eocc(1:Nup,1)) + sum(Eocc(1:Ndn,2))
100
101         Nitr = Nitr +1
102       end do ! end of LDA-SIC loop -----
```

Nsigma(beta) is a function which returns the number of occupied orbitals for each spin. Subroutines hartree(), LDA() and SIC() return the Hartree potential VH, the LDA potential Vxc and SIC correction Vsic, respectively. The subroutine

`LDA0()` is specifically for the LDA potential `Vxc0`, which is a functional of the total spin electron density, as opposed to the orbital spin electron density. As the iteration progresses, the sum of individual eigenvalues `Enet` converges to a unique value, which is used in an exit condition of the iteration loop (see L12 above). In order to accelerate the convergence process, the current potential (`V`) is weighted more than the previous potential (`Vold`); this is a numerical method called the damping scheme (see L71-72 above).

In the above code, `Enet` is the sum of individual eigenvalues $\varepsilon_{i\sigma}$ in Eq. 3.13. It should be noted that the total energy E in Eq. 3.19 is not `Enet`. Rather, it is[g]

[g] Using `Enet` has an advantage that one avoids computing the expectation value of the kinetic energy operator.

$$E = \sum_{\sigma} \sum_{i=1}^{N_\sigma} \varepsilon_{i\sigma} - J[n] - U_{\mathrm{xc}}[n] + E_{\mathrm{xc}}[n_\uparrow, n_\downarrow], \tag{3.30}$$

where

$$U_{\mathrm{xc}}[n] = \sum_{\sigma} \int d^3\mathbf{r}\; v_\sigma^{\mathrm{xc}}[n_\uparrow, n_\downarrow](\mathbf{r}) n_\sigma(\mathbf{r}). \tag{3.31}$$

The total energies given by Eqs. 3.19 and 3.30 must be equal to each other, which gives a good test to see if one implements the iteration properly. Table 3.1 lists the ground-state energy of atoms calculated with the pseudo-spectral method and various approximations for the exchange-correlation functionals [Murakami et al. (2018)]. The Fortran code for the DFT of many-electron atoms discussed in this subsection is provided in the following webpage: https://sites.google.com/site/mitsukomurakami02/codes/dft.

3.2.3 Adiabatic approximation

In 1984, Runge and Gross generalized the Hohenberg-Kohn theorem for time-dependent density functional theory (TDDFT), showing that there is a one-to-one correspondence between a time-dependent, external potential and an electron density evolving from its ground state [Runge and Gross (1984)]. At present, most TDDFT calculations [Bauer et al. (2001); Tong and Chu (2001); Telnov et al. (2013); Gao et al. (2016)] rely on the *adiabatic approximation* given by

$$v_\sigma^{\mathrm{KS}}[n_\uparrow, n_\downarrow](\mathbf{r}, t) := v_\sigma^{\mathrm{KS}}[n_\uparrow, n_\downarrow](\mathbf{r})|_{n_\sigma = n_\sigma(\mathbf{r},t)}. \tag{3.32}$$

That is, one feeds the time-dependent density into ground-state functionals $V_{\mathrm{H}}[n]$ and $v_{\mathrm{xc}}[n_\uparrow, n_\downarrow]$ assuming that this gives a reasonable description of the dynamics. In reality, electrons experience a sudden change in the exchange-correlation potential, known as the derivative discontinuity after one of them completely ionizes [Lein and Kümmel (2005)], which is not incorporated under the adiabatic approximation.

Table 3.1
The total energy of atoms ($Z \leq 18$) in atomic units (a.u.), obtained from Kohn-Sham wavefunctions using local density approximation (LDA), local-density approximation with self-interaction correction (LDA-SIC), Krieger-Li-Iafrate (KLI) approximation, Becke-Lee-Yang-Parr (BLYP) approximation. Also shown are the calculation based on the Hartree-Fock (HF) approximation [Bunge et al. (1992)].

Z	Atom	LDA	LDA-SIC	KLI	BLYP	HF
2	He	-2.724	-2.862	-2.862	-2.908	-2.862
3	Li	-7.193	-7.434	-7.432	-7.484	-7.433
4	Be	-14.223	-14.578	-14.572	-14.663	-14.573
5	B	-24.064	-24.549	-24.411	-24.649	-24.529
6	C	-37.112	-37.745	-37.529	-37.847	-37.688
7	N	-53.709	-54.506	-54.403	-54.599	-54.401
8	O	-73.992	-74.962	-74.617	-75.085	-74.809
9	F	-98.474	-99.635	-99.161	-99.766	-99.409
10	Ne	-127.491	-128.859	-128.545	-128.982	-128.547
11	Na	-160.644	-162.217	-161.844	-162.300	-161.859
12	Mg	-198.249	-200.027	-199.590	-200.102	-199.615
13	Al	-240.356	-242.341	-241.760	-242.390	-241.877
14	Si	-287.182	-289.377	-288.721	-289.397	-288.854
15	P	-338.889	-341.299	-340.689	-341.286	-340.719
16	S	-395.519	-398.148	-397.335	-398.138	-397.505
17	Cl	-457.343	-460.196	-459.295	-460.175	-459.482
18	Ar	-524.517	-527.599	-526.781	-527.566	-526.818

3.2.4 Example: TDDFT of atoms in a linearly-polarized field

Consider an atom exposed to a linearly-polarized, strong ($\sim 10^{14}$ W/cm^2) laser field $\mathbf{E}(t)$. We restrict ourselves to the non-relativistic regime and use the dipole approximation, i.e., the magnetic field of the laser pulse is neglected, so that the laser acts as an electric field which is time dependent but homogeneous in space. The time-dependent interaction potential due to the laser field in the limit of the dipole approximation is given by [Kulander (1987)]

$$V_{\text{laser}}(\mathbf{r},t) = \mathbf{r}\cdot\mathbf{E}(t). \tag{3.33}$$

Accordingly, we must solve a set of time-dependent Kohn-Sham equations

$$i\hbar\frac{\partial}{\partial t}\psi_{i\sigma}(\mathbf{r},t) = \left[\hat{h}^{\text{KS}}(t) + V_{\text{laser}}(\mathbf{r},t)\right]\psi_{i\sigma}(\mathbf{r},t), \tag{3.34}$$

where

$$\hat{h}^{\text{KS}}(t) = -\frac{\hbar^2}{2m}\nabla^2 - \frac{Z}{r} + v_\sigma^{\text{KS}}[n_\uparrow, n_\downarrow](\mathbf{r},t), \tag{3.35}$$

with

$$v_\sigma^{\rm KS}[n_\uparrow, n_\downarrow](\mathbf{r},t) = V_{\rm H}[n](\mathbf{r},t) + v_\sigma^{\rm xc}[n_{i\sigma}](\mathbf{r},t)\,. \qquad (3.36)$$

Under the adiabatic approximation (Eq. 3.32), the time-dependent Kohn-Sham potential (Eq. 3.36) is evaluated with the same functionals as in ground-state calculation, but using the time-dependent spin electron densities in each time step, namely,

$$n_\sigma(\mathbf{r},t) = \sum_{i=1}^{N_\sigma} n_{i\sigma}(\mathbf{r},t) = \sum_{i=1}^{N_\sigma} |\psi_{i\sigma}(\mathbf{r},t)|^2, \qquad (3.37)$$

and the time-dependent total electron density

$$n(\mathbf{r},t) = \sum_\sigma n_\sigma(\mathbf{r},t). \qquad (3.38)$$

The time dependence in the Kohn-Sham potential reflects the time evolution of the Kohn-Sham wavefunctions $\psi_{i\sigma}(\mathbf{r},t)$ in Eq. 3.34. This self-consistent feedback accounts for Coulomb polarization effects on the electronic response of the system. In this example, we shall adopt the LDA-SIC approximation [Tong and Chu (1998)] for the exchange-correlation potential $v_\sigma^{\rm xc}[n_{i\sigma}](\mathbf{r},t)$.

Without a loss of generality, we may assume that the laser polarization of a linearly-polarized field is along the z-axis, such that

$$\mathbf{E}(t) = E_o(t)\sin\omega_o t\ \hat{z}, \qquad (3.39)$$

where ω_o is a laser frequency, and $E_o(t)$ is a pulse envelope function. Then, the dipole approximation (Eq. 3.33) gives

$$V_{\rm laser}(\mathbf{r},t) = E_o(t)\, r\cos\theta\sin\omega_o t. \qquad (3.40)$$

Because of the azimuthal symmetry in $V_{\rm laser}(\mathbf{r},t)$ given by Eq. 3.40, the magnetic quantum number m_i of the i-th orbital is conserved during the time evolution in a linearly-polarized field, so that we can assume the solution of form

$$\psi_{i\sigma}(\mathbf{r},t) = \sum_\ell \frac{R_{i\sigma}^\ell(r,t)}{r} Y_\ell^{m_i}(\theta,\phi). \qquad (3.41)$$

In other words, the time-dependent wavefunction is expanded with the spherical harmonics $|\ell, m_i\rangle$ as a basis set, while the radial wavefunction $R_{i\sigma}^\ell(r,t)$ for each ℓ evolves in time on a one-dimensional grid.

For the stability of the evolution, the time-evolution operator must be split into three parts [Feit et al. (1982)] and sequentially applied to wavefunctions (Eq. 3.41) in each time step Δt [Murakami et al. (2017)],

$$\begin{aligned}\psi_{i\sigma}(\mathbf{r},t+\Delta t) &= e^{-i\hat{h}_0^{\rm KS}(\ell)\Delta t/2}\mathcal{L}^{-1}(\ell)e^{-iV(\mathbf{r},t+\Delta t/2)\Delta t} \\ &\quad\times\mathcal{L}(\theta)e^{-i\hat{h}_0^{\rm KS}(\ell)\Delta t/2}\ \psi_{i\sigma}(\mathbf{r},t)\,, \qquad (3.42)\end{aligned}$$

where $\hat{h}_0^{\mathrm{KS}}(\ell)$ is the stationary Hamiltonian (Eq. 3.29), and $V(\mathbf{r},t)$ is the time-dependent part of the Hamiltonian [Tong and Chu (1997a)],

$$V(\mathbf{r},t) = v_\sigma^{\mathrm{KS}}[n_\uparrow, n_\downarrow](\mathbf{r},t) - v_\sigma^{\mathrm{KS}}[\bar{n}_\uparrow, \bar{n}_\downarrow](r,0) + V_{\mathrm{laser}}(\mathbf{r},t). \tag{3.43}$$

Here, $v_\sigma^{\mathrm{KS}}[\bar{n}_\uparrow, \bar{n}_\downarrow](r,0)$ is the Kohn-Sham potential of the initial state, i.e., the ground state of an atom found by solving the initial value problem (Eq. 3.13) in the previous subsection. $\mathcal{L}$ denotes the Legendre transform defined by

$$\begin{aligned}\mathcal{L}\left\{R_{i\sigma}^\ell(r,t)\right\}(\theta) &\equiv \sum_\ell P_\ell^{m_i}(\theta,\phi)\, R_{i\sigma}^\ell(r,t) \\ &= \sum_\ell \sqrt{\frac{(2\ell+1)(\ell-m_i)!}{2(\ell+m_i)!}}\, P_\ell^{m_i}(\cos\theta)\, R_{i\sigma}^\ell(r,t) \\ &= R_{i\sigma}(r,\theta,t),\end{aligned} \tag{3.44}$$

which provides the change of representation between ℓ and θ. The operator $e^{-iV(\mathbf{r},t+\Delta t/2)\Delta t}$ is a complex-valued function of the electron coordinate $\mathbf{r}$, which can readily be multiplied onto the wavefunction in the θ-representation. On the other hand, the stationary Hamiltonian operator can be applied very accurately by using the ℓ-representation [Tong and Chu (1997b)],

$$e^{-i\hat{h}_0^{\mathrm{KS}}(\ell)\Delta t/2} R_{i\sigma}^\ell(r_j,t) = \sum_{jj'} (S_\ell)_{jj'} R_{i\sigma}^\ell(r_{j'},t), \tag{3.45}$$

where the evolution matrix $(S_\ell)_{jj'}$ is constructed from the set of eigenstates $\left\{R_{i\sigma}^{\ell k}(r_j)\right\}$ and their eigenvalues $\left\{\varepsilon_{i\sigma}^{\ell k}\right\}$ of the field-free Hamiltonian (Eq. 3.29) as

$$(S_\ell)_{jj'} = \sum_k R_{i\sigma}^{\ell k}(r_j) R_{i\sigma}^{\ell k}(r_{j'}) e^{-i\varepsilon_{i\sigma}^{\ell k}\Delta t/2}, \tag{3.46}$$

where the superscript k designates different energy levels ($k = 1, 2, ...$). The following is a code segment which utilizes the LAPACK subroutine ZGEMM for matrix multiplication in Eq. 3.46:

```
1        complex*16 SO(Nb,Nb,0:lmax,2)
2      ! local arrays for ZGEMM
3        complex*16, allocatable, dimension(:,:) :: P, Q
4
5      ! Set up the propagation matrix SO(l)
6        do beta=1,2 ! ----------------------------
7        SO(:,:,:,beta)=zero
8        if(Nsigma(beta).ne.0)then
9        if(beta.ne.2 .or. Nup.ne.Ndn)then
10         do l=0,lmax ! *************************
11
12         allocate( P(Nb,Nk(l,beta)), Q(Nb,Nk(l,beta)) )
13
14         do j=1,Nb
```

```
15          do k=1,Nk(l,beta)
16            P(j,k) = cdexp(-ai*E0(k,l,beta)*dt/2d0)
17     &        *U0(j,k,l,beta)
18            Q(j,k) = one*U0(j,k,l,beta)
19          end do
20          end do
21          call ZGEMM('N','T',
22     &         Nb,Nb,Nk(l,beta),one,P,Nb,Q,Nb,zero,
23     &         S0(:,:,l,beta),Nb)
24          deallocate(P,Q)
25          end do ! ******************************
26        end if
27        end if
28        end do ! ---------------------------------
```

Moreover, to eliminate the reflection from the boundary during the time evolution, we split the wave function at a given time t as

$$\begin{aligned}\psi(\mathbf{r},t) &= f(r)\psi(\mathbf{r},t) + [1-f(r)]\psi(\mathbf{r},t) \\ &= \psi_{(\mathrm{in})}(\mathbf{r},t) + \psi_{(\mathrm{out})}(\mathbf{r},t),\end{aligned} \tag{3.47}$$

where $f(r)$ is an absorbing function (called gobbler) that is 1 in the inner region ($0 \leq r \leq R_b$) and smoothly decreases to zero in the outer region ($R_b < r < r_{\max}$). For the evaluation of high-harmonic spectra (see Chapter 4), only the inner wave function is necessary. The following subroutine returns an array `rgob(1:Nb)` which contains a commonly-used absorbing function of $\cos^{1/8}$ for an array of radial coordinates `ri(1:Nb)`:

```
1         subroutine gobbler(Nb,rgob,ri)
2         implicit none
3         integer Nb
4         real*8 rgob(Nb), ri(Nb)
5         ! local variables
6         integer j
7         real*8 buff
8         buff = ri(Nb) - Rb ! i.e., buff = rmax - Rb
9         ! gobbler for radial direction
10        do j=1,Nb-1
11          if (ri(j).lt.Rb) then
12            rgob(j)=1d0
13          else
14            rgob(j)=(dcos((ri(j)-Rb)*pi/2d0/buff))**0.125
15          end if
16        end do
17        rgob(Nb)=0d0
18        return
19        end subroutine gobbler
```

The zero-force theorem of TDDFT [Vignale (1995)] states that the KS potential $v_\sigma^{\mathrm{KS}}[n_\uparrow, n_\downarrow](\mathbf{r},t)$, which accounts for the electron-electron interaction, does not contribute to the dipole acceleration only if it is exact. Although the exchange-correlation potential we use is approximate, we shall assume that the zero-force theorem holds when evaluating the dipole acceleration. It then follows that [Murakami et al. (2013)]

$$\langle a_z(t)\rangle = Z\sum_{\sigma}\sum_{i=1}^{N_\sigma}\sum_{\ell,m}\int dr\left(\frac{c_\ell^m}{r^2}\right)\mathbf{Re}\left[R_{i\sigma}^{\ell+1,m}(r,t)R_{i\sigma}^{\ell m}(r,t)\right], \tag{3.48}$$

where

$$c_\ell^m = \sqrt{\frac{(\ell+m+1)(\ell-m+1)}{(2\ell+1)(2\ell+3)}}. \tag{3.49}$$

The following is an excerpt of the TDDFT code which shows the time-loop:

```
1       ! opening output files
2       open(14, file='accl1.dat', status='replace')
3       open(15, file='accl2.dat', status='replace')
4       do n=1,Nt  ! Time loop ----------------------------
5         t = tmin + dfloat(n-1)*dt
6         ! Re-evaluate KS potential
7         do beta=1,2 ! spin
8           if(beta.eq.2 .and. Nup.eq.Ndn)then!closed system
9             VH(:,:,:,2) = VH(:,:,:,1)
10            Vxc(:,:,:,2) = Vxc(:,:,:,1)
11            Vxc0(:,:,2) = Vxc0(:,:,1)
12            Vsic(:,:,2) = Vsic(:,:,1)
13          else if (Z.eq.1) then      ! Hydrogen atom
14            VH(:,:,:,beta) = 0d0
15            Vxc(:,:,:,beta) = 0d0
16            Vxc0(:,:,beta) = 0d0
17            Vsic(:,:,beta) = 0d0
18          else
19            do s=1,Nsigma(beta)
20         m0=iabs(emm(s)) ! magnetic quantum number of each orbital
21              call calcDens(U(:,:,s,beta),Npsi(:,:,s),m0,xi,wy)
22              call hartree(VH(:,:,s,beta),Npsi(:,:,s),xi,wi,wy)
23              call LDA(Vxc(:,:,s,beta),Npsi(:,:,s)) ! orbital
24            end do
25            call LDA0(Vxc0(:,:,beta),Npsi,beta) ! total
26            call SIC(Vsic(:,:,beta),VH(:,:,:,beta),Vxc(:,:,:,beta),
27      &                 Npsi,xi,wi,wy,beta)
28          end if
29        end do
30        do k=0,lmax
31        do j=1,Nb
32        VH0(j,k)=sum(VH(j,k,1:Nup,1))+sum(VH(j,k,1:Ndn,2))
33        do beta=1,2
34          Vks(j,k,beta)=VH0(j,k)+Vxc0(j,k,beta)-Vsic(j,k,beta)
35         end do
36        end do
37        end do
38        ! loop over each electron ***********************
39        do beta=1,2 ! spin up/down
40          if(beta.eq.2 .and. Nup.eq.Ndn)then ! closed system
41            U(:,:,:,2) = U(:,:,:,1)
42            Da(:,2) = Da(:,1)
```

```
43          else
44          do s=1,Nsigma(beta)
45            m0=iabs(emm(s))
46            if(( Z.le.4 ) .or. ( Z.gt.4 .and. s.gt.1 )) then
47            ! evolve TDSE by using spectral representation
48            call evolve(U(:,:,s,beta),S0(:,:,:,beta),t,ri,
49      &                  rgob,y,wy,
50      &                  Vks(:,:,beta),Vks0(:,:,beta),m0)
51            end if
52            ! Find <a>(t)
53            Da(s,beta) = force(U(:,:,s,beta),ri,c(:,m0))
54          end do ! s
55          end if
56        end do ! beta
57        ! ********************************************
58        ! Output a(t) and z(t) to accl{1,2}.dat
59        write(14,330) t,(Da(s,1),s=1,Nsigma(1))
60        write(15,330) t,(Da(s,2),s=1,Nsigma(2))
61 330     format (1f12.5,2x,10e13.5)
62      end do ! end of time loop --------
63      close(15)
64      close(14)
```

VH and Vxc are arrays for the Hartree potential and the LDA potential, respectively, for orbital spin densities, and VH0 and Vxc0 are their corresponding arrays for the total electron density and for the spin electron densities. Vsic is the SIC term, and the KS potential for the current and the next time steps are stored in Vks0 and Vks, respectively. Because of the azimuthal symmetry of the linearly-polarized laser potential, both the wavefunction U and KS potentials are two-dimensional (r, θ), but they also depend on the orbital index (s=1,...,Nsigma(beta)) and spins (beta=1,2). Accordingly, arrays are defined as follows:

```
1 complex*16, dimension(1:Nb,0:lmax,Nsigmax,1:2)::U
2 real*8, dimension(1:Nb,0:lmax,Nsigmax,1:2) :: VH,Vxc
3 real*8, dimension(1:Nb,0:lmax)::VH0
4 real*8, dimension(1:Nb,0:lmax,1:2)::Vsic,Vxc0,Vks0,Vks
```

emm(s) is a function which returns the magnetic quantum number of the s-th orbital (s<Nsigmax), and subroutines calcDens() and force() return the spin orbital electron density Npsi(1:Nb,0:lmax,1:Nsigma) for each spin and the dipole acceleration matrix element. The time evolution is done by evolve():

```
1        subroutine evolve(U,S0,t,ri,rgob,y,wy,Vks,Vks0,m0)
2        use gridvars
3        use cvars
4        implicit none
5        complex*16, intent(inout) :: U(Nb,0:lmax)
6        complex*16, intent(in) :: S0(Nb,Nb,0:lmax)
7        real*8 t, ri(Nb), rgob(Nb)
8        real*8, dimension(0:lmax) :: y, wy
9        real*8, dimension(Nb,0:lmax) :: Vks, Vks0
10       integer m0
```

```
11         ! Local variables
12         complex*16 U2(Nb,0:lmax)
13         integer j, l
14         real*8 Vt
15         real*8 :: E
16      ! exp(-iH0dt/2d0)
17      !
18         do l=0,lmax
19         call ZGEMM('N','N',Nb,1,Nb,one,S0(:,:,l),Nb,
20       &            U(:,l),Nb,zero,U2(:,l),Nb)
21         end do
22      ! exp(-i[Vks(t)-Vks(0)+E(t)*r]dt)
23      !
24         call trans(U2,Nb,1,wy,m0) ! Legendre Transform
25         do l=0,lmax
26         do j=1,Nb
27           Vt = Vks(j,l) - Vks0(j,l)  ! TDDFT
28           !Vt = 0d0                   ! frozen-core
29           Vt = Vt + E(t+dt/2d0)*ri(j)*y(l) ! y=cos(theta)
30           U2(j,l) = U2(j,l) * cdexp(-ai*Vt*dt)
31         end do
32         end do
33         call trans(U2,Nb,-1,wy,m0!Inverse Legendre Transform
34      ! exp(-iH0dt/2d0)
35      !
36         do l=0,lmax
37         call ZGEMM('N','N',Nb,1,Nb,one,S0(:,:,l),Nb,
38       &            U2(:,l),Nb,zero,U(:,l),Nb)
39         end do
40         ! multiply gobblers
41         U2(:,:) = U(:,:) !copy U to U2
42         do j=1,Nb
43           U(j,:) = U(j,:)*dcmplx(rgob(j),0.0d0)
44         end do
45         return
46         end subroutine evolve
```

The Fortran code for the HHG of many-electron atoms driven by a linearly-polarized laser field discussed in this section is given in the following webpage:
https://sites.google.com/site/mitsukomurakami02/codes/tddft.

3.3 Quantum chemistry tools

The quantum chemistry method follows a different path, starting from the Hartree-Fock method, configuration interaction (single excitation and double excitation), all the way to coupled cluster methods, with increase in accuracy and efforts.

3.3.1 Basis functions

We know from the solutions of the hydrogen atom that the eigenstates of the Hamiltonian are even or odd functions of $\mathbf{r}$: the s orbitals are even functions, the p orbitals are odd functions, the d orbitals are even functions, etc. In fact in quantum chemistry one mostly uses linear combinations of the so-called Gaussian primitives

$$g(l_x, l_y, l_z, \zeta) = x^{l_x} y^{l_y} z^{l_z} e^{-\zeta r^2}, \tag{3.50}$$

to represent the atomic orbitals

$$\begin{aligned}\chi &= \sum_j c_j g(l_x, l_y, l_z, \zeta_j) = x^{l_x} y^{l_y} z^{l_z} \sum_j c_j e^{-\zeta_j r^2} \\ &= x^{l_x} y^{l_y} z^{l_z} R(r)\end{aligned} \tag{3.51}$$

Here parameters ζ_j decide how much the wavefunction extends into space, and the coefficients c_j are called contraction coefficients. The function $R(r) = \sum_j c_j e^{-\zeta_j r^2}$ is radially symmetric, while the exponents l_x, l_y, and l_z describe the angular part of the wavefunction. We only combine primitive Gaussians with the *same* l_x, l_y, and l_z in every atomic orbital χ. So, for example, for $l_x = l_y = l_z = 0$ there is no angle dependence, and we get a spherically symmetric s orbital. For $l_x = 1$ and $l_y = l_z = 0$ we get a p_x orbital, while for $l_x = l_y = 1$ and $l_z = 0$ we get a d_{xy} orbital.

To ease our presentation, we use the Dirac notation, where $|\chi\rangle$ represents the same basis function (atomic orbital) as Eq. 3.51, $\langle\chi|$ is the conjugation, and $\langle\chi|\hat{O}|\chi'\rangle$ refers to an integration with two basis functions χ and χ' and an operator $\hat{O}$. Consider one of the most frequently used integrals, which describes the interaction of a quantum system with an electric field along the x-, y-, and z-directions respectively,

$$\langle\chi|\hat{\mathbf{r}}|\chi'\rangle = \begin{pmatrix} \iiint_{-\infty}^{\infty} x^{l_x+l'_x+1} y^{l_y+l'_y} z^{l_z+l'_z} R(r)R'(r)\, dx\, dy\, dz \\ \iiint_{-\infty}^{\infty} x^{l_x+l'_x} y^{l_y+l'_y+1} z^{l_z+l'_z} R(r)R'(r)\, dx\, dy\, dz \\ \iiint_{-\infty}^{\infty} x^{l_x+l'_x} y^{l_y+l'_y} z^{l_z+l'_z+1} R(r)R'(r)\, dx\, dy\, dz \end{pmatrix},$$

for the special case in which χ' and χ reside on the same atom. We see that it is nonzero if and only if at least one of the exponents $l_x + l'_x + 1$, $l_y + l'_y + 1$, or $l_z + l'_z + 1$ is an even number (since we integrate over the whole space). Hence, we derive the next rule. Only transitions between s and p, between p and d, between d and f orbitals, and so on, are allowed. In fact we can even see what light polarization is needed. For instance, with linearly polarized light along the x-direction, we can induce a transition between s and p_x, or between p_y and d_{xy}, but neither a transition between s and p_y nor between d_{xy} and d_{yz}. These rules (called selection rules) can be summarized as follows:

$$\Delta s = 0, \Delta m_s = 0, \Delta l = \pm 1, \Delta m_l = 0, \pm 1, \tag{3.52}$$

where s denotes the spin momentum quantum number, m_s denotes the magnetic spin quantum number, l denotes the orbital, and m_l denotes the magnetic orbital quantum number.

Although the selection rules are mathematically rigorous for the atomic case, we must caution that in molecules the molecular orbitals are linear combinations of s, p etc., orbitals, and therefore more transitions are possible. It is the subject of point group theory to study the selection rules in molecular systems, by exploiting their exact symmetry [McWeeny (1963); Bishop (1973)].

Without giving too much importance to the mathematical details we only mention the two physical prerequisites of these selection rules, which in fact are valid for the cases considered here. The one is that the intensity of the laser pulse is weak enough not to destroy the material or excite it strongly enough that single-photon counting is not necessary, and the other is that we consider time scales at which spontaneous emission is not relevant.

Finally, we should add that Gaussian primitives are only one kind of all possible basis functions and they are more frequently used in quantum chemistry. The real space grid mesh is also used, where one replaces the differential operators using a finite difference method. In this case single-particle wavefunctions are three-dimensional functions. Computer storage is the limiting factor as to how big a grid mesh one can have.

3.3.2 Hartree-Fock approximation

In typical quantum chemical calculations the wavefunction $|\Psi\rangle$ is represented in real space. Usually the calculations are performed at least in two steps. In the first step the Hamiltonian of the system is solved with the **Hartree-Fock (HF) method**, which starts with a set of atomic orbitals (AO) $|\chi_i(\mathbf{r})\rangle$ called a *basis set*, typically localized on each atom, which resemble the hydrogenic solutions. Here, $\mathbf{r}$ is the position vector of the electron. In other words they consist of a radial part $R(r)$ and an angular part $Y(\theta, \phi)$, giving them the form of the well-known s, p, d, etc. orbitals of the hydrogen atom (r is the magnitude of $\mathbf{r}$). The Hartree-Fock procedure yields a set of molecular orbitals (MO), that is, one-body wavefunctions, out of which some are occupied and some remain empty (usually called virtual orbitals). The MOs $|\psi_a(\mathbf{r})\rangle$ are expressed as linear combinations of AOs (LCAO),

$$|\psi_a(\mathbf{r})\rangle = \sum_i c_{ai}\,|\chi_i(\mathbf{r})\rangle. \tag{3.53}$$

So we get two important characteristics of quantum chemistry: delocalization and hybridization. The first character refers to the fact that the electron is not necessarily orbiting just one atom but can spread across several atoms, while the second one means that the orbital does not necessarily have the exact shape of an s, p, d, etc. hydrogenic orbital (similar to the multipole expansion of an electrostatic potential: any potential can be expressed as a linear combination of a monopole, a dipole, a quadrupole and so on). It is important that within the HF approximation, each electron feels the potential of the rest in a static, averaged way: we say that HF is a mean-field theory.

3.3.3 Configuration interaction method

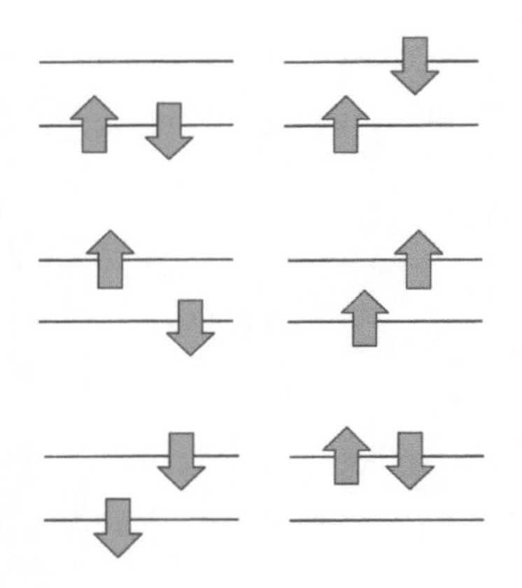

Figure 3.1
All possible configurations for two orbitals and two electrons.

A particular set of n occupied MOs is called a **configuration** (Fig. 3.1), and is a simple attempt to approximately describe a many-body electronic state $\Psi(\mathbf{r}_1, \mathbf{r}_2, \cdots, \mathbf{r}_N)$. Mathematically we put all the orbitals together into a so-called Slater determinant (which at least makes sure that the total wavefunction is antisymmetric, since the electrons are Fermions, and thus accounts for an important many-electron effect). A typical Hartree-Fock many-body wavefunction for N electrons in N number of orbitals is

$$\Psi_{\mathrm{HF}} = \frac{1}{\sqrt{N!}} \begin{vmatrix} \psi_1(\mathbf{r_1}) & |\psi_1(\mathbf{r_2}) & \cdots & \psi_1(\mathbf{r_N}) \\ \psi_2(\mathbf{r_1}) & |\psi_2(\mathbf{r_2}) & \cdots & \psi_2(\mathbf{r_N}) \\ \vdots & \vdots & \ddots & \vdots \\ |\psi_N(\mathbf{r_1}) & |\psi_N(\mathbf{r_2}) & \cdots & \psi_N(\mathbf{r_N}) \end{vmatrix}, \tag{3.54}$$

where Ψ_{HF} is an approximation to $\Psi(\mathbf{r}_1, \mathbf{r}_2, \cdots, \mathbf{r}_N)$.[2] We can also fill N electrons in M orbitals with M larger than N. Then we will have many more possible determinants or state configurations.

For instance, for two electrons with three orbitals ψ_a, ψ_b and ψ_c, we can get altogether three different configurations,

[2]Often when writing complicated wavefunctions, such as Slater determinants, it is customary to refrain from using the complete Dirac notation. In other words we simply write Ψ rather than $|\Psi\rangle$. Strictly speaking, $\Psi(\mathbf{r}) = \langle \mathbf{r}|\Psi\rangle$.

namely

$$\Psi_1 = \frac{1}{\sqrt{2!}} \begin{vmatrix} \psi_a(\mathbf{r_1}) & \psi_a(\mathbf{r_2}) \\ \psi_b(\mathbf{r_1}) & \psi_b(\mathbf{r_2}) \end{vmatrix}, \tag{3.55}$$

$$\Psi_2 = \frac{1}{\sqrt{2!}} \begin{vmatrix} \psi_b(\mathbf{r_1}) & \psi_b(\mathbf{r_2}) \\ \psi_c(\mathbf{r_1}) & \psi_c(\mathbf{r_2}) \end{vmatrix}, \tag{3.56}$$

$$\text{and } \Psi_3 = \frac{1}{\sqrt{2!}} \begin{vmatrix} \psi_a(\mathbf{r_1}) & \psi_a(\mathbf{r_2}) \\ \psi_c(\mathbf{r_1}) & \psi_c(\mathbf{r_2}) \end{vmatrix}. \tag{3.57}$$

If we populate starting from the energetically lowest MOs (called the *aufbau principle*), we get the HF ground state, which we call the reference state. Replacing one or more occupied ones with virtual orbitals[h] in Eq. 3.54 yields excited-state configurations. Each excited-state configuration physically corresponds to the excitation of one electron from an occupied to an empty state.

[h]Virtual orbitals refer to those orbitals that are unoccupied in the Hartree-Fock ground state. They are also called empty orbitals.

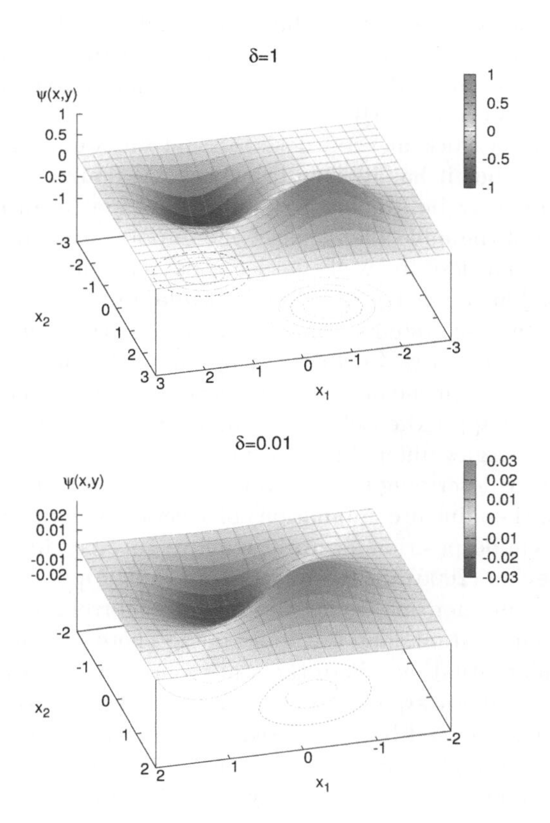

Figure 3.2
(Top) Two-electron wavefunction with the basis function centers separated by 2. The x-axis refers to the coordinate of electron 1, and the y-axis to that of electron 2. (Bottom) Two-electron wavefunction with the basis function centers separated by 0.02. The x-axis refers to the coordinate of electron 1, and y-axis refers to that of electron 2.

The Slater determinants are interesting. Consider two one-dimensional unnormalized basis functions, $\psi_a(x) = e^{-(x-\delta)^2}$ and $\psi_b(x) = e^{-(x+\delta)^2}$, centered at positions $+\delta$ and $-\delta$. The two-electron wavefunction is $\Psi(x_1, x_2) = [\psi_a(x_1)\psi_b(x_2) - \psi_b(x_1)\psi_a(x_2)]/\sqrt{2}$. Suppose $\delta = 1$ (dimensionless), so that the distance between the centers is 2. We then plot $\Psi(x_1, x_2)$ in the three-dimensional space in the top figure of Fig. 3.2, with the coordinate of electron 1 along the x-axis and that of electron 2 along the y-axis. We can see that the wavefunction has a peak at (-1,1) and a trough at (1,-1).

However, if we change δ to 0.01, the peak and the trough move closer as expected, but the entire magnitude of the wavefunction is sharply reduced (see the bottom of Fig. 3.2). This phenomenon is the manifestation of the Pauli exclusion principle due to the antisymmetry of the wavefunction. These two basis functions are not orthogonal to each other, but the physics does not change. This simple example demonstrates a key concept that there is an exchange-correlation hole around every electron, so another electron cannot occupy the same place of the electron. Missing the exchange hole leads to the self-interaction errors in DFT.

The Hartree-Fock method does take into account the exchange effect, but it has its own drawback. Although a single configuration may be a good first approximation to an electronic state of the molecule, it lacks one important characteristic: it does not describe well the electrons' tendency to avoid each other. This tendency is termed **correlations**,[i] and especially for magnetic systems it can represent a major contribution to the total energy. Correlations are divided into two categories: static, when the one-electron wavefunctions of the two electrons have approximately the same energy, and dynamic, when their energies differ. In quantum chemistry they are accounted for by describing the electronic state as a coherent superposition (i.e., linear combination) of several configurations. We call this superposition a **configuration interaction** (CI) method [Levine (2000); Szabo and Ostlund (1996)].

[i] Correlations are the tendency of the electrons in a many-body wavefunction to avoid each other due to the Pauli exclusion principle.

In most common molecules the dynamic correlations are by far more important than the static ones. There is, however, one particular situation where the static ones are much more important: magnetic systems. The reason is that magnetic systems are characterized by non-closed d (or f) shells, meaning that not all d (or f) orbitals are occupied. Since the d orbitals have approximately the same energy, the mixing of configurations with different d occupations corresponds per definition to static correlations which, unfortunately, are more difficult to describe. This difficulty, in fact, stems from the HF procedure itself. Remember that HF yields a set of occupied molecular orbitals and a set of virtual (empty) molecular orbitals.

The mathematical procedure of deriving those orbitals is that only the occupied ones are optimized. So if we have to accommodate only eight electrons in altogether ten d orbitals (which is the case for nickel), then we must face a rather arbitrary choice as to which ones to leave empty. Once selected, the occupied orbitals get optimized (their energy is lowered), while the virtual ones are left to their own fate, since they do not contribute to the HF energy. Thus we treat the d orbitals differently, although we shouldn't! The only way to treat them equally is to create a linear combination of all 45 (!) possible configurations (there are 45 (C_{10}^2) ways to put eight electrons into ten orbitals - there are ten orbitals, because each of the five d orbitals can be either spin-up or spin-down).

There are several computational procedures to tackle the correlation problem, ranging from simple ones, such as configuration interaction with single excitations (CIS) or with single and double excitations (CISD), up to complicated ones, in which all possible configurations are built (full CI).[j] Obviously, the more configurations accounted for, the more complicated the computation becomes; in fact, full CI is possible only for extremely small molecules. The common characteristic of these methods is that they describe the wavefunction as a linear combination of several configurations Ψ_i in the Dirac notation, the coefficients c_i of which are optimized: $\Psi_{\text{total}} = \sum_i c_i \Psi_i$.

[j] Typical state-of-the-art correlational methods for quantum chemistry include CI, CISD, CC, SAC-CI, EOM-CCSD, and CAS-SCF (see their definitions below).

Very often in quantum chemistry the name of the configuration is given relative to the Hartree-Fock configuration (typically named Ψ_{HF} or Ψ_0). The configuration Ψ_a^r means that an electron gets excited from the occupied molecular orbital a to the empty molecular orbital r. This is called a single virtual excitation. A double virtual excitation gives rise to the configuration Ψ_{ab}^{rs} (where $a \to r$ and $b \to s$), a triple to Ψ_{abc}^{rst} (where $a \to r$, $b \to s$ and $c \to t$), and so on. So a CIS wavefunction is

$$\Psi_{\text{CIS}} = c_0 \Psi_0 + \sum_{a,r} c_a^r \Psi_a^r, \tag{3.58}$$

and the CISD wavefunction, analogously is

$$\Psi_{\text{CISD}} = c_0 \Psi_0 + \sum_{a,r} c_a^r \Psi_a^r + \sum_{a,b,r,c} c_{ab}^{rs} \Psi_{ab}^{rs}. \tag{3.59}$$

Full CI means that *all* possible configurations are taken into account. Obviously, this can easily become a gigantic number of Slater determinants! In general, CI calculations, with the exception of full CI, are not size consistent, i.e., the energy of a many-particle system is not proportional to the number N of particles in the limit $N \to \infty$, as one would expect.

3.3.4 Coupled-cluster method

Another particularly interesting family of correlated calculations is the coupled-cluster (CC) method. Here the excitations mathematically are described by an exponential operator, which allows for a combination of multiple excitations to a single mathematical degree of freedom. If, as an example, we collectively denote as $\hat{R}_1$ all the single-virtual-excitation operators[3] and $\hat{R}_2$ all the double-virtual-excitation, then the CCSD wavefunction (coupled-cluster with single and double excitations) becomes

$$\begin{aligned}\Psi_{\text{CCSD}} &= e^{(\hat{R}_1+\hat{R}_2)}\,\Psi_0 \\ &= (1+\hat{R}_1+\hat{R}_2+\hat{R}_1^2+\hat{R}_2^2+\cdots)\Psi_0. \end{aligned} \tag{3.60}$$

The interesting aspect about the CC methods is that with lower excitation operators one achieves multiple excitations through the combination of the operators. So here the $\hat{R}_2^2$ term generates quadruple virtual excitations (of course, only a part of them), thus increasing the correlational level while at the same time keeping the calculation mathematically tractable. One additional, extremely important feature is the so-called *size consistency.* One can show that if the size of the system changes, the total energy changes in an almost linear way, thus making these methods suitable, among others, for dissociation and vibration analyses [Krylov (2008)].

This way it is possible to account for a higher correlational level with less computational effort. Here we find methods such as the symmetry-adapted-cluster configuration interaction (SAC-CI) [Nakatsuji (1979)], which additionally takes advantage of the molecule's symmetry, and the equation-of-motion coupled-cluster with single and double excitations (EOM-CCSD) [Krylov (2008)], well suited for the description of excited states.

One method which is well suited for static correlations (but limited only to a small number of excited states) is the complete-active-state self-consistent field (CAS-SCF) [Olsen (2011)]. The idea here is that although we restrict ourselves to a small number of configurations, we reoptimize the molecular orbitals for those configurations as well. This rectifies the above mentioned problem of the HF method of arbitrarily deciding which d orbitals to occupy and which to leave empty.

All in all, the quantum chemical calculation is tricky to carry out: one must manually identify the relevant configurations and run the correlational computation taking those into account. At the same time, their energy contributions must be estimated and a cut-off limit has to be established; otherwise the calculation quickly becomes numerically intractable. It can be helpful to visualize the MOs after a HF calculation and compare with some chemical intuition (which admittedly comes with time and experience)! A systematic study also includes inspection of several basis sets, which can be better suited for certain phenomena. Here again, too large a basis set is not a panacea, since it can lead to convergence problems.

---

[3] In second quantization, i.e., in terms of creation $a^\dagger$ and annihilation a operators, the single-excitation operator $\hat{R}_1 = a_\mu^\dagger a_\nu$ removes an electron from orbital ν and creates an electron in orbital μ.

3.4 Solid state physics: Essentials

Solids are among the most investigated materials within ultrafast phenomena. This requires some basic knowledge of solid state physics. However, it is difficult to review even a small portion of solid state physics here. The following materials serve as a working material for a novice, and the reader is encouraged to consult the solid state physics textbook such as [Kittel (1996); Ashcroft and Mermin (1976); Callaway (1976)]. The goal is to familiarize the reader with the basic knowledge that will be used in later chapters.

3.4.1 Crystal structure = Bravais lattice + basis

Solids contain lots of ions and electrons, and they come with different forms. Some are amorphous, some contain disorders and impurities, and some are crystallines. In this short introduction, we are only interested in a perfect crystal without any defects. We even ignore surfaces of solids. This approximation is not appropriate if we address surface properties.

A perfect crystal means that identical structural motifs repeat themselves infinitely. Therefore, we can introduce two crucial concepts for a crystal structure: (a) lattice (whose position is denoted by $\mathbf{T}$) and (b) basis which contains atoms (whose position is denoted by $\mathbf{p}$ referenced with the lattice).[k] In solid state physics, "lattice" is reserved for a unique grid of infinitely repeating motifs. There are only 14 types of distinctive and non-mutually transferable lattices, called Bravais lattices. These Bravais lattices have special names such as face-centered cubic lattice, which is why we purposely use "Bravais" here. Bravais lattices must catch the key features (such as symmetry of a crystal) of a structure. For this reason, conventional unit cells are used to represent Bravais lattices, instead of the primitive cell.[l] For instance, when we say that a solid has a face-centered cubic structure, we refer to its Bravais lattice, not its primitive cell.

[k] This basis has nothing to do with the basis functions in Section 3.3.

[l] Primitive cells contain only one lattice point.

"Lattice points" refer to those vertices on the lattice grid. All the lattice points must have the exact same environment as their other lattice points. This means that the angles with neighboring points are the same, the distances are the same, and the number of neighboring lattice points is the same. This ensures that a simple translation position vector[m] of lattice points $\mathbf{T}$ can carry one lattice point to another. This position vector must be integer multiples of primitive (minimum) lattice vectors $\mathbf{a}_1$, $\mathbf{a}_2$, and $\mathbf{a}_3$, or the sum of these integer multiple primitive lattice vectors, i. e., $\mathbf{T}(n_1, n_2, n_3) = n_1\mathbf{a}_1 + n_2\mathbf{a}_2 + n_3\mathbf{a}_3$, where n_1, n_2, and n_3 are integers. All the lattice points are linked with position vectors $\mathbf{T}$, and are completely equivalent when we view them from any directions. The space group symmetry of a lattice is determined from these lattice points (see below).

[m] The vectors are allowed to translate only. No other symmetry operations, such as inversion and mirror operations, are permitted.

Lattices provide a geometrical framework, somewhat abstract, but the basis adds more physical contents since it allows us to include ions, so we can have a real material. In other words, the basis is attached to lattice points. Since a basis can consist of a single atom or a group of atoms, we can describe any crystals. Therefore, a crystal structure = Bravais lattice (14 types) + basis (which may contain infinite types of atoms).

The concepts in solid state physics can be confusing in the beginning. How to identify a lattice structure? How to choose a basis? How to find a primitive lattice vector? There is no standard procedure to do so. Most of the time, one has to consult an expert. Fortunately, several very good tools are available nowadays to allow one to construct and view these structures. These include `xcrysden` and `vesta`. Another source is `Material projects`. `Youtube` also has some good videos on Bravais lattices.

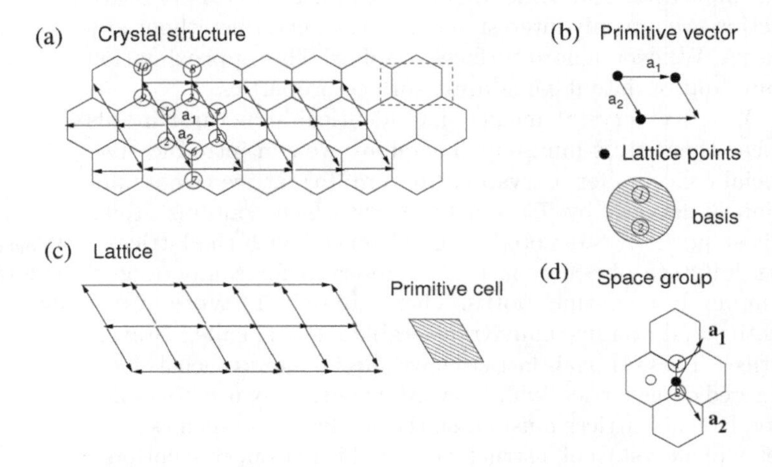

Figure 3.3
(a) Crystal structure of graphene. (b) Primitive vectors $\mathbf{a}_1$ and $\mathbf{a}_1$, lattice points, unit cell, and basis. (c) Lattice structure without basis. (d) To get a space group symmetry, one has to choose an appropriate origin of the cell and primitive lattice vectors. The origin must be at the center of hexagons.

In the following, we will make a moderate attempt to explain the key idea behind lattice and basis. It is not as simple as one might think. We take a graphene sheet as an example, which has a honeycomb structure.[n] Figure 3.3(a) shows a small portion of a graphene sheet. Each vertex around the hexagonal honeycombs is occupied by a carbon atom. The numbers attached to those vertices are added for easy presentation. These atoms help us figure out what the lattice is. Let's start with atom 1 in Fig. 3.3(a) to see whether it is qualified to be a basis that is attached to a lattice point. If the basis with atom 1 is a lattice point, then all the atoms should be reproduced from atom 1 by two lattice vectors (because we have a two-

[n] The honeycomb is not a Bravais lattice. The underlying lattice is a triangular Bravais lattice.

dimensional system). It is indeed true that atoms like 4 and 6, including their equivalent atoms, can be reproduced using two lattice vectors $\mathbf{a}_1$ and $\mathbf{a}_2$ (see Fig. 3.3(a)). However, there is no way to reproduce atoms 2 and 5, using $\mathbf{a}_1$ and $\mathbf{a}_2$.[o] Therefore, we can conclude that atom 1 alone cannot be a basis that can be attached to a lattice point. This also means the basis for our system cannot contain only one atom. The basis must include two atoms, but these two atoms are not any arbitrary two atoms, and they must be of different types such as 1 and 2, but not 1 and 4 except when one wants to build a supercell for some reason. Another way to see why the basis must contain two atoms is to note that atoms 1 and 2 have a different environment: Above atom 1, we have atoms 3 and 9, but above atom 2, we have atom 1 only. This equivalency is realized through the basis with two atoms that we choose. In other words, when we discuss the lattice of a crystal, how we choose a basis and how many atoms the basis consists of affect the lattice choice. Lattice and basis are intertwined.

[o]The reason why we are only allowed to use lattice vectors is because only the translational symmetry should be used to construct a lattice. No proper or improper (inversion and mirror) rotations are allowed when we construct a lattice.

If we agree to choose atoms 1 and 2 as our basis (Fig. 3.3(b)), then atom pairs such as (10,9), (8,3), (4,5), and (6,7) are all bases. If we draw arrows from one basis to another basis (see one example in Fig. 3.3(b)), these arrows are an example of lattice vectors.[p] Figure 3.3(b) reproduces these two primitive lattice vectors, $\mathbf{a}_1$ and $\mathbf{a}_2$. If we attach a basis to each lattice point, the crystal structure can be reproduced. Figure 3.3(c) removes the original honeycomb and only shows the lattice. The entire lattice can be reproduced by additions or subtraction of two primitive lattice vectors $\mathbf{a}_1$ and $\mathbf{a}_2$. We note in passing that if we take atoms 1 and 3 as our basis, then a new set of primitive lattice vectors must be chosen. Figure 3.3(c) also shows an example of the primitive cell, which is the minimum unit of the lattice and contains only one lattice point (each corner is shared by four neighboring cells). One is free to choose a different primitive cell, as far as the cell reflects the highest possible symmetry of the lattice.

[p]There are in principle infinitely possible ways to draw lattice vectors from one basis to another basis not necessarily between neighbors. However, these lattice vectors are not suitable for space group symmetry determination.

Once we find a lattice, then we can figure out its symmetry by checking whether a particular symmetry operation leaves the lattice unchanged. These operations form a group. Since the lattice obeys the spatial translational symmetry (space) and additional symmetry operations, this symmetry group is called a space group.[q] A proper choice of lattice vectors is essential to this. For instance, if one chooses four atoms or two lattice points (see the dashed box on the right side of Fig. 3.3(a)), then this cell contains more than one lattice point and is not a primitive cell. Since two lattice vectors have different lengths, a 60° rotation symmetry is artificially lost. Figure 3.3(b) has two primitive lattice vectors, $\mathbf{a}_1 = (a, 0, 0)$ and $\mathbf{a}_2 = (\frac{a}{2}, -\frac{\sqrt{3}a}{2}, 0)$, with the same length $|\mathbf{a}_1| = |\mathbf{a}_2|$, so this choice keeps the rotation symmetry of the crystal. Here a is the lattice constant[r] which is the side length of the Bravais lattice.

[q]In atoms and molecules, there is no such translation symmetry. We call thus a point group because the operation is carried out with respect to a point.

Next, we must determine the atom positions of a basis

[r]We do not use the primitive cell to define the lattice constant. There are three lattice constants in a 3-dimensional crystal.

within the primitive cell. Atom positions $\boldsymbol{\tau}$ are mapped onto primitive vectors to allow us to see the symmetry of these atoms. Because the atoms are situated within the basis, their position vectors are written in terms of primitive vectors with fractional multipliers,[s]

[s]Compare: Lattice points are integral multiples of primitive vectors.

$$\boldsymbol{\tau} = w_1\mathbf{a}_1 + w_2\mathbf{a}_2 + w_3\mathbf{a}_3, \tag{3.61}$$

where fractional numbers w_1, w_2, and w_3 are called Wyckoff positions.[t] To find the Wyckoff positions, we much choose the origin of the lattice. If we are not interested in space group symmetry, the choice of origin is arbitrary. But if we consider the space group, the choice is not arbitrary. Suppose we choose the middle point between atoms 1 and 2 as the origin; then atom 1 has the position vector $\boldsymbol{\tau}_1 = (0, \frac{b}{2}, 0)$, and atom 2 has $\boldsymbol{\tau}_2 = (0, -\frac{b}{2}, 0)$, where b is the side length of the hexagon and is equal to $a/\sqrt{3}$.

[t]The Wyckoff positions are often denoted by x, y, and z (see Table 3.2), but here to avoid confusion, w_1, w_2 and w_3 are used.

To find w_1, w_2, and w_3 from $\boldsymbol{\tau}$ is difficult if the primitive vectors $\mathbf{a}_1$, $\mathbf{a}_2$, and $\mathbf{a}_3$ are not orthogonal to each other, as in our case. Many textbooks do not have this information. Here we present a generic method to do this. We first dot-product Eq. 3.61 by $\mathbf{a}_1$, $\mathbf{a}_2$, and $\mathbf{a}_3$, and for atom 1 we get

$$\boldsymbol{\tau}_1 \cdot \mathbf{a}_1 = w_1\mathbf{a}_1 \cdot \mathbf{a}_1 + w_2\mathbf{a}_2 \cdot \mathbf{a}_1 + w_3\mathbf{a}_3 \cdot \mathbf{a}_1, \tag{3.62}$$

$$\boldsymbol{\tau}_1 \cdot \mathbf{a}_2 = w_1\mathbf{a}_1 \cdot \mathbf{a}_2 + w_2\mathbf{a}_2 \cdot \mathbf{a}_2 + w_3\mathbf{a}_3 \cdot \mathbf{a}_2, \tag{3.63}$$

$$\boldsymbol{\tau}_1 \cdot \mathbf{a}_3 = w_1\mathbf{a}_1 \cdot \mathbf{a}_3 + w_2\mathbf{a}_2 \cdot \mathbf{a}_3 + w_3\mathbf{a}_3 \cdot \mathbf{a}_3. \tag{3.64}$$

Once we know the atom positions and basis vectors, all the dot products are known. Solving these three equations gives us w_1, w_2 and w_3. For atom 1 we have $\boldsymbol{\tau}_1 = \frac{1}{6}\mathbf{a}_1 - \frac{1}{3}\mathbf{a}_2$, i.e., the Wyckoff position is (1/6, -1/3, 0), and for atom 2 we have $\boldsymbol{\tau}_2 = -\frac{1}{6}\mathbf{a}_1 + \frac{1}{3}\mathbf{a}_2$, (-1/6, 1/3,0) However, if we compare the above Wyckoff positions with those listed under group No. 191 space group $P6/mmm$ or No. 17 plane group $P6mm$ in the International Tables for Crystallography, we find the latter has positions at ($2b$) (1/3, 2/3, 0) and (2/3, 1/3, 0).[u] There are two reasons for this difference. (i) The International Tables for Crystallography chooses the origin at the center of the hexagon (see the empty circle in Fig. 3.3(d)), so two carbon atoms' positions in the basis are at $(a/2, -b/2, 0)$ and $(a/2, b/2, 0)$, where $b = a/\sqrt{3}$. (ii) The lattice vectors in the International Tables for Crystallography are illustrated in Fig. 3.3(d), where $\mathbf{a}_1 = (\frac{a}{2}, \frac{\sqrt{3}a}{2}, 0)$ and $\mathbf{a}_2 = (\frac{a}{2}, -\frac{\sqrt{3}a}{2}, 0)$. If one uses these lattice vectors, one can reproduce the same Wyckoff positions. The proof is left as an exercise.

[u]2 refers to the multiplicity, and b denotes the site symmetry with $3m.$, meaning threefold rotation and a mirror image. We set z to zero in these positions.

To reinforce the above message, we want to show the reader another example of how she/he starts to compute a solid. Most textbooks use some very simple examples to demonstrate the concept of Bravais lattices, but in research, one often encounters more complicated structures. We are going to use C_{60} solid as an example. You must first search the literature for its

Table 3.2
Solid C_{60} structure parameters. Lattice constant is $a = b = c = 14.255$Å [Dorset and McCourt (1994)] at room temperature due to the rapid spinning of C_{60}. No. 202 is the number listed in the International Tables of Crystallography.

Space Group	No. 202 $Fm\bar{3}$		
Atom	x	y	z
C_1	0.052	0	0.249
C_2	0.105	0.085	0.220
C_3	0.185	0.052	0.165

structure, and if you are lucky, you may find its space group symmetry. Table 3.2 shows the information that you need to start the investigation.

A single C_{60} molecule has icosahedral (I_h) point group symmetry,[v] with 120 symmetry operations. When they form solids, most of these operations are lost because solids must obey translational symmetry, and solid C_{60} has space group (No. 202, $Fm\bar{3}$) [David et al. (1991)]. Its basis at each lattice point has 60 carbon atoms whose Wyckoff positions are given below.

[v] All the symmetry operations such as rotation, reflection, and inversion are defined with respect to a single point. In C_{60}, this is the center of C_{60}. We change from point group symmetry for a single molecule to space group symmetry for a solid. No. 202 denotes the number in the International Tables for Crystallography. In total, there are 230 groups. F refers to face-centered cubic (FCC). The Bravais lattice for solid C_{60} is an FCC structure.

In principle one needs to find all 60 positions, but because of the space symmetry, we can generate them from only three coordinates. Carbon atom 1 C_1 (with multiplicity of 12) is at ($x = 0.04913$, $y = 0$, $z = 0.249$), C_2 (with multiplicity of 24) at ($x = 0.105$, $y = 0.085$, $z = 0.220$), and C_3 (with multiplicity of 24) at ($x = 0.185$, $y = 0.052$, $z = 0.165$). All the other carbon atom positions are generated using 24 symmetry operations in the $Fm\bar{3}m$ group such as

$$\begin{pmatrix} 0 & 0 & -1 \\ -1 & 0 & 0 \\ 0 & -1 & 0 \end{pmatrix}.$$

The sum of the multiplicities is 12+24+24=60. One can see the space group operation greatly simplifies calculations.

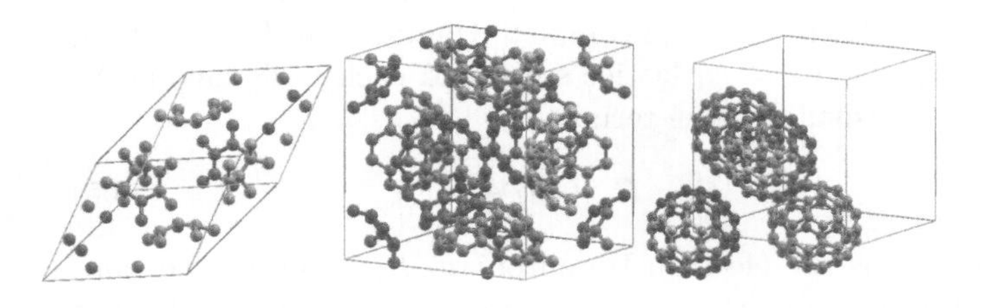

Figure 3.4
The same structure of solid C_{60} in a primitive cell (left), in a conventional fcc unit cell (middle), and in a displaced cell (right).

Figure 3.4 shows the primitive cell, conventional cell, and displaced structures. The primitive cell is the smallest cell, but it is hard to see each individual buckyball. When we adopt the conventional cell (four times larger), the FCC structure reveals itself. If we properly displace some of those atoms, we can recover those buckyballs. These three figures are generated from the same structure, but they show different perspectives.

3.4.2 Band structure: How electronic energy disperses with crystal momentum

It is necessary to introduce the band structure to appreciate why we introduce the lattice and basis in the first place.

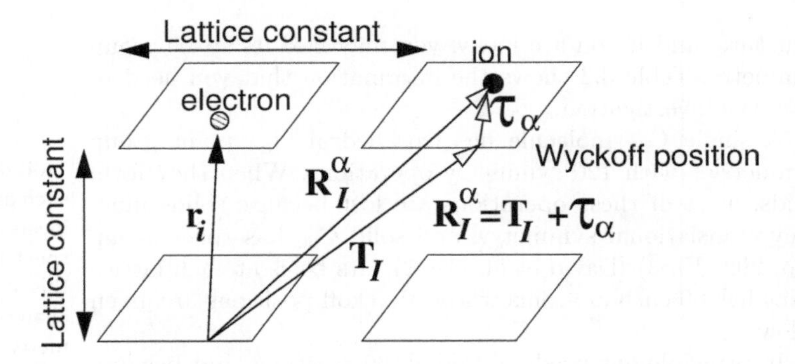

Figure 3.5
A crystal consists of electrons and ions. The position of the electron is denoted by $\mathbf{r}_i$. The ion's position $\mathbf{R}_I^\alpha = \mathbf{T}_I + \tau_\alpha$ has two indices, I and α, where I denotes a particular lattice point and α denotes the ion in the basis. $\mathbf{T}_I$ is the translation vector. τ_α is the position vector of the α ion, referenced with respect to a high symmetry point in the basis. This symmetry point is used to construct a specific space symmetry group. The Wyckoff positions are the coefficients when τ_α is projected onto three primitive vectors $\mathbf{a}_1$, $\mathbf{a}_2$, and $\mathbf{a}_3$.

Solids contain both electrons and ions. Figure 3.5 shows four primitive cells. $\mathbf{r}_i$ denotes the i-th electron position, and $\mathbf{R}_I^\alpha$ denotes the position of the α-th ion at the I-th lattice point where the basis sits. The complete Hamiltonian for a system, with N_e electrons, N_l lattice sites, and each basis having N_a atoms, consists of three terms,

$$H = H_e + H_l + H_{e-l}, \tag{3.65}$$

where H_e is the electron Hamiltonian, H_l is the lattice (ion) Hamiltonian, and H_{e-l} is the interaction between the electron and lattice. The electron Hamiltonian is

$$H_e = \sum_{i=1}^{N_e} \frac{\mathbf{p}_i^2}{2m_e} + \frac{1}{2}\sum_{i,j=1}^{N_e} V^{e-e}(\mathbf{r}_i - \mathbf{r}_j), \tag{3.66}$$

where the first term is the kinetic energy of the electron and is summed over all the electrons, and the second term is the electron-electron interactions and the summation is taken over the pairs of electrons. 1/2 is used to avoid double counting. The electron Hamiltonian alone is the same for any system. The periodicity of lattice is not present. This feature is employed to develop the density functional theory, where the most complicated interaction (electron-electron interaction) term does not depend on the lattice directly.

The lattice Hamiltonian of ions is

$$H_l = \underbrace{\sum_{I=1}^{N_l}}_{\text{lattice}} \underbrace{\sum_{\alpha^I=1}^{N_a}}_{\text{basis}} \frac{\mathbf{P}_{\alpha^I}^2}{2M_{\alpha^I}} + \frac{1}{2} \underbrace{\sum_{I,J=1}^{N_l}}_{\text{lattice}} \underbrace{\sum_{\alpha^I,\beta^J=1}^{N_a}}_{\text{basis}} V^{\text{ion}-\text{ion}}(\mathbf{R}_I^{\alpha^I} - \mathbf{R}_J^{\beta^J}), \tag{3.67}$$

where the first term is the kinetic energy of an ion α in a basis situated at lattice site I. Thus the ion has two indices (I and α). Its momentum is $\mathbf{P}_{\alpha^I}$ and the mass is M_{α^I}. The summation over I runs from 1 to the total number of lattice sites in the system N_l, while the summation over α^I runs from 1 to N_a (the total number of ions in a basis). N_a does not depend on a particular lattice site since all the bases have the exact same number of atoms and the same type of atoms with the same coordinates. This explains why when we choose lattice points, we must make sure these lattice points are exactly equivalent. This is a direct manifestation of the concept of crystal structure: lattice+basis. In Eq. 3.67, the second term, $V^{\text{ion}-\text{ion}}$, is the repulsion between ions. Note that each ion in a basis must have two indices. The interaction can be between one ion at lattice site I and another ion at lattice site J, or between ions at the same lattice site in which case $I = J$.

The electron-lattice interaction Hamiltonian is

$$H_{e-l} = \underbrace{\sum_{I=1}^{N_l}}_{\text{lattice}} \underbrace{\sum_{\alpha^I=1}^{N_\alpha}}_{\text{basis}} \underbrace{\sum_{i=1}^{N_e}}_{\text{electron}} V^{\text{electron}-\text{ion}}(\mathbf{r}_i - \mathbf{R}_I^{\alpha^I}), \tag{3.68}$$

where $V^{\text{electron}-\text{ion}}$ is the electron-ion interaction. If we directly solve the eigenvalue problem for the combined Hamiltonian, the dimension grows with the lattice size N_l, even if we completely ignore the degrees of freedom of electrons. So the key to solve this problem is to use the lattice periodicity as illustrated in Eqs. 3.67 and 3.68, because ions at lattice site I and at site J are exactly the same. The only exception is at the beginning or at the end of a crystal where the repetition of lattices ends. In real materials, N_l is extremely large, and one often takes the limit to infinity. Or one can adopt a periodic boundary condition, where the first lattice site is connected to the last lattice site. In the one-dimensional case, this is just a ring.

This is where the crystal momentum[4] comes in. We work in the crystal momentum space, or the reciprocal lattice space $\mathbf{k}$. The translational symmetry allows us to solve the problem for one $\mathbf{k}$ at a time, instead of solving the problem for total N_l (close to infinity) lattice sites.[5] For a three-dimensional system, $N_l = N_1 \times N_2 \times N_3$, and the system dimension is defined by three vectors $N_1\mathbf{a}_1$, $N_2\mathbf{a}_2$, and $N_3\mathbf{a}_3$, where N_1 is the number of lattice points (or primitive cells) along the primitive vector $\mathbf{a}_1$ and the same for others (Fig. 3.7).

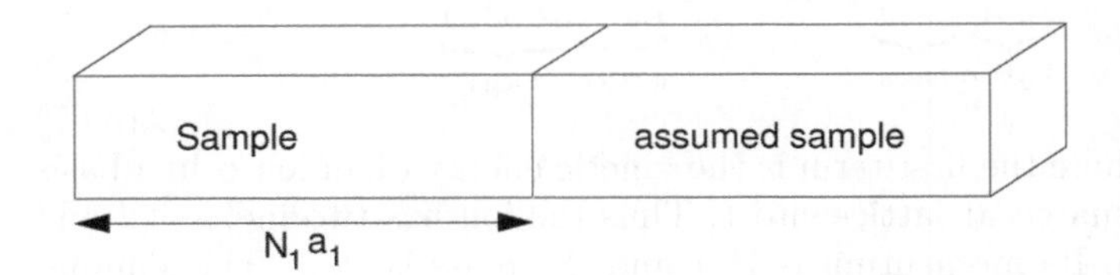

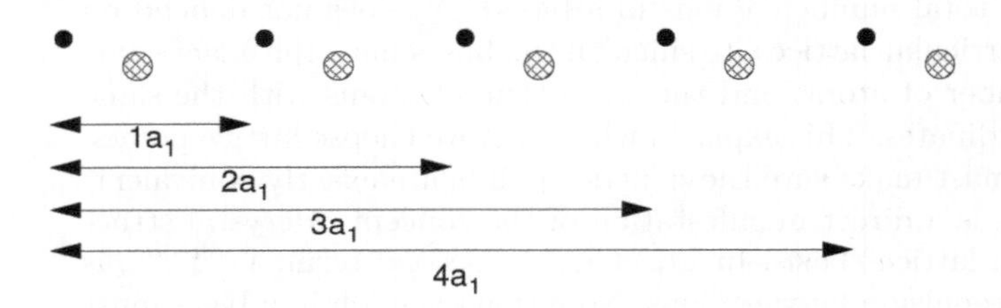

Figure 3.6
(a) The Born-von Karman boundary condition is applied to the wavefunction even if the sample has no translational symmetry. The allowed wavelength is the length of the sample, in this case N_1a_1. The crystal momentum has one point, $2\pi/(N_1a_1)$. (b) Once the translational symmetry exists, there are more crystal momentum points allowed. In the figure, they are $2\pi/a_1$, $2\pi/(2a_1)$, $2\pi/(3a_1)$, and $2\pi/(4a_1)$.

To find out what value $\mathbf{k}$ takes, we impose the Born-von Karman boundary condition to the electron wave function $\psi(\mathbf{r})$ at location $\mathbf{r}$. Here $\mathbf{r}$ is the electron coordinates, not to be confused with the ion positions above. The boundary condition is

$$\psi(\mathbf{r} + N_i\mathbf{a}_i) = \psi(\mathbf{r}), \qquad i = 1, 2, 3, \qquad (3.69)$$

[4]In free space, crystal momentum is the same as the momentum of an electron, but in solids the crystal momentum and the momentum of an electron are not the same thing. This is because the wavefunction in solids consists of two parts, the plane wave part and the periodic wavefunction part. The crystal momentum $\mathbf{k}$ is a wavevector of a wave.

[5]We solve N_K small $M \times M$ matrices, instead of a large matrix of $(N_kM) \times (N_kM)$, where N_k is the number of $\mathbf{k}$ points and M is the number of basis functions used for each unit cell.

which says that the wavefunction repeats itself (see Fig. 3.6(a)) with a period of $N_1\mathbf{a}_1$, $N_2\mathbf{a}_2$, and $N_3\mathbf{a}_3$, This periodicity has nothing to do with the translational symmetry of the lattices in the sample. We assume that a fictitious sample follows the sample repetitively. In other words, even if the sample has no translational symmetry, we still can write down the same equation, but one has an important restriction on $\mathbf{k}$. Take a one-dimensional material (which has no translational symmetry) as an example. The length along the x-axis is N_1a_1, so the wavelength λ of this wave, which satisfies the Born-von Karman boundary condition, is N_1a_1. As a result, the wavevector, which is $2\pi/\lambda$, is $2\pi/N_1a_1$. This is the only value that k_1 takes, just one point. This explains why if one carries out the band structure calculation for a system such as a water molecule or DNA chain that does not have a translational symmetry, one needs one $\mathbf{k}$ point. In this case, the primitive cell is the entire system itself.

If a crystal has translational symmetry, then there will be many more wavelengths that match the above boundary condition, which results in many more $\mathbf{k}$ points. Figure 3.6(b) shows such an example with four repeating units, so we have four wavelengths and four wavevectors, $2\pi/a_1$, $2\pi/(2a_1)$, $2\pi/(3a_1)$, and $2\pi/(4a_1)$.

We can make this more quantitative. According to the Bloch theorem, the electron wavefunction of a solid is a Bloch function, which is a product of a plane wave and a periodic function,

$$\psi_{n\mathbf{k}}(\mathbf{r}) = e^{i\mathbf{k}\cdot\mathbf{r}}u_{n\mathbf{k}}(\mathbf{r}), \tag{3.70}$$

where n is the band index[w] and $u_{n\mathbf{k}}(\mathbf{r})$ is a periodic function, i.e., $u_{n\mathbf{k}}(\mathbf{r}+\mathbf{T}) = u_{n\mathbf{k}}(\mathbf{r})$. Here the generic lattice translational vector is $\mathbf{T} = j_1\mathbf{a}_1 + j_2\mathbf{a}_2 + j_3\mathbf{a}_3$, where j_1, j_2, and j_3 are integers. $e^{i\mathbf{k}\cdot\mathbf{r}}$ is an eigenstate of translational symmetry, and in general it is not a periodic function of $\mathbf{a}_i$ for an arbitrary $\mathbf{k}$, i.e., $e^{i\mathbf{k}\cdot(\mathbf{r}+\mathbf{j}_\mathbf{i}\mathbf{a}_\mathbf{i})} \neq e^{i\mathbf{k}\cdot\mathbf{r}}$. Instead, $e^{i\mathbf{k}\cdot j_i\mathbf{a}_i}$ gives a phase factor at lattice site $j_i\mathbf{a}_i$.

[w] Atoms have many energy levels, such as $1s$, $2s$ $2p$, $\cdots$. When they form a solid, orbitals overlap and form bands, in particular valence orbitals. n is used to denote those bands states.

If we apply the Born-von Karman boundary condition to the Bloch wave function, we have

$$\psi_{nk}(\mathbf{r} + N_i\mathbf{a}_i) = e^{i\mathbf{k}\cdot(\mathbf{r}+N_i\mathbf{a}_i)}u_{nk}(\mathbf{r} + N_i\mathbf{a}_i) \tag{3.71}$$

$$= e^{i\mathbf{k}\cdot(\mathbf{r}+N_i\mathbf{a}_i)}u_{nk}(\mathbf{r}) = \psi_{nk}(\mathbf{r})e^{iN_i\mathbf{k}\cdot\mathbf{a}_i} \tag{3.72}$$

$$= \psi_{nk}(\mathbf{r}). \tag{3.73}$$

Equation 3.73 uses the Born-von Karman boundary condition. Comparing Eqs. 3.72 and 3.73 leads to

$$N_i\mathbf{k}\cdot\mathbf{a}_i = N_ik_ia_i = 2\pi m_i, \rightarrow k_i = \frac{2\pi}{a_i}\frac{m_i}{N_i} = b_i\frac{m_i}{N_i}, \tag{3.74}$$

where b_i represents the primitive reciprocal lattice vectors. In general, they can be computed from

$$\mathbf{b}_i = 2\pi\frac{\mathbf{a}_j \times \mathbf{a}_k}{\mathbf{a}_1\cdot(\mathbf{a}_2 \times \mathbf{a}_3)}, \tag{3.75}$$

where i, j, k are cyclic, with 1, 2, and 3. Here $m_i = 1, 2, 3, ..., N_i$. The total number of $\mathbf{k}$ points along $\mathbf{b}_i$ is N_i, which is exactly the same as the number of lattice sites along $\mathbf{a}_i$. Note that m_i cannot be zero, because this would imply an infinitely long wavelength. $\mathbf{k}$ can approach to zero only through N_i (if N_i goes to infinity), but not through m_i. If we consider a three-dimensional system, the total number of $\mathbf{k}$ points is $N_1 N_2 N_3$, or N_l.

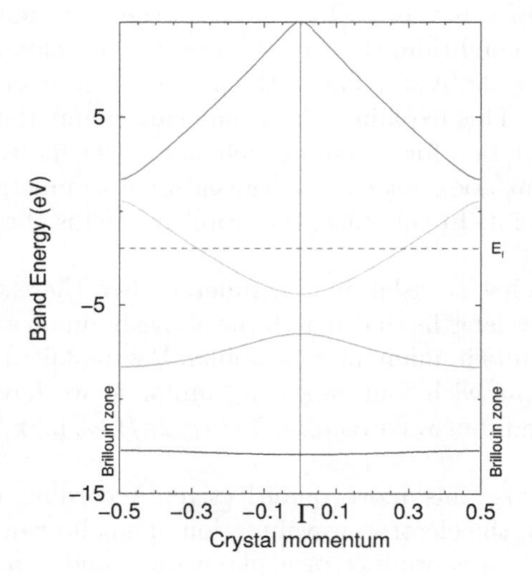

Figure 3.7
Band structure of the Kronig-Penney model for the first Brillouin zone. The Γ point at $\mathbf{k} = 0$ is added to show the location in a band structure.

Equation 3.74 shows the largest crystal momentum is $\mathbf{k}_i = \mathbf{b}_i$, while the minimum crystal momentum is $\mathbf{k}_i^{min} = \mathbf{b}_i / N_i$. The minimum volume formed by three minimum crystal momenta is

$$V_{\text{min}}^{\text{reciprocal}} = \mathbf{k}_1^{\text{min}} \cdot (\mathbf{k}_2^{\text{min}} \times \mathbf{k}_3^{\text{min}}) = \frac{1}{N_1 N_2 N_3} \mathbf{b}_1 \cdot (\mathbf{b}_2 \times \mathbf{b}_3), \tag{3.76}$$

or

$$V_{\text{min}}^{\text{reciprocal}} = (2\pi)^3 / V_{\text{sample}}, \tag{3.77}$$

where V_{sample} is the volume of the sample. Here all the $\mathbf{k}$ are positive. Using positive $\mathbf{k}$ has an advantage as it allows us to compute the wavelength by $\lambda = 2\pi/k$. However, in some cases, a negative $\mathbf{k}$ is also useful as it represents a wave traveling along a negative direction. For instance, if we discuss the band gap opening at the Brillouin zone,[6] it is useful to introduce

[6]Brillouin zones are a region in reciprocal lattice spaces that is enclosed by planes that are perpendicular to the reciprocal lattice vectors and bisect the reciprocal lattice vectors. In the one-dimensional case, the first

both the positive and negative $\mathbf{k}$, so a simple perturbation treatment can yield an energy gap [Kittel (1996)].

Figure 3.7 shows an example of a band structure for the Kronig-Penney model. The horizontal dotted line denotes the Fermi energy. For real materials, one often uses first-principles methods. The most popular one is to solve the Kohn-Sham equation for a solid. At the end of this chapter, we list some of the popular codes.

In the one-dimensional case, the coordinates of the special $\mathbf{k}$ point are easy, but in a three-dimensional system, the coordinates of special $\mathbf{k}$ points, with special designations, are not given. Many solid state physics books do not provide their coordinates. Only two books [Lax (1974); Meijer and Bauer (1962)] that we are aware of have all fourteen Bravais lattices labeled with special symbols, but only one book [Lax (1974)] has the actual coordinates. For convenience, we reproduce two tables in the Appendices (see A.11).

3.5 Two special features in ultrafast dynamics

To investigate ultrafast dynamics, there are several features that need special attention. One is spin-orbit coupling, and the other is the interaction between laser radiation and matter.

3.5.1 Spin-orbit coupling

Spin-orbit coupling (SOC) is an important interaction in ultrafast phenomena, and is directly responsible for intersystem crossing, spin-orbit torques in spintronics and demagnetization. SOC is a relativistic effect, and microscopically can be derived from the Dirac equation,

$$i\hbar \sum_{\mu=0,1,2,3} \gamma^{\mu}\partial_{\mu}\Psi - mc\Psi = 0, \tag{3.78}$$

where Ψ is an eigenfunction of an electron with four components (not to be confused with the above many-body electron wavefunction). c is the speed of light in vacuum, m is the mass of the electron, and $\partial_0 = \frac{\partial}{c\partial t}$, $\partial_1 = \frac{\partial}{\partial x}$, $\partial_2 = \frac{\partial}{\partial y}$, and $\partial_3 = \frac{\partial}{\partial z}$, where t is the time, x, y, and z are Cartesian coordinates. γ's are 4×4 dimensionless matrices, given as

$$\begin{pmatrix} I_2 & 0 \\ 0 & -I_2 \end{pmatrix}, \begin{pmatrix} 0 & \sigma_x \\ -\sigma_x & 0 \end{pmatrix}, \begin{pmatrix} 0 & \sigma_y \\ -\sigma_y & 0 \end{pmatrix}, \begin{pmatrix} 0 & \sigma_z \\ -\sigma_z & 0 \end{pmatrix}. \tag{3.79}$$

Brillouin zone is between $-b_1/2$ and $b_1/2$. These Brillouin zones have a special role in solid state physics as they represent the strongest Bragg reflections on electrons by the crystal lattice, so the electrons cannot survive there. A band gap is open.

Here σ_x, σ_y, and σ_z are 2×2 Pauli matrices (see Sec. 3.6), and I_2 is

$$\begin{pmatrix} 1 & 0 \\ 0 & 1 \end{pmatrix}, \tag{3.80}$$

so the entire matrix is a 4×4 matrix.

The above Dirac equation is free of any external field. If we introduce a crystal potential to the Dirac equation, and if we then take a non-relativistic limit for the Dirac equation, we have a term $H_{\rm SOC}$

$$H_{\rm SOC} = \frac{\hbar}{4m^2c^2}(\nabla V \times \frac{\hbar}{i}\nabla) \cdot \sigma, \tag{3.81}$$

where V is the crystal potential. This is the spin-orbit coupling. Its most interesting feature for our relevant first-row transition metals (we basically study Fe, Co, and Ni) is that it couples a spatial degree of freedom (crystal potential) to the otherwise isotropic spin degree of freedom. Hence, in combination with the exchange interaction, it yields a preferential magnetization axis and the magneto-optical Kerr effect (MOKE). Furthermore, it allows for femtosecond, laser-induced spin dynamics, because it indirectly couples the spins to the electric field of a laser. Since we are dealing with rather light atoms, we restrict ourselves to the one-electron contributions of the spin-orbit-coupling operator, while the two-electron contributions are accounted for via an effective nuclear charge ($Z_a^{\rm eff}$) [Koseki et al. (1998)]:

$$\hat{H} = \sum_{i=1}^{N_{\rm el}} \frac{Z_a^{\rm eff}}{2c^2 R_i^3} \hat{\mathbf{L}} \cdot \hat{\mathbf{S}}, \tag{3.82}$$

R_i is the position vector of the i-th electron, and $\hat{\mathbf{L}}$ and $\hat{\mathbf{S}}$ are the angular[x] and the spin orbital momentum operators. If we compare Eqs. 3.81 and 3.82, we see that the terms in parentheses in Eq. 3.81 become $\hat{\mathbf{L}}$. This is only possible if one has an atomic system with spherical symmetry. The proof is left as an exercise.

[x] This is only possible if we have a spherical potential.

3.5.2 Interaction between laser radiation and matter

Before we continue, let us shortly review how one calculates the transition matrix elements. We start from the so-called minimal coupling Hamiltonian, in which we substitute the operator of the canonical momentum of a particle in a vector potential $\mathbf{A}$ for the simple mechanical momentum $\mathbf{p} \mapsto \mathbf{p} - e\mathbf{A}$. We work in standard units (SI) (if we use Gaussian units, we would have $\hbar = 1$ in atomic units) in the Coulomb gauge ($\mathbf{A} \cdot \mathbf{p} = \mathbf{p} \cdot \mathbf{A}$). The Hamiltonian of the free electron with mass m_e and charge $-e$ becomes

$$\hat{H} = \frac{1}{2m_e}(\mathbf{p} - e\mathbf{A})^2 = \frac{\mathbf{p}^2}{2m_e} - \frac{e\mathbf{p} \cdot \mathbf{A}}{m_e} + \frac{e^2\mathbf{A}^2}{2m_e}. \tag{3.83}$$

Out of the three resulting terms, the first one is of the unperturbed system, the second one oscillates with frequency ω, while the third one oscillates with frequency 2ω and can thus be neglected. If we assume a sinusoidal vector potential $\mathbf{A} = \mathbf{A}_0 e^{-i(\frac{\omega}{c}\mathbf{n}\cdot\mathbf{r}-\omega t)}$ and perform a Taylor expansion of it up to the linear term, we can derive the transition matrix elements of the perturbative Hamiltonian H' between states $|\alpha\rangle$ and $|\beta\rangle$.

$$H'_{\alpha\beta} = \langle\alpha|\hat{H}'|\beta\rangle = -\frac{e}{m_e}\langle\alpha|\mathbf{p}\cdot\mathbf{A}|\beta\rangle + \underbrace{\frac{e^2}{2m_e}\langle\alpha|\mathbf{A}^2|\beta\rangle}_{\approx 0}\,. \quad (3.84)$$

The vectorial electric-dipole transition matrix element is

$$\mathbf{d}_{\alpha\beta} = \langle\alpha|\mathbf{r}|\beta\rangle. \quad (3.85)$$

The vectorial magnetic-dipole transition matrix element is

$$\mathbf{L}_{\alpha\beta} = \langle\alpha|p_a r_n - p_n r_a|\beta\rangle, \quad (3.86)$$

and the rank-two-tensorial electric-quadrupole transition matrix element is

$$\overset{\leftrightarrow}{Q}_{\alpha\beta} = \langle\alpha|p_a r_n + p_n r_a|\beta\rangle. \quad (3.87)$$

Here p_a and r_n are abbreviations for the projections of $\mathbf{p}$ on $\mathbf{a}$ and of $\mathbf{r}$ on $\mathbf{n}$, respectively (see Appendix A.5). The tensorial product in the perturbation element of the electric dipole is defined as

$$\overset{\leftrightarrow}{Q}_{\alpha\beta}\!:\!(\nabla\mathbf{E}) = \sum_{i=x,y,z}\sum_{j=x,y,z}\left(\overset{\leftrightarrow}{Q}_{\alpha\beta}\right)_{i,j}\frac{\partial E_i}{\partial r_j}. \quad (3.88)$$

The details are given in Appendix A.5.

For the actual calculation and depending on our preference, we can either compute the integrals $\langle i|\hat{\mathbf{p}}|j\rangle$ directly or use the position operator $\hat{\mathbf{r}}$ in combination with the commutation relation $[\hat{\mathbf{r}}, \hat{H}] = \frac{i\hbar}{m}\hat{\mathbf{p}}$, yielding $\langle i|\hat{\mathbf{p}}|j\rangle = im\,(E_i - E_j)\,\langle i|\hat{\mathbf{r}}|j\rangle/\hbar$, if $|i\rangle$ and $|j\rangle$ are eigenstates of the unperturbed Hamiltonian. Here $\hbar$ is the reduced Planck constant, $h/2\pi$.

One final aspect, which also comes out of this derivation, is the importance the relative magnitude of the transition-matrix elements. The electric-dipole transition element contributes to the zero-th order term of the Taylor expansion, and hence it is the most important one. The magnetic-dipole and electric-quadrupole *both* stem from the linear term of the expansion and are therefore of comparable importance. In other words, strictly speaking taking magnetic-dipole transitions into account should be always be done *together* with the electric-quadrupole transitions.

3.5.3 Further notes on the vector potential

Ultrafast phenomena introduce some special features to vector potentials. Because they are auxiliary fields, whether the results depend on our choice of these fields should be tested. We mention two features that are not often covered in standard textbooks: one is the spatial feature and the other is the temporal feature.

According to the Maxwell equations, in terms of the scalar potential V and vector potential $\mathbf{A}$, the electric field $\mathbf{E}$ and the magnetic field $\mathbf{B}$ can be written as

$$\mathbf{E} = -\nabla V - \frac{\partial \mathbf{A}}{\partial t}, \tag{3.89}$$

$$\mathbf{B} = \nabla \times \mathbf{A}. \tag{3.90}$$

If we take the dipole approximation, we only keep the leading spatial term of $\mathbf{A}(\mathbf{r},t) = \mathbf{A}_0 e^{i(\mathbf{k}\cdot\mathbf{r}-\omega t)}$, i.e., $\mathbf{A}(\mathbf{r},t) = \mathbf{A}_0 e^{-i\omega t}$. According to Eq. 3.90, $\mathbf{B} = 0$. This means that the above dipole approximation zeros out the magnetic field. Because for radiation the ratio of the magnitude of the electric field to that of the magnetic field is $|\mathbf{E}|/|\mathbf{B}| = c$, the speed of light c goes to infinity. For this reason, the dipole approximation to the vector potential should be used with great caution.[y]

[y] Within the dipole approximation, an electron simply follows the electric field and does not precess as otherwise one would expect from the Lorentz force $-e\mathbf{E} - e\mathbf{v} \times \mathbf{B}$, where $\mathbf{v}$ is the velocity of the electron.

Secondly, to simulate an ultrafast laser pulse, one often uses an envelope function for the electric field such as

$$\mathbf{E}(\mathbf{r},t) = \mathbf{E}_0 \underbrace{e^{-t^2/\tau^2}}_{\text{envelope}} e^{i(\mathbf{k}\cdot\mathbf{r}-\omega t)}, \tag{3.91}$$

where $\mathbf{E}_0$ denotes the magnitude and direction of the field. To see this clearly, we assume $V = 0$. If we directly substitute this into Eq. 3.89, $\mathbf{A}$ becomes

$$\mathbf{A}(\mathbf{r},t) = -\int_{-\infty}^{t} \mathbf{E}(\mathbf{r},t')dt'. \tag{3.92}$$

However, this $\mathbf{A}(t)$ is incorrect. The magnetic field computed from $\mathbf{A}(t)$ through Eq. 3.90 does not match the electric field's time profile in $|\mathbf{E}|/|\mathbf{B}| = c$. This in addition ignores the fact of how the pulse laser is generated. From Chapter 2, we know that a laser pulse consists of a series of planewaves whose phases are locked (see Sec. 2.4.3). Therefore, the vector potential should have exactly the same envelope as the electric field. And its amplitude $|\mathbf{A}_0|$ is $|\mathbf{E}_0|/\omega$,

$$\mathbf{A}(\mathbf{r},t) = \mathbf{A}_0 e^{-t^2/\tau^2} e^{i(\mathbf{k}\cdot\mathbf{r}-\omega t)}. \tag{3.93}$$

Then the magnetic and electric fields have the same time dependence.

3.6 Rotation matrices for spins

To rotate a regular polar vector **A**, one simply multiplies **A** by a rotation matrix by $\mathcal{R}$, i.e., $\mathcal{R}\mathbf{A}$, where $\mathcal{R}$ is the rotation matrix. One sees that three components of **A** are transformed simultaneously. Spin is an axial vector, so its rotation matrices are quite different from those for polar vectors. It belongs to a special unitary group with degree 2, or SU(2). Its three Pauli matrices

$$\sigma_x = \begin{pmatrix} 0 & 1 \\ 1 & 0 \end{pmatrix}, \sigma_y = \begin{pmatrix} 0 & -i \\ i & 0 \end{pmatrix}, \sigma_z = \begin{pmatrix} 1 & 0 \\ 0 & -1 \end{pmatrix} \quad (3.94)$$

represent three components of spin σ along the x-, y-, and z-axes, respectively. This greatly increases the complexity of transformation. For spin, one has to multiply each component by 2×2 matrices U as $U\sigma U^+$. If we use the Euler angles α, β, and γ,[z] then the $U(\alpha, \beta, \gamma)$ matrix is [Xie et al. (1984)]

[z] Eulerian angles are defined as follows: One first rotates an object around the z-axis by γ, then rotates around the y-axis by β, and finally rotates around the z-axis by α.

$$U(\alpha, \beta, \gamma) = \begin{pmatrix} e^{-i(\alpha+\gamma)/2}\cos(\beta/2) & -e^{i(\gamma-\alpha)/2}\sin(\beta/2) \\ e^{-i(\gamma-\alpha)/2}\sin(\beta/2) & e^{i(\alpha+\gamma)/2}\cos(\beta/2) \end{pmatrix}. \quad (3.95)$$

(Many textbooks end here.) If we carry out a complete investigation for each component, then we find that the transformed Pauli matrices are simply the linear combination of the original Pauli matrices. For σ_z, it becomes

$$\sigma_z' = (\cos\alpha \sin\beta)\sigma_x + (\sin\alpha \sin\beta)\sigma_y + (\cos\beta)\sigma_z. \quad (3.96)$$

The details of derivation are left as a homework assignment.

For the same three Eulerian angles (α, β, γ), we have three corresponding rotation matrices for the spatial rotation.

$$R_z(\gamma) = \begin{pmatrix} \cos\gamma & -\sin\gamma & 0 \\ \sin\gamma & \cos\gamma & 0 \\ 0 & 0 & 1 \end{pmatrix}, \quad (3.97)$$

$$R_y(\beta) = \begin{pmatrix} \cos\beta & \sin\beta & 0 \\ 0 & 1 & 0 \\ -\sin\beta & \cos\beta & 0 \end{pmatrix}, \quad (3.98)$$

$$R_z(\alpha) = \begin{pmatrix} \cos\alpha & -\sin\alpha & 0 \\ \sin\alpha & \cos\alpha & 0 \\ 0 & 0 & 1 \end{pmatrix}. \quad (3.99)$$

We compare the coefficients in Eq. 3.96 with the rotation matrix $\mathcal{R}$ for the same Eulerian angles, α, β, and γ,

$$\mathcal{R}(\alpha, \beta, \gamma) = R_z(\alpha)R_y(\beta)R_z(\gamma) = \begin{pmatrix} \mathcal{R}_{11} & \mathcal{R}_{12} & \mathcal{R}_{13} \\ \mathcal{R}_{21} & \mathcal{R}_{22} & \mathcal{R}_{23} \\ \mathcal{R}_{31} & \mathcal{R}_{32} & \mathcal{R}_{33} \end{pmatrix}, \quad (3.100)$$

where $\mathcal{R}_{11} = \cos\alpha\cos\beta\cos\gamma - \sin\alpha\sin\gamma$, $\mathcal{R}_{12} = -\cos\alpha\cos\beta\sin\gamma - \sin\alpha\cos\gamma$, $\mathcal{R}_{13} = \cos\alpha\sin\beta$, $\mathcal{R}_{21} = \sin\alpha\cos\beta\cos\gamma +$

$\cos\alpha\sin\gamma$, $\mathcal{R}_{22} = -\sin\alpha\cos\beta\sin\gamma + \cos\alpha\cos\gamma$, $\mathcal{R}_{23} = \sin\alpha\sin\beta$, $\mathcal{R}_{31} = -\sin\beta\cos\gamma$, $\mathcal{R}_{32} = \sin\beta\sin\gamma$, and $\mathcal{R}_{33} = \cos\beta$. It is amazing that the coefficients in Eq. 3.96 are just the third column, i.e., $\mathcal{R}_{13}$, $\mathcal{R}_{23}$, and $\mathcal{R}_{33}$. The same is true for σ_x and σ_y,

$$\sigma_x' = (\cos\alpha\cos\beta\cos\gamma - \sin\alpha\sin\gamma)\sigma_x + (\cos\alpha\sin\gamma + \sin\alpha\cos\beta\cos\gamma)\sigma_y + (-\sin\beta\cos\gamma)\sigma_z \quad (3.101)$$

$$\sigma_y' = (-\cos\alpha\cos\beta\sin\gamma - \sin\alpha\cos\gamma)\sigma_x + (\cos\alpha\cos\gamma - \sin\alpha\cos\beta\sin\gamma)\sigma_y + \sin\beta\sin\gamma\sigma_z \quad (3.102)$$

This shows that one can directly use $\mathcal{R}(\alpha,\beta,\gamma)$ to rotate spins. The only difference is that when there is an improper rotation, after rotation one must multiply it with the determinant of the rotation matrix [Birss (1964)]. Note that for proper rotations, the determinant is always one, so the difficulty is in improper rotations. This method should be used more extensively.

We show one example below. Suppose $\mathcal{R}$ is an inversion,

$$\mathcal{R}(\alpha,\beta,\gamma) = \begin{pmatrix} -1 & 0 & 0 \\ 0 & -1 & 0 \\ 0 & 0 & -1 \end{pmatrix}. \quad (3.103)$$

We also assume the spin is along the $+z$-axis.

$$\begin{pmatrix} S_x' \\ S_y' \\ S_z' \end{pmatrix} = \mathrm{Det}(\mathcal{R})\mathcal{R}(\alpha,\beta,\gamma)\mathbf{S} = (-1)\begin{pmatrix} -1 & 0 & 0 \\ 0 & -1 & 0 \\ 0 & 0 & -1 \end{pmatrix}\begin{pmatrix} 0 \\ 0 \\ S_z \end{pmatrix}. \quad (3.104)$$

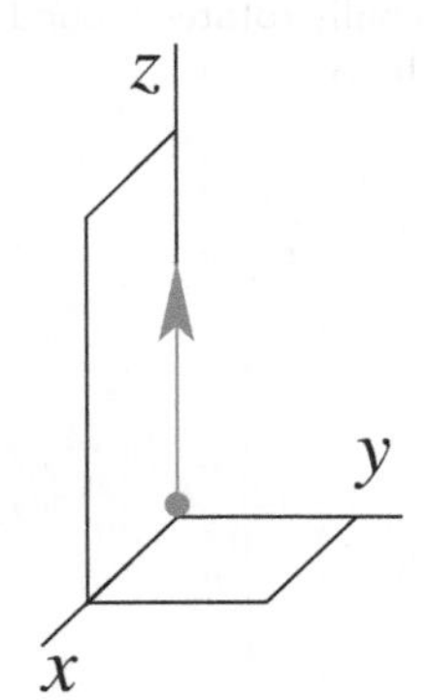

Figure 3.8
Rotation of a spin. The arrow represents the spin. Two mirror planes in the xy and xz planes are highlighted.

After the rotation, the spin is along the $-z$-axis. Since the determinant of $\mathcal{R}$ is -1, the final spin is still along the $+z$-axis. This is consistent with our expected result. If we have a mirror symmetry operation with respect to the xy plane (see Fig. 3.8), the spin direction remains unchanged. On the other hand, if the mirror plane is within the xz plane,

$$\mathcal{R}(\alpha,\beta,\gamma) = \begin{pmatrix} 1 & 0 & 0 \\ 0 & -1 & 0 \\ 0 & 0 & 1 \end{pmatrix}, \quad (3.105)$$

then the spin rotates to the $-z$-axis. For more complicated cases, there is no major difficulty. This greatly eases spin rotations.

3.7 Software packages

Table 3.3 lists some of the most popular software that is in use for the first-principles calculations, both commercial and free.

Table 3.3
Some of free and low-cost first-principles packages. PW: planewave. LAPW/APW: Linearized augmented plane wave/augmented plane wave. TD: Time-dependent.

Software	Basis function	System	TD	Cost
`Abinit`	PW	Solids	Yes	Free
`CP2K`	Gaussian/PW	Molecules/solids	Yes	Free
`ELK`	LAPW/APW	Solids	Yes	Free
`Octopus`	Grid mesh	Molecules/solids	Yes	Free
`SIESTA`	Atomic orbitals	Molecules/solids	No	Free
`Wien2k`	LAPW/APW	Solids	No	Low
`Gaussian`	Gaussian	Molecules	Yes	High
`Vasp`	PW	Solids	No	High

Useful information on materials of different kinds is available on `materialsproject.org`. Although there are some errors in the database, they are a good starting point to work on a new material.

3.8 Exercises

1. If we apply the Born-von Karman boundary condition to a DNA chain which does not have translational symmetry, how many **k** points do we have? And why?

2. A Bloch wavefunction's periodic function in a one-dimensional system is a δ function which is approximated by a Gaussian function. Sketch the wavefunction with $\mathbf{k} = 0.1(2\pi/a)$, where the lattice constant $a = 2.0$Å.

3. Prove that if we use Figure 3.3(d)'s lattice vectors $\mathbf{a}_1 = (\frac{a}{2}, \frac{\sqrt{3}a}{2}, 0)$ and $\mathbf{a}_2 = (\frac{a}{2}, -\frac{\sqrt{3}a}{2}, 0)$, and the atoms' positions in the basis are at $(a/2, -b/2, 0)$ and $(a/2, b/2, 0)$, where $b = a/\sqrt{3}$, then the Wyckoff positions for carbon atoms in the basis are (1/3,2/3,0) and (2/3,1/3,0), respectively.

4. Derive Eqs. 3.96-3.102. Hints: (a) First express all the Pauli matrices in their matrix form. (b) Then perform the matrix multiplication $U\sigma U^+$. (c) All the diagonal elements are the coefficients of σ_z. All the real off-diagonal elements are the coefficients of σ_x, and all the imaginary parts are the coefficients of σ_y.

5. The following matrices are from a real calculation.

$$\mathcal{R}_1 = \begin{pmatrix} 0 & 1 & 0 \\ 1 & 0 & 0 \\ 0 & 0 & -1 \end{pmatrix}, \mathcal{R}_2 = \begin{pmatrix} 0 & -1 & 0 \\ 1 & 0 & 0 \\ 0 & 0 & -1 \end{pmatrix}. \tag{3.106}$$

(a) Suppose the spin is originally along the $+z$-axis. Which matrix(ces) keep the spin in its original direction? (b) Consider what happens if a spin is along the y-axis. Does $\mathcal{R}_1$ change its axis? What about $\mathcal{R}_2$?

6. Explain why when we compute the vector potential from the electric field which is not a planewave, we cannot directly use $\mathbf{E} = -\frac{\partial \mathbf{A}}{\partial t}$.

7. If we compare Eqs. 3.81 and 3.82, we see that the terms in the parenthesis in Eq. 3.81 become $\hat{\mathbf{L}}$. This is only possible if one has an atomic system with spherical symmetry. Prove this.

Part II

Applications

4 High-harmonic generation

4.1 Brief history and key features of high-harmonic generation
4.2 Working principles of HHG
4.3 Applications
4.4 Experimental demonstration of high-harmonic generation in C_{60}
4.5 High-harmonic generation in solids
4.6 Exercises

Harmonic generation is a special case of the more general sum-frequency generation, where a group of photons with the same energy ($\hbar\omega$) is converted to a photon with multiple $\hbar\omega$, $n\hbar\omega$, where n is the harmonic order. Harmonic generation can occur in acoustics. In optics, even low-order harmonic generation is useful as it converts two low-frequency photons to a single high-frequency photon. A green laser pointer uses this principle, where a nonlinear optical crystal, called potassium titanyl phosphate (KTP), with chemical formula $KTiOPO_4$, converts infrared light with wavelength of 1064 nm to green light at 532 nm, i.e., doubling the laser frequency, or second-order harmonic generation (see discussions in Chapter 2)). After frequency doubling, the output power decreases from the incident power of about 100 mW to 5 mW, about a 20 times reduction. Such a reduction is expected from traditional nonlinear optics, where harmonic generation is treated as *a perturbation process*.

This chapter describes a new type of high-order harmonic generation or high-harmonic generation (HHG). It is nonperturbative and has a very different generating mechanism from traditional nonlinear optics. HHG is directly responsible for multiple new frontiers in ultrafast phenomena such as attosecond physics. Its impact has far exceeded the original scientific curiosity and rapidly expands into materials science and engineering. We consider HHG as a vital venue for future scientific breakthroughs, and thus we introduce it to the reader here. This chapter will serve as an introduction to more advanced topics in this field.

4.1 Brief history and key features of high-harmonic generation

High-harmonic generation was first observed in gaseous and noble gas atoms. A typical experiment consists of a gas jet and a strong laser pulse. Much work was done with Ne, Ar, Kr, and Xe very early [McPherson et al. (1987); Ferray et al. (1988)]. The generated 33rd harmonic in the XUV region (32.2 nm) surpassed the perturbative record of the 11th-harmonic generation [Wildenauer (1987)]. Research was intensified.

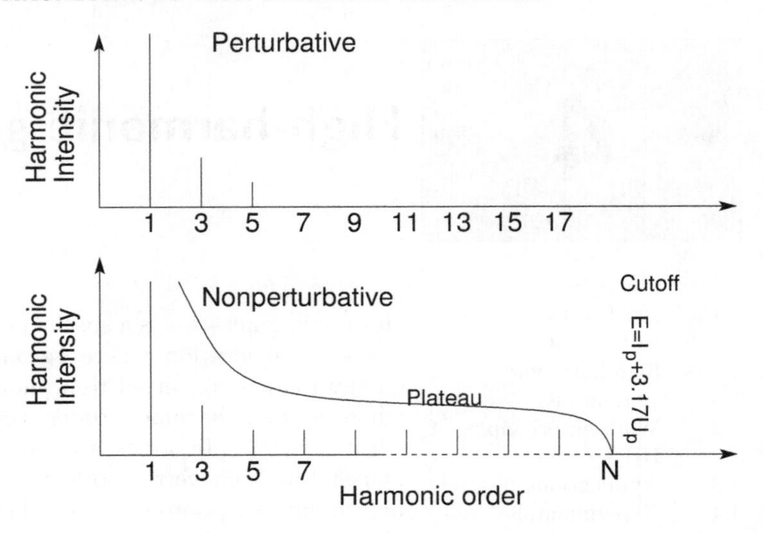

Figure 4.1
(Top) Perturbative harmonic generation and (bottom) nonperturbative harmonic generation. They are similar to each other in the lower-order region, but in the high-order regions they differ. There is a plateau region where the harmonic intensity does not drop as predicted by perturbation theory. Then a steep cutoff at $I_p + 3.17U_p$ appears in the nonperturbative region. I_p and U_p are the ionization potential and the pondermotive energy, respectively.

The key features of high-harmonic generation in gaseous atoms can be summarized as follows. (1) The entire spectrum has a plateau region where the harmonic intensity does not drop as expected from perturbation theory. Figure 4.1 schematically compares the traditional harmonic generation (top half) with the HHG (bottom half). (2) There is an abrupt cutoff after some orders (see the bottom half). The energy at the cutoff is the summation of the ionization energy I_p of the atom and 3.17 times the electron ponderomotive energy, $3.17U_P$ [Krause et al. (1992)]. Corkum [Corkum (1993)] explained the reason behind the cutoff $I_p + 3.17U_p$ observed experimentally, where I_p is the ionization potential and U_p is the ponderomotive energy, and Lewenstein et al. [Lewenstein et al. (1994)] presented a quantum-mechanical formula for the harmonic cutoff. But for over a decade, HHG remained "a solution looking for a problem" [L'Huillier (2017)].

In 1992, Farkas et al. [Farkas et al. (1992)] first demonstrated the 5th-order harmonics on a gold surface. von der Linde et al. [von der Linde et al. (1995)] reported up to 15th-order harmonics in an Al film and 14th order in glass. The main theoretical work was done by Plaja and Roso-Franco [Plaja and Roso-Franco (1992)] who examined the mechanism of harmonic generation in silicon, while Faisal et al. [Faisal and Kamiński (1996, 1997)] developed a nonperturbative Floquet-

Bloch theory and Varro and Ehlotzky [Varro and Ehotzky (1994)] computed HHG from a metal surface. We should also mention that Bandrauk's group was the first to investigate HHG in a one-dimensional chain [Pronin et al. (1994)]. Kálmán and Brabec investigated hard-x-ray emission from a thin crystal [Kálmán and Brabec (1995); Faisal and Kamiński (1996)].

In 2005, HHG in C_{60} was predicted theoretically [Zhang (2005)], which was confirmed experimentally [Ganeev et al. (2009a,b)]. HHG in ZnO was reported [Ghimire et al. (2011)], but these investigators possibly were unaware of those prior investigations in solids and nanostructures. A recent flurry of investigations touched many fronts that are difficult to cover in this book [Mücke (2011); Schubert et al. (2014b); Krausz and Stockman (2014); Hommelhoff and Higuchi (2015); Hohenleutner et al. (2015); Luu et al. (2015); Garg et al. (2016); Ndabashimiye et al. (2016); Liu et al. (2017); Langer et al. (2017)]. There are several recent reviews [Ciappina et al. (2017); Kruchinin et al. (2018)].

The physical understanding and the power of HHG in solids is unfolding, with its underlying mechanism that is likely to be different from that in atoms and small molecules.

4.2 Working principles of HHG

High-order harmonic generation has several unique features that are not common in traditional nonlinear optics. We feel that it is worthwhile to review what is known in nonlinear optics. High-order harmonic generation in nonlinear optics results from multiple transitions among eigenstates. These eigenstates are weakly perturbed by external fields, so they do not change a lot. It is usually the case that for each subsequent higher order, the harmonic signal decreases sharply by several orders of magnitude. These perturbative high-order harmonics, which are developed by a series of expansions in terms of an external laser field (see Section 2.9), form the basis of traditional nonlinear optics [Shen (1984); Mukamel (1995)].

High-order harmonic generation in atoms is built upon a different paradigm. It relies on a much stronger laser field. The system is significantly altered, so that the time evolution of the system cannot be described by a superposition of field-free eigenstates. The entire process is more like ionization or transport, where the system loses and gains electrons during laser excitation. Figure 4.2 illustrates the entire process. In the first-half cycle of the laser field, the electron overcomes the Coulombic potential of the ion core and appears in the vacuum, the rebirth of the electron. Before the electron can move too far, in the second-half cycle the laser field reverses its direction, and the ionized electron, under the joint force of the laser and Coulomb interactions, rushes toward the core. The accelerated charge radiates light with very high photon energies. In essence, HHG converts the large laser field "amplitude" to the high energy photons.

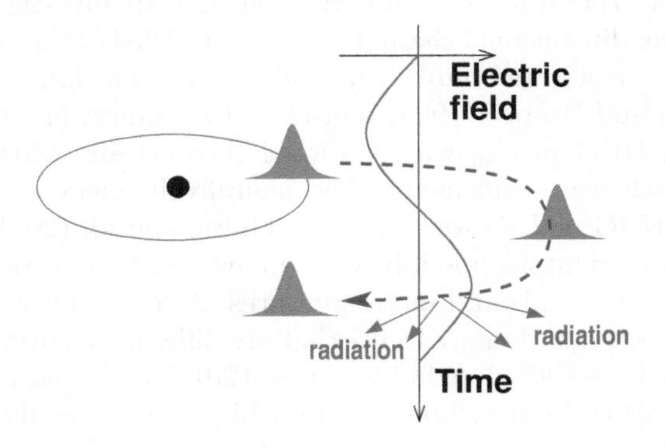

Figure 4.2
High-harmonic generation starts with the initial excitation of electrons by an intense laser pulse. An electron first escapes from the Coulombic potential of the atom in the first half cycle of the laser field. When the next half cycle comes, with the electric field direction reversed, the electron is accelerated toward the ion core under influence of the laser and the Coulomb interaction. The accelerated electron radiates light with frequencies 10-100 times higher.

4.2.1 Laser electric field strength and Coulomb potential in an atom

To develop a conceptual picture, we compare the magnitude of the laser field with the Coulomb potential in a model atom. If the laser field potential is weaker than the Coulomb potential, the electron is perturbed very slightly. On the other hand, if the opposite is true, the electron will be strongly affected.

Let us repeat some basic concepts concerning the laser intensity[a]

[a] The SI unit of the laser intensity is W/m^2. Experimentally, the common unit is W/cm^2. The SI unit of fluence is J/m^2.

$$I = \frac{1}{2}\epsilon_0 c A_0^2, \tag{4.1}$$

where 1/2 is from the time average of a cosine wave, ϵ_0 is the permittivity in vacuum, c is the speed of light, and A_0 is the laser electric field amplitude. We use the same experimental parameters from [Varjú et al. (2009)], with pulse duration of 40 fs, wavelength of 800 nm, and pulse energy of 1 mJ. The peak power of such a laser is P = energy/duration = 1 mJ/40 fs = 25×10^9 W or 25 GW. If the laser focuses on an area of radius of 100 μm, the intensity is

$$I = 25 \times 10^9 \text{ W}/(\pi(100 \times 10^{-4} \text{ cm})^2) = 7.96 \times 10^{13} \text{ W/cm}^2. \tag{4.2}$$

From Eq. 4.1 we find the electric field amplitude A_0 at 2.45 V/Å. This means that if we move along a particular direction by 1 Å, the potential changes by 2.45 V. This potential is going to be compared with the Coulomb potential experienced by the electron.

In a hydrogen-like atom, the Coulomb potential is

$$V = -\frac{Ze}{4\pi\epsilon_0 r}, \tag{4.3}$$

so the electric field of the Coulomb field, $\mathbf{E} = -\nabla V$,

$$\mathbf{E} = -\frac{Ze}{4\pi\epsilon_0 r^2}\hat{r}, \tag{4.4}$$

where Z is the atomic number, r is the radius, and $\hat{r}$ is the unit vector along the radial direction. If we take $Z = 1$ and r to be the Bohr radius of 0.529 Å, we find $|\mathbf{E}| = 3.17$ V/Å. This shows that the laser-induced potential change is about 22.8% of the Coulomb electric field even for the $1s$ orbital of a hydrogen atom. This percentage is way too high for any perturbative scheme to work. For higher-lying orbitals, the percentage is higher. Therefore, we enter a nonperturbative regime where we cannot simply treat the laser electric field as a perturbation.

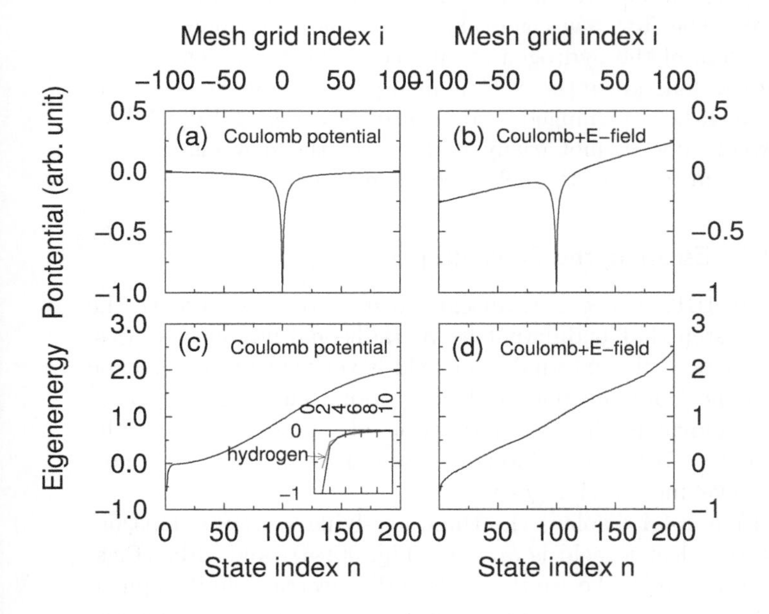

Figure 4.3
Impact of an external electric field on the one-dimensional model system. (a) Bare Coulomb potential. (b) Coulomb plus external electric field potential. (c) Eigenenergy versus state index. Inset: Comparison of the lowest eigenvalues between the 1D system and the hydrogen atom. (d) Eigenenergy versus state index with the presence of both a Coulomb potential and laser E-field potential.

To see what really happens, we adopt a simple one-dimensional model. The total potential is the sum of the

Coulomb potential (Eq. 4.3) and the electric field potential,

$$V_{tot} = -\frac{Ze}{4\pi\epsilon_0(|x|+a)} + A_0(x - x_{center}), \tag{4.5}$$

where a is a small positive number to avoid the singularity, A_0 is the external field amplitude (Eq. 4.1), and x_{center} is the center of the system. We have a code in Appendix A.7 to compute the eigenvalues and eigenvectors. The Hamiltonian includes both the kinetic energy and potential energy. Figure 4.3(a) shows the original Coulomb potential being negative everywhere. Here we choose $a = 1$ and 200 mesh grid points, half of which are on the positive x axis. Figure 4.3(b) shows the original Coulomb potential plus the laser field potential. Here we choose a static electric field. One sees that the new potential is tilted and asymmetric. On the left half, the potential becomes more negative, but on the right half, the potential passes through zero and becomes positive, where the electron is unbounded and can be liberated.

We diagonalize the Hamiltonian matrix (which includes the kinetic and potential energy terms) and find the eigenstates. We plot the eigenvalues as a function of state index in Fig. 4.3(c). The first few eigenvalues resemble the atomic energy spectrum of the hydrogen atom (see the inset in Fig. 4.3(c)). Once we introduce the electric field potential, we see that the eigenvalues are changed significantly (see Fig. 4.3(d)).[b] This is where one cannot easily understand the new eigenstates' change in terms of the field-free eigenstates.

[b] A much stronger change is in the wavefunction (see below).

4.2.2 Escaping the Coulomb potential

Before HHG can occur, an excited electron has to leave the ion, a step that is different from traditional nonlinear optics. Electrons, if the external laser field is very strong, will escape from the Coulomb potential, ionization. The above eigenenergy change is the first indication of electrons leaving the ion core. To visualize electrons escaping spatially, we examine wavefunction changes in space.

Figure 4.4 displays the three low-lying wavefunctions before the electric field is applied (Fig. 4.4(a)) and after (Figs. 4.4(b) and (c)). We see that before the electric field is applied, all three states ϕ_1, ϕ_2 and ϕ_3 are centered at zero, since our Coulomb potential is centered in the system. They are bound states and have the shape resembling s, p, and d waves in a hydrogen atom. When we apply an electric field, the potential is changed. In Fig. 4.4(b) we apply a potential that is higher on the left and lower on the right, so we expect that the electron tends to escape from the right side of the potential. This is exactly what happens: ϕ_2 and ϕ_3 are shifted to the right.[c] They escape as far as the simulation box can go. We find that if we increase the simulation box from 200 to 400 mesh points, ϕ_2 and ϕ_3 still appear at a similar location of the simulation

[c] Due to the symmetry reason, ϕ_1 is not affected.

box. The reader can use our own code in Appendix A.7 to verify this. This proves that they are not bound states any more. If we reverse the field direction, Fig. 4.4(c) shows that these waves escape in the opposite direction. In contrast to the common belief, at least from our simple model, the wavefunctions do not retain their original shapes when they escape from the Coulomb potential – compare ϕ_2 in Fig. 4.4(a) and Fig. 4.4(b). This is similar to quantum mechanical tunneling.[1]

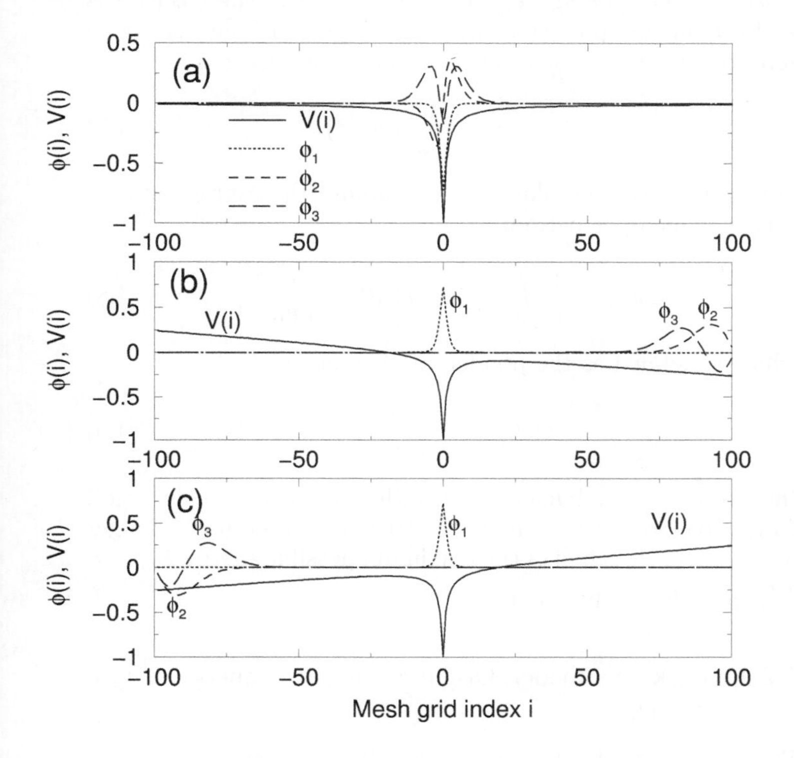

Figure 4.4
Wavefunctions of the three lowest eigenstates ϕ_1, ϕ_2, and ϕ_3 in the one-dimensional system under three different conditions. (a) Field-free wavefunctions. (b) Wavefunctions after an electric field is applied. (c) Same as (b) but the electric field is applied in the opposite direction. All the results are computed using the code in Appendix A.7.

4.2.3 Ponderomotive energy

Now we have an electron that just escaped the Coulomb potential into vacuum. In the literature, this is called the rebirth of the electron. What matters to HHG is how much energy

[1] In quantum tunneling, wavefunctions are complex. Some portion of the waves tunnel through the barrier, while the rest scatter back.

the electron can gain from the laser electric field alone. We introduce an unfamiliar concept: ponderomotive energy.

Consider again a cw electric field along the x-direction,

$$E(t) = A_0 \cos(\omega t), \tag{4.6}$$

where A_0 is the field amplitude and the laser period is $T = 2\pi/\omega$. An electron is treated as a classical particle with mass m_e and charge $-e$. The acceleration of the electron is $-eE(t)/m_e$. Suppose that the initial velocity of the electron is zero, so that the velocity of the electron at time t is

$$v(t) = -\frac{eA_0}{m_e\omega}\sin(\omega t). \tag{4.7}$$

Since the field is oscillatory, we compute the average kinetic energy gained per period,

$$\bar{E}_{kin} = \frac{1}{T}\int_0^T \frac{1}{2}m_e v(t)^2 dt = \frac{1}{4}\frac{e^2A_0^2}{m_e\omega^2}, \tag{4.8}$$

which is defined as the ponderomotive energy U_p,

$$U_p = \frac{1}{4}\frac{e^2A_0^2}{m_e\omega^2}. \tag{4.9}$$

Interestingly, for a longer wavelength or smaller ω, U_p is larger. This is part of the reason why in HHG one often uses a longer wavelength. U_p is half of the maximum possible energy $E_{max} = \frac{1}{2}\frac{e^2A_0^2}{m_e\omega^2} = 2U_p$ from Eq. 4.7.

4.2.4 Corkum's theory: Origin of the cutoff energy of $I_p + 3.17U_p$

What we have discussed so far is critical to our understanding of HHG, but this is not the entire picture. From the last subsection, we see the time-averaged kinetic energy gain is U_p. However, in high-harmonic generation, we are not interested in the time-averaged kinetic energy, but rather the instantaneous kinetic energy. Second, if the ionized electron does not come back to the ion core and becomes a true free electron, there is no harmonic generated in this way, because a true plane wave does not absorb or emit photons. What is important to HHG is the collision of the electron with the core. In other words, we are interested in how the electron under the influence of the laser field accelerates back toward the ion core and how much energy is gained. This energy determines the maximum energy that the electron can emit, the energy cutoff E_{cut}. This is the single most important concept in HHG, at least in atoms.

The first part of E_{cut} is the ionization potential energy of the atom, or I_p. This is easy to understand since the electron must have this amount of energy to enter the vacuum. The higher the ionization energy is, the larger the cutoff energy is. The second part of the energy is $3.17U_p$. This strange term was discovered in experiments first and has a simple theoretical explanation [Corkum (1993)]. The entire derivation is similar to the above subsection, with one exception: We do not set the initial velocity of the electron to zero. Instead, we have (compare with Eq. 4.7)

$$v(t) = -\frac{eA_0}{m\omega}\left[\sin(\omega t) - \sin(\omega t_i)\right], \tag{4.10}$$

where t_i is the initial time that the electron appears in vacuum.[d] The kinetic energy is

$$E_{kin}(t) = 2U_p(\sin(\omega t) - \sin(\omega t_i))^2. \tag{4.11}$$

This equation shows that the maximum kinetic energy $E_{kin}^{\max}(t)$ is at time t when two phases ωt and ωt_i behind sine functions differ more than 180°, so the two sine functions ($\sin(\omega t)$ and $\sin(\omega t_i)$) have opposite signs.

[d] This initial time, which often appears nonessential to many physical phenomena, contains the most important physics for HHG in atoms.

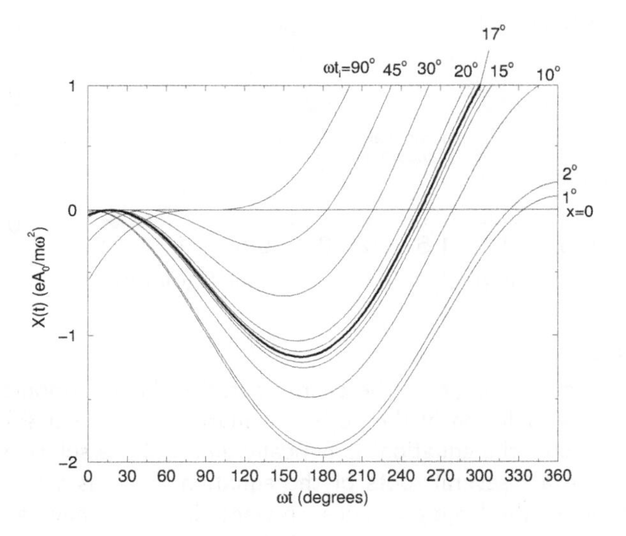

Figure 4.5
Position $x(t)$ versus ωt for a few selected ωt_i. The horizontal line is $x = 0$. The crossing point between the position and the horizontal line is the solution of the equation $x(t) = 0$. The thick solid line is at $\omega t_i = 17°$, where the maximum velocity emerges. Note that the unit of the position is $eA_0/(m\omega^2)$.

The underlying physics would be simple if the above equation were the only one to satisfy. But we have one more constraint. Since the electron must return to the core to generate the harmonic radiation, the position of the electron must be zero at time t, $x(t) = 0$. We integrate Eq. 4.10 and find the electron position is

$$x(t) = \frac{eA_0}{m\omega^2}(\cos(\omega t) - \cos(\omega t_i) + (\omega t - \omega t_i)\sin(\omega t_i)). \tag{4.12}$$

This equation has some unusual features. We see that the first term oscillates with frequency ω and passes zero at $t = n\pi/\omega$, where $n = 0, 1, 2...$, but the last term increases/decreases with a slope $\sin(\omega t_i)$ determined by the initial time t_i and ω. To get some feeling as to how the position changes, we plot the position $x(t)$ versus ωt for a few selected ωt_i in Fig. 4.5. We see that when t_i is earlier or smaller (such as $\omega t_i = 1°$ and $2°$), there are many points crossing the $x = 0$ line, mainly due to the first term in Eq. 4.12. The figure is plotted up to $\omega t = 2\pi$ (360°). This appears promising, but Eq. 4.10 shows that the velocity depends on the difference between two sine functions, so it does not lead to a larger velocity. But if we increase ωt_i to 90°, then the root is $\omega t = 2\pi$, so we get zero velocity. A calculation and discussion for the case with (ωt_i) larger than 90° is left as the exercises at the end of this chapter.

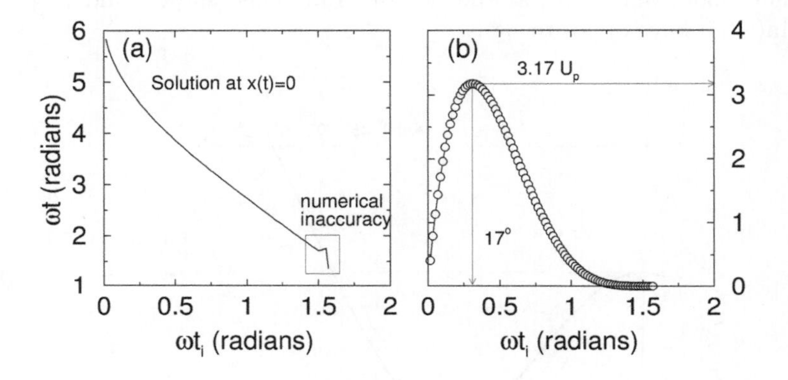

Figure 4.6
Explanation of the origin of the energy cutoff in high-harmonic generation. (a) Solution of the position equation (4.12). For every ωt_i, we solve the equation numerically and find the solution for ωt. There are several roots of the equation if ωt_i is small. We only choose the large one since we want $\sin(\omega t)$ to have an opposite sign of $\sin(\omega t_i)$ where the velocity is maximized. ωt decreases as ωt_i. At $\omega t_i = \pi/2$, there is no root. The small box denotes a few points at large ωt_i, which are inaccurate due to our numerical method where the position deviates strongly from zero. These points should be dropped. (b) Kinetic energy versus ωt. The velocity is calculated by Eq. 4.10, where $\sin(\omega t_i)$ is from (a). The maximum kinetic energy is precisely at $3.17U_p$.

Figure 4.5 provides guidance as to where to find the solution. Numerically, for every ωt_i, we solve Eq. 4.12 for ωt.[e] In Appendix A.8, we provide a simple code to compute the roots, together with a more accurate but tedious Matlab code. Among all the solutions, we plot ωt versus ωt_i in Fig. 4.6(a). It is clear that ωt decreases as ωt_i increases. Note again that the velocity relies on the difference between two sine functions (see

[e]We first set $x(t)$ to zero, so we have $\cos(\omega t) - cos(\omega t_i) + (\omega t - \omega t_i)\sin(\omega t_i) = 0$. ω must be in radians. For every ωt_i, we find ωt.

Eq. 4.10); to maximize the kinetic energy (Eq. 4.11), we look for a solution that these two terms ($\sin(\omega t)$ and $\sin(\omega t_i)$) have an opposite sign. Figure 4.6(b) has a peak at ωt_i around 17° or 0.2967 radian. Then from Fig. 4.6(a) we use $\omega t_i = 0.2967$ radian to find $\omega t = 4.45$ radian. Plugging these numbers into Eq. 4.11, we find the kinetic energy $E_{kin} = mv(t)^2/2 = 3.17U_p$.[f] This is the reason behind the energy cutoff.

[f] With $\omega t_i = 0.2967$ and $\omega t = 4.45$. $[\sin(\omega t) - \sin(\omega t_i)]^2 = 1.5829$, so $E_{kin} = 3.165U_p \approx 3.17U_p$.

4.3 Applications

4.3.1 Applications to hydrogen and neon atoms

Experiments on hydrogen atoms are extremely difficult to carry out, and there have been few comparisons of exact numerical calculations to experimental data [Dörr et al. (1990); Stodolna et al. (2013); Torlina et al. (2015)]. HHG experiments for practical applications are typically done with noble-gas atoms (He, Ne, Ar, Kr, and Xe). In this section, we employ two different methods for hydrogen and neon atoms. We will first explain how the actual high-harmonic generation is computed for the hydrogen atom in a strong laser field, which is the simplest system to consider as an example of high-harmonic generation. Then we move on to the neon atom.

It is unnecessary to make the *single active electron approximation*[g] [Kulander and Rescigno (1991)] to a hydrogen atom. We numerically solve the time-dependent Schrödinger equation (TDSE). We shall use atomic units, i.e., $e = m_e = \hbar = 1$, throughout this section, unless specified otherwise.

[g] We assume that only one electron in an atom is active, while the other electrons are frozen at the core, with a suitable choice of an effective nuclear-binding potential which includes the electrostatic screening due to the frozen-core electrons.

The TDSE of the hydrogen atom in a strong laser field is given by

$$i\frac{\partial}{\partial t}\psi(\mathbf{r},t) = \left[\mathcal{H}^0 + V(\mathbf{r},t)\right]\psi(\mathbf{r},t), \tag{4.13}$$

where $\mathcal{H}^0$ is the stationary-state Hamiltonian

$$\mathcal{H}^0 = -\frac{1}{2}\nabla^2 - \frac{1}{r}. \tag{4.14}$$

The first term is the kinetic energy, and the second term is the nuclear potential energy.[2] $V(\mathbf{r},t)$ in Eq. 4.13 is the laser potential.[h] In the length gauge (LG) under the dipole approximation, we replace $V(\mathbf{r},t)$ by $\mathbf{r}\cdot\hat{\mu}$, so Eq. 4.13 becomes [Chu and Cooper (1985); Kulander (1987)]

[h] The laser vector potential $\mathbf{A}(\mathbf{r},t)$ depends on both space and time. Under the dipole approximation, $\mathbf{A}(\mathbf{r},t) \approx \mathbf{A}(t)$, so spatial dependence is lost. From Chapter 3, we know this is equivalent to ignoring the magnetic field.

$$i\frac{\partial}{\partial t}\psi_{\mathrm{LG}}(\mathbf{r},t) = \left[\frac{-1}{2}\nabla^2 - \frac{1}{r} + \mathbf{r}\cdot\hat{\mu}E(t)\right]\psi_{\mathrm{LG}}(\mathbf{r},t), \tag{4.15}$$

where $E(t)$ is the laser field, and $\hat{\mu}$ is its polarization vector

[2] For the single active electron approximation of a many-electron atom, the binding potential $-1/r$ in Eq. 4.14 is replaced with a pseudopotential whose groundstate energy matches the experimental single-ionization potential of the atom.

which is constant for linear polarization but varies with time for elliptic and circular polarizations.

For a linearly-polarized laser field along the z-axis in particular, $\mathbf{r} \cdot \hat{\mu} = r\cos\theta$, and we can write

$$E(t) = \sqrt{I_o} f(t) \sin(\omega_o t), \tag{4.16}$$

where $\sqrt{I_o}$ and ω_o are the peak electric-field strength and the frequency[3] of a driving laser, respectively, and $f(t)$ is an envelope function given by

$$f(t) = \begin{cases} \cos^2\left(\dfrac{\omega_o t}{2n}\right) & \text{if } |t| < \dfrac{n\pi}{\omega_o}, \\ 0 & \text{otherwise}, \end{cases} \tag{4.17}$$

with n being the number of optical cycles ($T = 2\pi/\omega_o$) per pulse of a driving laser.

An elliptically-polarized laser field on the xz-plane is given by [Murakami and Chu (2016)]

$$\hat{\mu} E(t) = \sqrt{I_o} f(t) \left[\frac{1}{\sqrt{\varepsilon^2+1}} \sin(\omega_o t)\, \hat{z} + \frac{\varepsilon}{\sqrt{\varepsilon^2+1}} \cos(\omega_o t)\, \hat{x} \right], \tag{4.18}$$

where ε is an ellipticity constant, ranging from 0 to 1. $\varepsilon = 0$ gives the linearly-polarized field, and $\varepsilon = 1$ gives the circularly-polarized field. In spherical coordinates, the interaction potential in LG is therefore

$$\mathbf{r} \cdot \hat{\mu} E(t) = \sqrt{I_o} f(t)\, r \left[\frac{1}{\sqrt{\varepsilon^2+1}} \cos\theta \sin(\omega_o t) + \frac{\varepsilon}{\sqrt{\varepsilon^2+1}} \sin\theta \cos\phi \cos(\omega_o t) \right]. \tag{4.19}$$

Alternatively, one could solve the TDSE in the velocity gauge (VG), given by [Han and Madsen (2010)],

$$i \frac{\partial}{\partial t} \psi_{\rm VG}(\mathbf{r}, t) = \left[\frac{-1}{2} \nabla^2 - \frac{1}{r} - i\mathbf{A}(t) \cdot \nabla \right] \psi_{\rm VG}(\mathbf{r}, t), \tag{4.20}$$

where $\mathbf{A}(t) = \int_t^\infty \hat{\mu} E(t')\, dt'$ is a vector potential[i] associated with the driving laser field. Here the dipole approximation makes $V(\mathbf{r}, t) = -i\mathbf{A} \cdot \nabla$. The LG and VG solutions are related according to the following gauge transformation under the dipole approximation:

[i]Caution must be taken if an envelope function is used since one cannot simply integrate $E(t')$ over time to get $\mathbf{A}$. See Chapter 3.

$$\psi_{\rm LG}(\mathbf{r}, t) = e^{i\mathbf{A}(t)\cdot\mathbf{r}}\, \psi_{\rm VG}(\mathbf{r}, t). \tag{4.21}$$

The choice of gauge has a significant influence on the spatial and time resolutions required for convergence of numerical

[3]The typical wavelength of a Ti:sapphire laser is $\lambda = 800$ nm, which is equivalent to the energy of $E = hc/\lambda = 1.55$ eV $= 0.057$ a.u., where h is Planck's constant and c is the speed of light. (The atomic unit of energy is $E_h = 27.21138602$ eV.) The corresponding frequency is $\omega_o = 0.057$ a.u. since $E = \hbar\omega_o$ where $\hbar = h/(2\pi)$=1 in atomic units.

calculations [Schafer (2008)]. In general, in atoms the length gauge is preferred near the binding potential, whereas the velocity gauge is more effective when evolving for the electron far from the core. For HHG calculations, one keeps track of the electron that recombines with the parent atom, so that a relatively small ($r_{\max} \leq 100$ a.u.[4]) size of a computational space is sufficient, and therefore the length gauge is typically used. On the other hand, the velocity gauge is more efficient when calculating photoelectron spectra which are obtained from the ionized wave function at large distances away from the nucleus.

Regardless of the gauges, we may assume the solution of the form[j]

$$\psi(\mathbf{r},t) = \sum_{\ell,m} \frac{R_\ell^m(r,t)}{r} Y_\ell^m(\theta,\phi), \tag{4.22}$$

where $Y_\ell^m(\theta,\phi)$ are the spherical harmonics,

$$Y_\ell^m(\theta,\phi) = \sqrt{\frac{(2\ell+1)}{4\pi}\frac{(\ell-m)!}{(\ell+m)!}}\, P_\ell^m(\cos\theta)\, e^{im\phi}, \tag{4.23}$$

and $P_\ell^m(\cos\theta)$ are the associated Legendre polynomials. Then, Eq. 4.15 for LG becomes

$$i\frac{\partial}{\partial t} R_\ell^m(r,t) = \left[\mathcal{H}_\ell^0 + \mathbf{r}\cdot\hat{\mu}E(t)\right] R_\ell^m(r,t), \tag{4.24}$$

where

$$\mathcal{H}_\ell^0 \equiv \frac{-1}{2}\frac{\partial^2}{\partial r^2} + \frac{\ell(\ell+1)}{2r^2} - \frac{1}{r}, \tag{4.25}$$

with the normalization condition:

$$\sum_{\ell,m}\int_0^\infty dr\, |R_\ell^m(r,t)|^2 = 1. \tag{4.26}$$

Similarly, Eq. 4.20 for VG becomes

$$i\frac{\partial}{\partial t} R_\ell^m(r,t) = \left[\mathcal{H}_\ell^0 - i\mathbf{A}(t)\cdot\left(\nabla - \frac{1}{r}\hat{r}\right)\right] R_\ell^m(r,t), \tag{4.27}$$

where $\hat{r}$ is a unit vector along $\mathbf{r}$.

The wavefunction is evolved in time using the split-operator scheme as [Feit et al. (1982)]

$$R_\ell^m(r,t+\Delta t) = e^{-i\mathcal{H}_\ell^0\Delta t/2} e^{-iV(\mathbf{r},t)\Delta t} e^{-i\mathcal{H}_\ell^0\Delta t/2} R_\ell^m(r,t), \tag{4.28}$$

where $V(\mathbf{r},t) = \mathbf{r}\cdot\hat{\mu}E(t)$ or $-i\mathbf{A}(t)\cdot\nabla$. If without an external electric field or $V(\mathbf{r},t)=0$, $R_\ell^m(r,t+\Delta t) = e^{-i\mathcal{H}_\ell^0\Delta t} R_\ell^m(r,t)$, as it should.

Discretization of the stationary Hamiltonian $\mathcal{H}_\ell^0$ based on the finite difference method is discussed for both LG and VG

[4]The atomic unit of length is the Bohr radius: $a_0 = 5.2917721067 \times 10^{-11}$ m.

[j]The separation of radial and angular variables is possible because the system has spherical symmetry. The time evolution of a function $R_\ell^m(r,t)$ is explicitly monitored on a radial grid $r \in [0, r_{\max}]$ for each of the quantum numbers ℓ and m, which are weighted by time-dependent expansion coefficients of spherical harmonics. Note that, before the laser field turns on, only the ground state of the hydrogen atom ($\ell = 0 = m$) is occupied, so $\langle R_\ell^m(r,t=0)\rangle = \delta_\ell^0\delta_m^0$. As the laser field intensifies, however, the electron population in the ground state decreases because some of them go to the excited states and eventually to the continuum and ionize to free space. When the quantum numbers ℓ and m are all summed up in Eq. 4.22, the probability amplitude of an electron remains as unity throughout the time evolution.

in Ref. [Schafer (2008)]. The propagator for $\mathcal{H}_\ell^0$ can be approximated by the Crank-Nicolson scheme as

$$e^{-i\mathcal{H}_\ell^0\Delta t/2} \simeq \left(1 + i\mathcal{H}_\ell^0\frac{\Delta t}{2}\right)^{-1}\left(1 - i\mathcal{H}_\ell^0\frac{\Delta t}{2}\right). \tag{4.29}$$

The propagator for the time-dependent potential $V(\mathbf{r}, t)$ can be applied using the spherical-harmonic transforms [Murakami and Chu (2016)], or by using the Wigner-3j coefficients. For a linearly-polarized driving laser field, in particular, the operator $e^{-iV(\mathbf{r},t)\Delta t}$ is 2×2 block diagonal[5] and can be applied exactly by using the following identity [Schafer (2008)]:

$$\exp\left\{-i\begin{pmatrix} 0 & c_\ell^m \\ c_\ell^m & 0 \end{pmatrix}\right\} = \begin{pmatrix} \cos c_\ell^m & -i\sin c_\ell^m \\ -i\sin c_\ell^m & \cos c_\ell^m \end{pmatrix}, \tag{4.30}$$

where

$$c_\ell^m = \sqrt{\frac{(\ell+m+1)(\ell-m+1)}{(2\ell+1)(2\ell+3)}}. \tag{4.31}$$

To eliminate the reflection from the boundary, we must divide the wave function into inner (in) and outer (out) wavefunctions at a given time t as

$$\begin{aligned} R_\ell^m(r,t) &= g(r)R_\ell^m(r,t) + [1 - g(r)]R_\ell^m(r,t) \\ &= R_{\ell(\text{in})}^m(r,t) + R_{\ell(\text{out})}^m(r,t), \end{aligned} \tag{4.32}$$

where $g(r)$ is an absorbing function that is 1 in the inner region $(0 \le r \le r_b)$ and smoothly decreases to zero in the outer region $(r_b < r < r_{\max})$, such as

$$g(r) = \cos^{\frac{1}{8}}\left(\frac{\pi(r - r_b)}{2(r_{\max} - r_b)}\right). \tag{4.33}$$

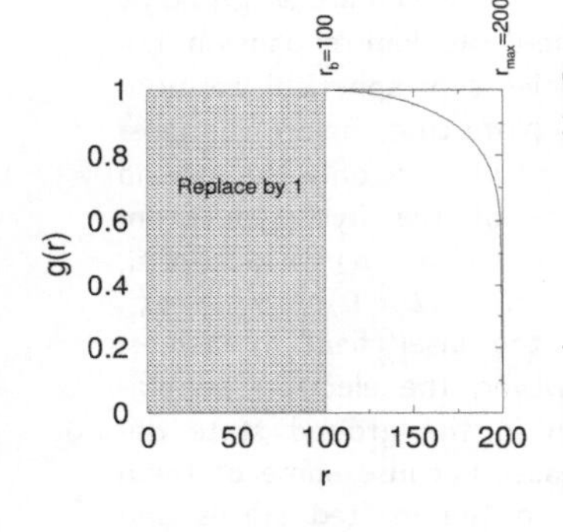

Figure 4.7
The absorbing function $g(r)$ has two parts stitched together. Within r_b, $g(r) = 1$; and for $r > r_b$, $g(r)$ obeys Eq. 4.33.

Figure 4.7 shows an example of absorbing functions. For the evaluation of high-harmonic spectra, only the inner wave function is necessary. It is used to obtain the expectation value of a dipole moment along the laser polarization axis as

$$d(t) = \sum_{\ell,m}\langle R_{\ell(\text{in})}^m(r,t)\mathbf{r}\cdot\hat{\mu}R_{\ell(\text{in})}^m(r,t)\rangle, \tag{4.34}$$

where $\langle\rangle$ denotes an integration over space. If the velocity gauge is used, $\mathbf{r}$ should be replaced by $-i\nabla$. High-harmonic

[5]The details can be found in Ref. [Schafer (2008)], but in the linearly-polarized laser field, laser potential $V(t)$ in the length gauge is diagonal in the radial variable. As we expand the Hamiltonian $\exp(-iV(t)dt)$ in the spherical harmonic basis, it is tri-diagonal in the angular-momentum quantum number ℓ, with each matrix element given by Wigner 3j-coefficients. The full derivation is beyond the scope of this book. This formalism that uses block-diagonal matrix is only applicable to a linearly-polarized field. For other 2D-fields, one must use a fully three-dimensional evolution matrix, which is done by using the spherical harmonic transforms.

spectra are proportional to the square modulus of its Fourier transform, i.e.,

$$D(\omega) = \left| \frac{1}{\tau} \int_0^{\tau} dt\ e^{-i\omega t} d(t) \right|^2, \tag{4.35}$$

where τ is the duration of a driving-laser pulse. Alternatively, one can use the acceleration of electron by applying Ehrenfest's theorem [Murakami et al. (2013, 2017)]. The z-component of an acceleration vector is given by

$$a_z(t) = -\sum_{\ell,m} \int_0^{\infty} dr\ \frac{c_\ell^m}{r^2} \mathbf{Re} \left[R_{\ell+1(\text{in})}^{m\,*}(r) R_{\ell(\text{in})}^{m}(r) \right], \tag{4.36}$$

where c_ℓ^m is given by Eq. 4.31. Similarly, the x-component of an acceleration vector is

$$a_x(t) = -\frac{1}{2} \sum_{\ell,m} \int_0^{\infty} dr\ \frac{1}{r^2} \mathbf{Re} \left[\kappa_\ell^m R_{\ell(\text{in})}^{m\,*}(r) R_{\ell+1(\text{in})}^{m+1}(r) + \kappa_\ell^{-m} R_{\ell(\text{in})}^{m\,*}(r) R_{\ell+1(\text{in})}^{m-1}(r) \right], \tag{4.37}$$

where

$$\kappa_\ell^m = \sqrt{\frac{(\ell+m+1)(\ell+m+2)}{(2\ell+1)(2\ell+3)}}. \tag{4.38}$$

Accordingly, the harmonic spectra are found as

$$D(\omega) = |\tilde{a}_z(\omega)|^2 + |\tilde{a}_x(\omega)|^2, \tag{4.39}$$

where the respective components of the acceleration expectation values are Fourier-transformed as

$$\tilde{a}_{z,x}(\omega) = \frac{1}{\tau} \frac{1}{\omega^2} \int_0^{\tau} dt\ e^{-i\omega t} a_{z,x}(t). \tag{4.40}$$

A Fortran code which calculates the high-harmonic spectra of a hydrogen atom driven by a linearly-polarized laser field using the solution of TDSE in the length gauge can be downloaded from the following website:
https://sites.google.com/site/mitsukomurakami02/codes/hhg

To compute the high harmonic spectrum in neon, we employ TDDFT, whose implementation can be found in Section 3.2.4. We choose $r_{\text{max}} = 5\alpha_0$, where $\alpha_0 = {E_0}^2/{\omega_0}^2$ is the classical oscillator radius of a driving laser field, and $R_b = 1.1273\alpha_0$ which is known to minimize the effect of long paths in high-harmonic spectra [Strelkov et al. (2012)].

Figure 4.8 shows the high-harmonic spectra of hydrogen (H) and neon (Ne) atoms, driven by an 800-nm ($\omega_o = 0.05696$ a.u.) laser pulse of peak intensity $I_o = 2\times10^{14}$ W/cm^2 and full-width-half-maximum duration of $\tau = 40$ fs. The corresponding classical oscillator radius and ponderomotive potential are $\alpha_0 = 23.30$ a.u. and $U_p = 0.44$ a.u., respectively. The cutoff

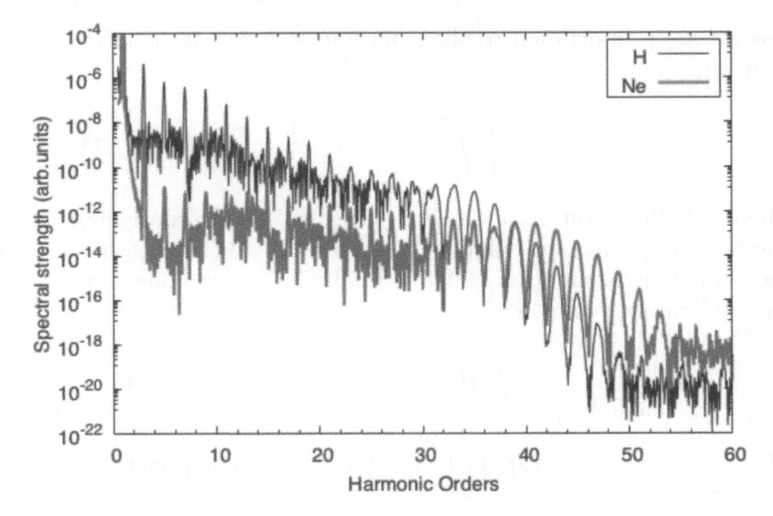

Figure 4.8
High-harmonic spectra of hydrogen (H) and neon (Ne) atoms, calculated with the time-dependent Schrödinger equation and TDDFT, respectively, driven by an 800-nm laser pulse of peak intensity $I_o = 2 \times 10^{14}$ W/cm^2.

energy predicted by the cutoff law is $33\omega_o$ for the H atom and $38\omega_o$ for valence electrons of the Ne atom, which agree with the cutoffs observed in Fig. 4.8. In both cases, the spectra consist of a series of equispaced peaks at odd multiples of the driving-laser frequency ω_o. Even harmonics are forbidden for high-harmonic generation from an atom because of the symmetry in electron dynamics along the laser polarization axis.

4.3.2 Applications to C_{60}

Gaseous atoms have dominated high-harmonic generation for several decades. Only a handful of investigations were on solids in the 1990s before the recent resurgence of studies of solids. There were very few investigations of nanostructures until 2005, when some of the authors of this book started to investigate HHG in C_{60}. Nanostructures have their disadvantages and advantages for harmonic generation. A number of noticeable disadvantages include the following: (1) They may be easily fragmented under intense laser radiation. C_{60} may be broken into smaller clusters. Fragmentation itself may not be necessarily detrimental to HHG, but may interfere with photon production. (2) They may have stronger multiple-electron emission and ionization. Electron emission and ionization tend to compete with HHG production.

On the other hand, nanostructures have unparalleled advantages: (i) Tailorability. Nanomaterials can be systematically engineered to maximize HHG yields. This is mainly due to their tunable material structures. For instance, silicon and gold clusters adopt varieties of shapes. The dimension and size can be controlled systematically. Small diamondoids such as adamantane and rimendanine are excellent candidates. (ii) Large cross section. HHG mainly results from the collision of an electron with the ion core. In nanostructures, parent cores are much larger than atoms, and recollision rates are much higher. Electronically, the electron density is very high. So the efficiency of HHG is higher than that in atoms. (iii) Due to a low ionization potential, a weaker laser should be enough to knock electrons out of nanomaterials and generate HHG signals.

In the following, we take C_{60} as an example to show how to investigate HHG in these nanostructures. C_{60} is a beautiful symmetric molecule with the highest I_h group symmetry. Sixty carbons form a cage structure. Among 360 electrons, 120 are core electrons and 240 valence electrons with mixed sp^2 and sp^3 hybridization. Among the 240 electrons, there are 180 σ electrons forming 90 bonds (see Fig. 4.9). The remaining 60 electrons are π electrons. The σ bonds are much stronger, so they form the carbon cage structure. The optical gap of C_{60} is about 2 eV. In Appendix A.9, we provide a code to compute the structure of C_{60}. Different from HHG in atoms, the number of electronic states in C_{60} is much larger, so it is necessary to limit ourselves to a small set of states. The calculation can be outlined as follows.

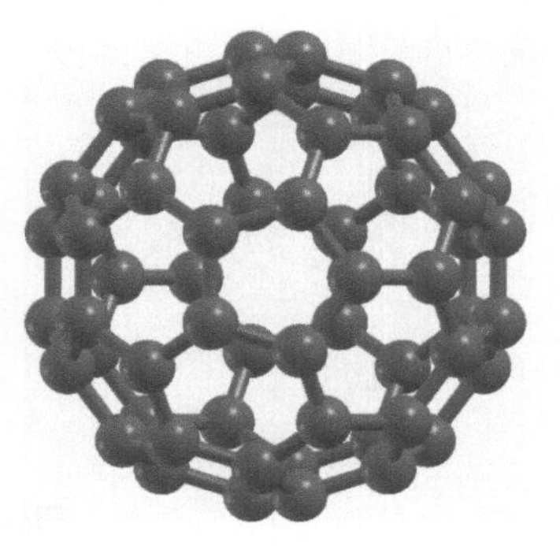

Figure 4.9
C_{60} structure. There are ninety bonds. Single bonds are between the pentagons. Double bonds are between the pentagons and hexagons. All the coordinates of carbon atoms can be generated using three symmetry operations from a generating point. See the code in Appendix A.9 at the end of this book.

4.3.2.1 Model

This step consists of structural optimization and ground-state property comparisons. Such calculations can be performed at either a simple model Hamiltonian level or first-principles level, but the key is that one has to include a sufficient number of states in the energy spectrum.

Our model system does not consider electron ionization, and thus we keep the laser intensity as low as possible to avoid fragmentation. Our calculation starts with a tight-binding model [Zhang (2005)],

$$\begin{aligned} H_0 = & - \sum_{\langle ij\rangle,\sigma} t_{ij}(c^\dagger_{i,\sigma}c_{j,\sigma} + h.c.) + \frac{K_1}{2}\sum_{\langle i,j\rangle}(r_{ij} - d_0)^2 \\ & + \frac{K_2}{2}\sum_i d\theta^2_{i,5} + \frac{K_3}{2}\sum_i(d\theta^2_{i,6,1} + d\theta^2_{i,6,2}), \quad (4.41) \end{aligned}$$

where the first term is the electron hopping term, and the remaining three terms are the lattice stretching, pentagon-hexagon, and hexagon-hexagon bending energy terms, respectively. $c^\dagger_{i,\sigma}$ is the electron creation operator at site i with spin σ $(=\uparrow\downarrow)$, $t_{ij} = t_0 - \alpha(r_{ij} - d_0)$ is the hopping integral between nearest-neighbor atoms at positions $\mathbf{r}_i$ and $\mathbf{r}_j$, $r_{ij} = |\mathbf{r}_i - \mathbf{r}_j|$, and the summation $\langle ij\rangle$ over $i(j)$ runs from 1 to 60 with

$i \neq j$. By fitting two bond lengths (one single and one double bond), optical energy gap and 174 normal mode vibrational frequencies to the respective experimental values, we obtain the average hopping integral $t_0 = 1.91$ eV, electron-lattice coupling $\alpha = 5.0$ eV/Å, spring constants for lattice stretching as $K_1 = 42$ eV/Å^2, for pentagon-hexagon bending as $K_2 = 8$ eV/rad^2, and for hexagon-hexagon bending as $K_3 = 7$ eV/rad^2, and $d_0 = 1.5532$ Å. This model includes 60 π electrons but ignores all other electrons.

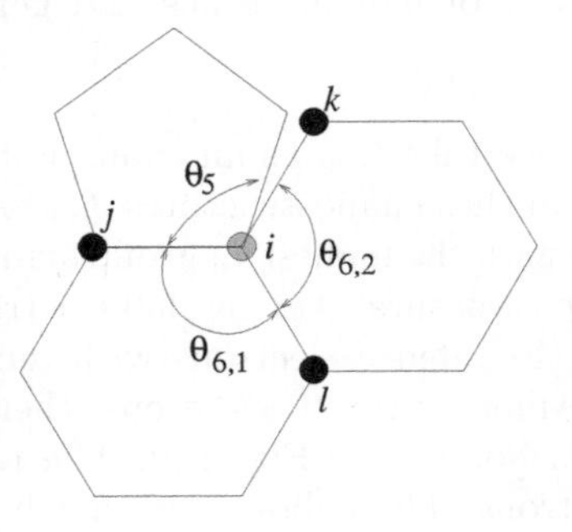

Figure 4.10
There are two kinds of bond angles: One is within the pentagons θ_5, and the other is in the hexagons. There are two possible choices for hexagons, $\theta_{6,1}$ and $\theta_{6,2}$. Note that two hexagons cannot connect with a pentagon seamlessly within a plane since the sum of all the angles is 348°, not 360°. This is the reason why C_{60} is a sphere.

Caution must be taken when we describe those three different vibration energy terms. Figure 4.10 shows an example of carbon atom i which is surrounded by atoms j, k and l. For each atom i, there is only one bond angle on pentagons, angle $\angle jik$, which is called θ_5, where 5 refers to the number of vertices in a pentagon. Since atom i has two angles in two hexagons, we label them $\theta_{i,6,1}$ and $\theta_{i,6,2}$. In Eq. 4.41, we need to find the angle differences $d\theta_{i,5}$ and $d\theta_{i,6,1(2)}$. $d\theta_{i,5}$ is the bond angle difference on the pentagons, $d\theta_{i,5} = \theta_{i,5} - \theta^o_{i,5}$, where $\theta^o_{i,5}$ is the initial angle at equilibrium. $d\theta_{i,6}$ is the difference between the angles on the pentagons, $d\theta_{i,6,1(2)} = \theta_{i,6,1(2)} - \theta^o_{i,6,1(2)}$, where $\theta^o_{i,6,1(2)}$ is the initial angle at equilibrium. The following code shows how these angles are computed:

```
1 c      mmm is the index of all neighboring atoms;
2 c      nm=60 (60 carbon atoms)
3 c       elastic energy angle part (pentagon)
4 c       write(*,*)'Checking angles'
5         do  i=1,nm
6             j1=mmm(i,1)
7             j2=mmm(i,2)
8 d1=((xp(i )-xp(j1))**2+(yp(i )-yp(j1))**2+(zp(i)-zp(j1))**2)
9 d2=((xp(i )-xp(j2))**2+(yp(i )-yp(j2))**2+(zp(i)-zp(j2))**2)
```

```
10 d3=((xp(j1)-xp(j2))**2+(yp(j1)-yp(j2))**2+(zp(j1)-zp(j2))**2)
11          temp=(d3-d1-d2)/2.d0/dsqrt(d1*d2)
12          ang=dacos(-temp)
13 c         write(*,*)i,ang,anp0
14          ep0=ak2/2.d0*(ang-anp0)**2+ep0
15 c     this is correct since there is no double counting here
16       enddo
17 c     elastic energy angle part (hexagon)
18       do 4 i=1,nm
19          do 5 ij=1,2
20         j55=mmm(i,ij)
21         j56=mmm(i,3)
22        d1=((xp(i)-xp(j55))**2+(yp(i)-yp(j55))**2
23      & +(zp(i)-zp(j55))**2)
24        d2=((xp(i)-xp(j56))**2+(yp(i)-yp(j56))**2
25      & +(zp(i)-zp(j56))**2)
26       d3=((xp(j56)-xp(j55))**2+(yp(j56)-yp(j55))**2
27      & +(zp(j56)-zp(j55))**2)
28         temp=(d3-d1-d2)/2.d0/dsqrt(d1*d2)
29         ang=dacos(-temp)
30         ep0=ak3/2.d0*(ang-anh0)**2+ep0
31 c         write(*,*)i,ang,anh0
32 c     since there are two different angles
33 c     for each i
34 5       continue
35 4       continue
```

4.3.2.2 Time-dependent Liouville equation

Before we present the details of our time-dependent Liouville equation, we want to comment on the time-dependent Schrödinger equation. In nanostructures, we have many electrons. In traditional quantum mechanics textbooks, the time-dependent Schrödinger equation is developed for a single electron,[k] not for many electrons. Applying the single-electron time-dependent Schrödinger equation to a many-electron problem may violate the Pauli exclusion principle. This occurs because electrons are Fermions, and even without interactions among themselves (such as Coulombic and exchange interactions), "interactions" are still present due to the Pauli exclusion principle.

To compute HHG signals, we adopt the Born-Oppenheimer approximation, where we treat the atomic motion classically and the electron dynamics quantum mechanically. The lattice dynamics is simulated classically by the Newton equation, so at each time step we can find the velocity and position of each atom. There is a complication for the force term for angles. The following code gives the details as to how this part is computed with the forces around the pentagons. A similar code can be written for hexagons.

[k] Take a two-electron system as an example. For electron 1 in a state ϕ_1, we have $i\hbar\dot{\phi}_1 = H_1\phi_1$; for electron 2 in ϕ_2, we have $i\hbar\dot{\phi}_2 = H_2\phi_2$. If ϕ_1 and ϕ_2 evolve with time coherently, $\langle\phi_1(t)|\phi_2(t)\rangle = \langle\phi_1(0)|\phi_2(0)\rangle = 0$. But if these states evolve incoherently (i.e. $H_1 \neq H_2$), which is the case for time-dependent density functional theory currently in use, $\langle\phi_1(t)|\phi_2(t)\rangle \neq 0$. Then the Pauli exclusion principle is violated.

```
1       do i=1,nm
2          fx(i)=0.d0
3          fy(i)=0.d0
```

```
4         fz(i)=0.d0
5      enddo
6      do 100 i=1,60
7         j=mmm(i,1)
8         k=mmm(i,2)
9      rij=dsqrt((x(i)-x(j))**2+(y(i)-y(j))**2+(z(i)-z(j))**2)
10     rik=dsqrt((x(i)-x(k))**2+(y(i)-y(k))**2+(z(i)-z(k))**2)
11     rjk=dsqrt((x(k)-x(j))**2+(y(k)-y(j))**2+(z(k)-z(j))**2)
12        cn=(rij**2+rik**2-rjk**2)/(2.d0*rij*rik)
13        theta=dacos(   cn   )
14        sn=dsqrt(1.d0-cn**2)
15        aa=(rik*cn-rij)/rij
16        bb=(rij*cn-rik)/rik
17        pxi=bb* (x(i)-x(k)) -aa* (x(j)-x(i))
18        pyi=bb* (y(i)-y(k)) -aa* (y(j)-y(i))
19        pzi=bb* (z(i)-z(k)) -aa* (z(j)-z(i))
20        pxj=(x(j)-x(k)) +aa* (x(j)-x(i))
21        pyj=(y(j)-y(k)) +aa* (y(j)-y(i))
22        pzj=(z(j)-z(k)) +aa* (z(j)-z(i))
23        pxk=(x(k)-x(j)) +bb* (x(k)-x(i))
24        pyk=(y(k)-y(j)) +bb* (y(k)-y(i))
25        pzk=(z(k)-z(j)) +bb* (z(k)-z(i))
26        temp=1.d0/(rij*rik*sn)*ak2*(theta-anp0)
27
28        fx(i)=fx(i)-pxi*temp
29        fx(j)=fx(j)-pxj*temp
30        fx(k)=fx(k)-pxk*temp
31
32        fy(i)=fy(i)-pyi*temp
33        fy(j)=fy(j)-pyj*temp
34        fy(k)=fy(k)-pyk*temp
35
36        fz(i)=fz(i)-pzi*temp
37        fz(j)=fz(j)-pzj*temp
38        fz(k)=fz(k)-pzk*temp
39 100    continue
```

The electron dynamics is simulated quantum mechanically by the Liouville equation,

$$i\hbar\frac{\partial\langle n|\rho|m\rangle}{\partial t}=\langle n|[H,\rho]|m\rangle, \tag{4.42}$$

where $\langle n|\rho|m\rangle$ is the electron density matrix between states $|n\rangle$ and state $|m\rangle$. H is the total Hamiltonian, $H=H_0+H_I$, where H_0 is the field Hamiltonian. The Liouville equation rigorously respects the Pauli exclusion principle, since any double occupancy is eliminated exactly through the density matrix term.Consider $[H_I,\rho]$ since this term is the only term that leads to transition between two states. Take two states $|n\rangle$ and $|m\rangle$ as an example. $i\hbar\frac{\partial\langle n|\rho|m\rangle}{\partial t}=\langle n|H_I|n\rangle\langle n|\rho|m\rangle-\langle n|\rho|m\rangle\langle m|H_I|m\rangle=(\langle m|\rho|m\rangle-\langle n|\rho|n\rangle)\langle n|H_I|m\rangle$, where we assume $\langle n|H_I|n\rangle=0$ and $\langle m|H_I|m\rangle=0$. $\langle m|\rho|m\rangle-\langle n|\rho|n\rangle$ ensures the Pauli principle. If both states are occupied with one electron, there is no transition. The interaction Hamiltonian is

$$H_I = -e \sum_{i\sigma} n_{i\sigma} \mathbf{E}(t) \cdot \mathbf{r}_i, \tag{4.43}$$

where $\mathbf{E} = \hat{n} A_0 F(t) \cos(\omega t)$. Here, $\hat{n}$, A_0, t, and ω are the laser polarization, amplitude, time, and frequency, respectively. We choose a sine squared function $F(t) = \sin^2(t/T)$. In Eq. 4.42, $|n(m)\rangle$ represents either the site index or state index. While these two representations are formally equivalent if the lattice vibration is not considered, when the vibration is included, it is much simpler to implement the site index, because in the state space[l] the transition matrix has to be recalculated at each time step, while in the site space the transition matrix is diagonal and is easily computed by a new electron/lattice position which is computed anyway during the lattice vibration.

[l] In the state representation, we have to diagonalize the instantaneous Hamiltonian to compute the transition matrix elements between eigenstates. It is not correct to use the field-free states to expand them, because once the lattice starts to vibrate, the state space changes.

4.3.2.3 Power spectrum

In classical electrodynamics in a nonrelativistic and single point charge limit, the power P radiated per unit solid angle is given by Larmor's formula [Jackson (1962)], page 469, in electrostatic units),

$$P = \frac{2}{3} \frac{e^2 \dot{v}^2}{c^3}, \tag{4.44}$$

where c is the speed of light, e is the elementary charge, and $\dot{v}$ is the acceleration. In quantum mechanics, the acceleration term is replaced by the dipole acceleration.

In our present case, our dipole acceleration is $\ddot{\mathbf{D}} = \sum_{i\sigma} \langle \ddot{n}_{i\sigma} \mathbf{r}_i + 2\dot{n}_i \dot{\mathbf{r}}_i + n_i \ddot{\mathbf{r}}_i \rangle$, where $\dot{n}_i$ and $\ddot{n}_i$ can be directly computed from Eq. 4.42. $\dot{\mathbf{r}}_i$ and $\ddot{\mathbf{r}}_i$ are the velocity and acceleration of the carbon atoms, which carry the information of the molecular vibrations. In order to compute the power spectrum, we Fourier-transform $\mathbf{D}(t)$ into the frequency domain [Zhang (2005)],

$$\ddot{\mathbf{D}}(\Omega) = \frac{1}{T} \int_0^T \ddot{\mathbf{D}}(t) \exp(i\Omega t) dt, \tag{4.45}$$

where $\ddot{\mathbf{D}}(\Omega)$ is a vector and Ω is the emitted photon energy.

We choose the laser photon energy of 0.4 eV. Since the energy gap for the first dipole-allowed transition between the lowest unoccupied molecular orbital plus one (LUMO+1) and the highest occupied molecular orbital (HOMO) is about 2.7 eV, this photon energy is away from the resonance. The pulse duration is $\tau = 32$ laser field cycles. Figure 4.11(a) shows the power spectrum, where the x-axis denotes the harmonic order and the y-axis represents the logarithmic of Fourier-transformed dipole accelerations. The highest harmonic order is 31. The 7th-order harmonic is where the first resonant peak appears. The spectrum is very interesting in that before the 7th order, all the harmonic orders are integers. But after the

7th order, the harmonic orders are no longer integers. Peak a is associated with the A_g, H_g, and T_{1u} normal modes, and does not appear when we switch off the lattice vibration.

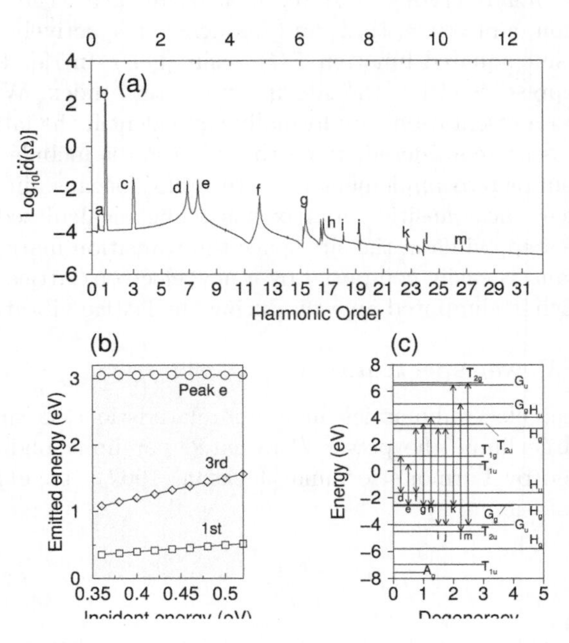

Figure 4.11
(a) Harmonic generations in C_{60}. The laser photon energy is 0.4 eV, pulse duration is 32 laser cycles, and field strength is 0.02 eV/Å. Peak a results from the lattice vibration. Peaks b, c, and d are at the 1st, 3rd, and 7th orders, respectively. Above the 7th order, harmonics mainly result from the intrinsic electronic excitations. (b) The emitted photon energy versus the incident energy. The circles, diamonds and squares denote peak e, the third and first harmonics, respectively. (c) Assignment of peaks from d to m (see letters below arrows) to their respective transitions (double-arrowed lines). The H_u (HOMO) is at -1.6158 eV, and the T_{1u} (LUMO) is at 0.5255 eV [Zhang (2005)]. Used with permission from the American Physical Society.

To understand the origin of other high-order harmonics, a method, that is very popular in the resonant inelastic x-ray scattering (RIXS) [Zhang et al. (2002a)], is very handy. In RIXS, one sends in x-ray photons and collects emitted photons. A spectrometer can measure the emitted photon energy. If we increase the incident photon energy, the emitted photon energy may or may not change. If the emitted photon energy does not change with respect to the incident photon energy, then the associated transition must correspond to the intrinsic electron states. Figure 4.11(b) shows that the energies of the first and third harmonics scale linearly with the incident energy, but peak e does not. This shows that there is a fundamental difference among those harmonic peaks. Peaks like peak e originate from those intrinsic transitions that involve the energy states. Figure 4.11(c) shows that nearly all the peaks can be attributed to unique transitions.

Finally we should point out that the vector nature of $\ddot{\mathbf{D}}(\Omega)$ [Zhang and George (2006)] enables one to compute the polarization of emitted photons [Zhang and George (2007b)].

4.4 Experimental demonstration of high-harmonic generation in C_{60}

Theoretical prediction needs an experimental verification. The first experimental study was carried out four years later. Ganeev *et al.* employed an intense femtosecond Ti:sapphire laser to demonstrate HHG from the plasma flumes generated from C_{60} [Ganeev et al. (2009a)].

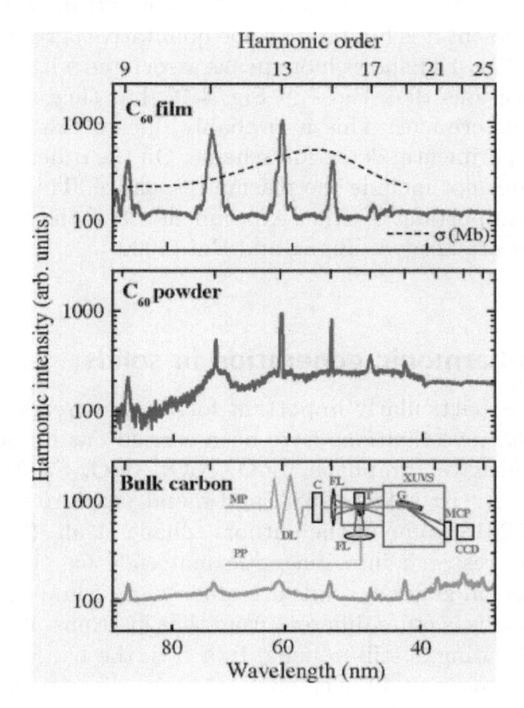

Figure 4.12
High-harmonic generation using plasma flumes. (Top) C_{60} film. (Middle) C_{60} powder. (Bottom) Bulk carbon [Ganeev et al. (2009a)]. The inset in the bottom panel is the experimental setup. MP: main pulse; PP: prepulse; DL: delay line; C: grating compressor; FL: focusing lenses; T: target; XUVS: XUV spectrometer; G: gold-coated grating; MCP: microchannel plate; CCD: charge-coupled device. Used with permission from the American Physical Society.

They first used a subnanosecond prepulse for the sample ablation. This is different from the commonly used gas jet. They found that the structure of ablated fullerenes is intact, where they kept the laser intensity low at 2×10^9 to 2×10^{10}W/cm^2. They then applied a stronger 35-fs pulse to the ablated fullerene to generate HHG. The 35-fs pulse intensity is higher, up to 7×10^{14}W/cm^2, with pulse energy of 8 to 20 mJ. The HHG signal was detected by a microchannel plate and phosphor screen. Figure 4.12 shows the HHG signals from three different samples, C_{60} film, C_{60} powder, and bulk carbon. One sees that the harmonic signals from C_{60} film are stronger than other samples. The peak centers around 20 eV, corresponding to the plasma frequency in the C_{60} film. In the powder sample, the signal is still there, but is weaker. Both powder and film data are 25 times stronger than those in bulk carbon.

This represents the first experiment effort. A direct comparison between the experiment and theory is certainly desirable, but at present it is limited to some qualitative agreement. For instance, Fig. 4.11 shows more peaks associated with many intrinsic transitions than those in Fig. 4.12, but they are not seen in the experiment. This is probably due to the limitation of the experimental detection scheme. On the other hand, the theory does not include the plasma resonance. This is expected to be important. Further experimental and theoretical investigations are needed [Zhang and Bai (2020)].

4.5 High-harmonic generation in solids

This section is particularly important for the new reader. To this end, HHG investigations have been carried out in monolayer and multilayer graphene, ZnO, NiO, MgO, Si, MoS_2, Bi_2Se_3, SiO_2, Ar/Kr solids, GaSe, and metal-sapphire nanostructures. In 2018, some of the authors [Zhang et al. (2018c)] expanded the research into magnetic materials for the first time. However, in general, solids are much more complex.

HHG in solids is quite different from that in atoms. A complete understanding is still missing. In solids, the band structure has an important impact on how harmonic signals are generated. Vampa et al. [Vampa et al. (2015)] even suggested reconstructing the band structure from the HHG spectrum. However, in general, this is very challenging, and HHG cannot be easily considered as a band mapping tool like photoemission.

Depending on the type of materials and the strength of a laser field, electrons react to the laser pulse differently. To explain the complication of HHG in solids, Fig. 4.13 shows the band structures across the Fermi energy E_f for four possible cases. The crystal momentum on either side does not touch the Brillouin zone boundary yet; otherwise, the band dispersion has to be perpendicular to the boundary. In metals, electrons fill the lower part of the band up to the Fermi level E_f (see the horizontal dashed line in Fig. 4.13(a)). Under the influence of a laser pulse, the short arrow in Fig. 4.13(a) shows that an electron can move within a single band beyond the Fermi level to fill an unoccupied state. Such a process is called an intraband transition. Once the electron gains enough energy, it comes back to fill the occupied states by releasing a photon, giving harmonic generation. HHG signals due to intraband transition have a lower energy. In metals, it is also possible to have interband transitions between different bands (the long vertical arrow).

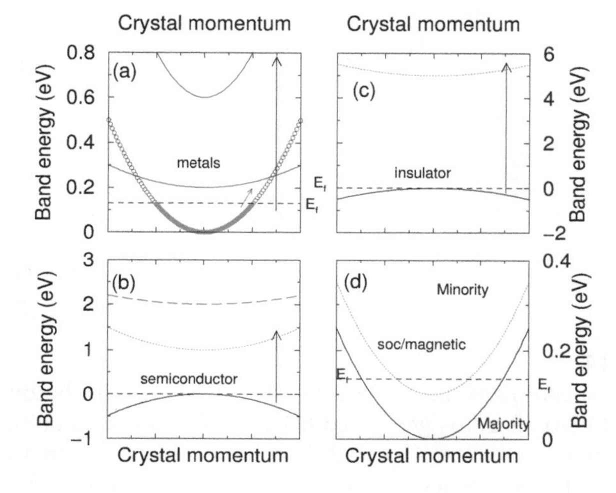

Figure 4.13
High-harmonic generation in solids for different kinds of materials. (a) Metals can have intraband and interband transitions. (b) In semiconductors, intraband effects are small since the lower bands are all filled. The system is dominated by interband transitions. (c) Insulators are dominated by interband transitions. (d) Magnetic materials with strong spin-orbit coupling have more channels open to laser excitation.

Different from metals, Fig. 4.13(b) shows a situation for semiconductors. There is a finite gap separating the valence and conduction bands. Although intraband transitions are possible, they often occur with participation of lattice vibrations, since the valence bands are completely filled and electrons have nowhere to go. Interband transitions should dominate HHG in semiconductors. Similar to semiconductors, insulators have a much stronger interband transition to generate HHG signals (see Fig. 4.13(c)). Figure 4.13(d) shows an example for a magnetic system, with the spin majority and minority bands shifted with respect to each other. Here, harmonic generations are quite different from those in nonmagnetic materials, since two spin channels contribute differently. We will provide an example below. But at first, we are going to discuss graphene.

4.5.1 Graphene

As briefly introduced in Sec. 3.4.1, graphene is a two-dimensional sheet, and strictly is one atomic layer thin. Carbon atoms form a honeycomb structure, with a triangular lattice. The distance between two neighboring atoms is 1.42 Å. Its band structure at the Γ point disperses with crystal momentum linearly, Dirac fermions. This allows the efficient electron transport.

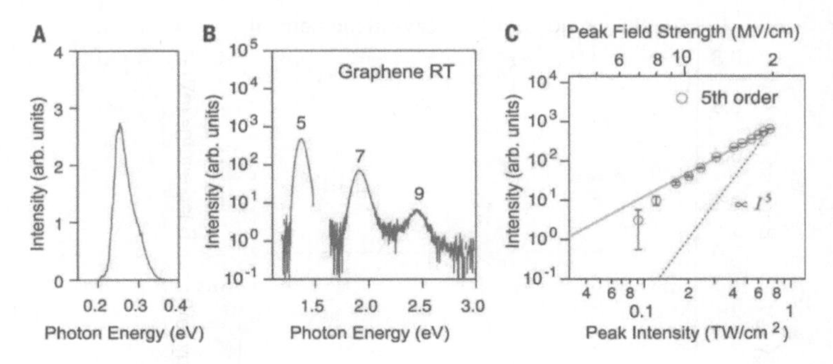

Figure 4.14
(a) Energy spectrum of the pump pulse. The pulse, whose photon energy is 0.23-0.29 eV, is generated from an 800-nm 35-fs pulse of 1 kHz repetition rate. (b) High-harmonic generation up to 9th order. (c) Dependence of harmonic intensity on the pump intensity [Yoshikawa et al. (2017)]. Used with permission from the American Association for the Advancement of Science.

Experimental and theoretical HHG investigations on graphene have been very active [Yoshikawa et al. (2017); Al-Naib et al. (2014); Bowlan et al. (2014); Cox et al. (2017); Chizhova et al. (2017); Dimitrovski et al. (2017)]. Figure 4.14 shows an experimental result, which is generated by a 25-fs mid-infrared pulse of photon energy 0.26 eV, whose energy spectrum is shown in Fig. 4.14(a). The laser electric field is quite strong, 30 MV/cm, or 0.3V/Å. Fig. 4.14(b) shows the harmonic signals with order up to 9 are observed. Figure 4.14(c) shows that the dependence of the 5th harmonic intensity on the laser peak intensity I does not follow the perturbative prediction of I^5. This is different from perturbative harmonics. Since the linear dispersion is the Γ point, using a THz pulse is proven to be more efficient [Hafez et al. (2018)].

However, despite these interesting results, none of these investigations is capable of revealing more information about graphene than other methods. In addition, the highest energy of HHG is only 2.5 eV (see Fig. 4.14(b)), too low to become competitive, and the THz driven HHG is even lower at 8.68 meV,[m] in comparison with HHG energies from atoms. Therefore, at this time, the future direction of HHG in solids is unclear, and further investigation is necessary [Zhang and Bai (2019)].

[m] The fundamental frequency is 0.3THz, or 1.24 meV. The highest order is 7, so we can get 8.68 meV.

4.5.2 Going to magnets

There has been no experimental investigation on magnetic materials of any kind, as far as we know. Ferromagnets have a distinctive feature, where the majority and minority electrons have different band structures (Fig. 4.13(d)). There are more electrons in majority bands. For this reason, we expect that HHG signals are spin dependent. Antiferromagnets and ferrimagnets are also interesting since each spin sublattice has different electronic and magnetic properties, so they are expected to yield different HHG signals.

Theoretically, we take one monolayer ferromagnetic Fe(110) as an example. Figure 4.15 shows its structure for a typical slab calculation, where a vacuum layer of 12 Å is inserted between layers so the interaction between them is small, while the lattices along the lateral directions remain periodic. The calculation employs the full potential linearized augmented planewave method [Blaha et al. (2018)], where one solves the Kohn-Sham equation using the dual basis function. The spherical harmonic basis is used for atoms with a radius called the Muffin-tin radius R_{MT}, and in the interstitial region between atoms, a planewave basis is used. These two basis functions are joined together at the atomic sphere boundary with the same value and same slope, so spherical harmonics acquire an index of the crystal momentum. The full potential refers to the fact that the potential used for atoms is no longer restricted to a potential only retaining $l = 0$ inside the sphere and outside sphere $K = 0$, where l and K refer to the orbital angular momentum quantum number and the reciprocal lattice vector, respectively. The linearized method means that the basis function is no longer dependent on eigenenergies, in contrast to the original augmented planewave method where the basis has an energy dependence which is extremely complicated if one wants to compute band energies.

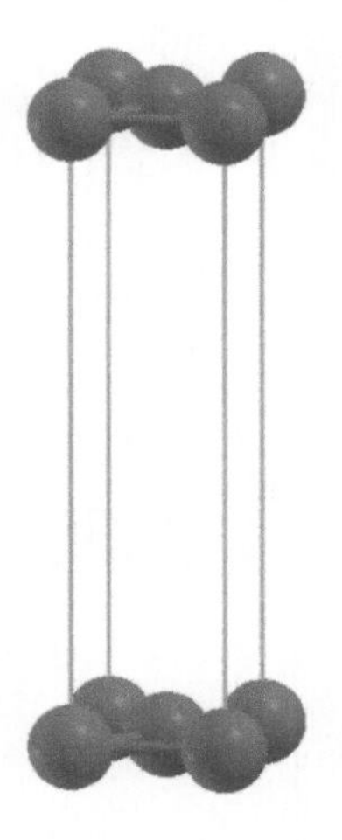

Figure 4.15
One monolayer Fe(110) in the supercell. The vacuum layer that separates from the layer to layer distance is 12 Å thick.

The laser electric field is along the y-axis, with photon energy of 2.0 eV and duration of 60 fs. To compute the harmonic signals, we solve the time-dependent Liouville equation,

$$i\hbar\frac{\partial \rho}{\partial t} = [H, \rho], \tag{4.46}$$

where H includes the system Hamiltonian H_0 and the interaction between the laser and system H_I. $H_0 = \sum_{nk} E_{nk}\rho_{\mathbf{k};n,n}$. $H_I = -\sum_{\mathbf{k};n,m} \rho_{\mathbf{k};n,m}\mathbf{P}_{\mathbf{k};m,n} \cdot \mathbf{A}(t)$, where $\mathbf{P}_{\mathbf{k};mn}$ is the momentum matrix elements between bands m and n at $\mathbf{k}$ point and $\mathbf{A}(t)$ is the vector potential of the laser field. We note that the momentum operator is computed from the Wien2k's optics program: outmat.f. We have modified the code so it exports the matrix elements with full accuracy. The following is our code.

```
1 !      Sept 8 2015
2 write(400)((OX(NB1,NB2),NB2=NB1,NEMAX),NB1=NEMIN,NEMAX)
3 write(400)((OY(NB1,NB2),NB2=NB1,NEMAX),NB1=NEMIN,NEMAX)
4 write(400)((OZ(NB1,NB2),NB2=NB1,NEMAX),NB1=NEMIN,NEMAX)
5 write(400)(E(NB1),NB1=NEMIN,NEMAX)
```

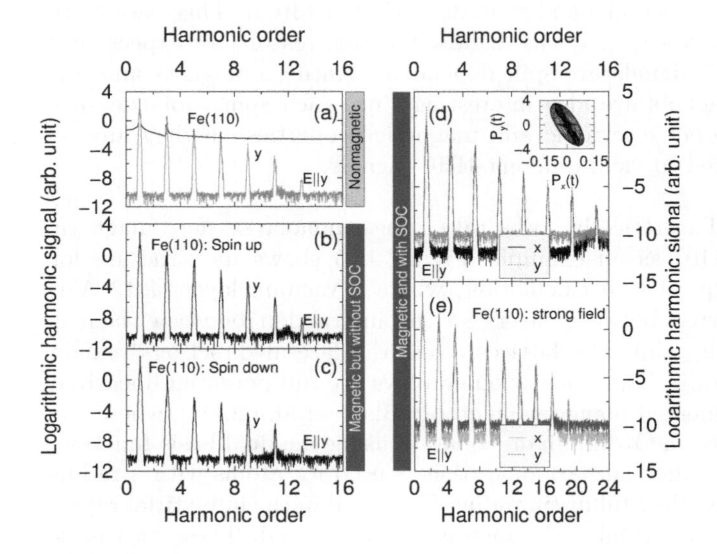

Figure 4.16
Logarithmic harmonic signals from a Fe(110) monolayer. The laser E-field is along the y-axis, with $\hbar\omega = 2.0$ eV, $\tau = 60$ fs and field amplitude $E_0 = 0.09\mathrm{V/\AA}$, for the results in this figure. (a) HHG signal from a nonmagnetic Fe monolayer. The top curve is obtained without using the window function, while the bottom is processed with the window function. (b) HHG from the spin-up channel in a magnetic Fe monolayer. The spin-orbit coupling is not included. (c) Similar to (b), but from the spin-down channel. (d) HHG signal with spin-polarized electrons and spin-orbit coupling. The solid and dashed lines denote the signals along the x- and y-axes, respectively. The spin is oriented perpendicular to the Fe(110) surface. Inset: Phase diagram of $P_x(t)$ versus $P_y(t)$. (e) Same as (d) but with $E_0 = 0.15\mathrm{V/\AA}$, where high harmonics up to 19th order are observed. The figure and caption are from [Zhang et al. (2018c)] with permission from Springer Nature.

Caution is necessary that these matrix elements cannot be used directly since it does not have a correct unit.[6] Its unit is 1/bohr, just $-i\nabla$, where i is the imaginary and the negative

[6] Thanks to Claudia Draxl for numerous communications.

sign is included in the matrix element. If our vector potential has a unit of Vfs/Å, the matrix element should be multiplied by $\mathrm{pa_{constant}} = 100 * \mathrm{h}/(2\pi * \mathrm{bohr} * \mathrm{emass})$, so $\mathbf{P_{k;m,n}} \cdot \mathbf{A(t)}$ has a unit of eV. Density functional calculation can be carried out at either a nonmagnetic or a magnetic level. Thus we can investigate how inclusion of spin affects our spectrum.

Figure 4.16(a) shows the high-harmonic signal on the logarithmic scale versus harmonic order for a nonmagnetic Fe(110) monolayer. We see that harmonics always appear at odd numbers since our system has an inversion symmetry, and the highest harmonic order is at 13th.

Next, we run a spin-polarized calculation, where the spin-up and spin-down channels are computed differently (see Figs. 4.16(b) and (c)). The spectrum shows some differences, but on the logarithmic scale, the difference is small. This indicates that HHG in solids is better plotted on a regular scale. We can also investigate the effect of spin-orbit coupling which couples the majority and minority spins. This has an important consequence. Even if the laser is polarized along the y-axis, the signal is also from the x-axis. Figure 4.16(d) shows both the x- and y-components have comparable magnitudes. What is even more interesting is that if we plot the x- and y-components against each other, an ellipse shape appears. This is the realization of a time-dependent magneto-optical effect in the time domain. If we increase the laser field amplitude to 0.15 V/Å, we see the harmonic order reaches 19.

Finally, we want to draw attention to two curves in Fig. 4.16(a). The one at the top is computed without using the window function, and the one at the bottom is computed with the window function.[7] This matters if $\mathbf{P}(t)$ at the final time has a long tail with a very small but nonzero oscillation. If we carry out the regular Fourier transform without window functions, these small oscillations will increase the overall baseline in the FFT transformed power spectrum so they will smear out harmonic peaks. The treated spectrum has many more peaks by suppressing the base line. To suppress the base line, one multiplies the original data by a window function $\mathcal{W}$ before feeding into the Fourier transform code.

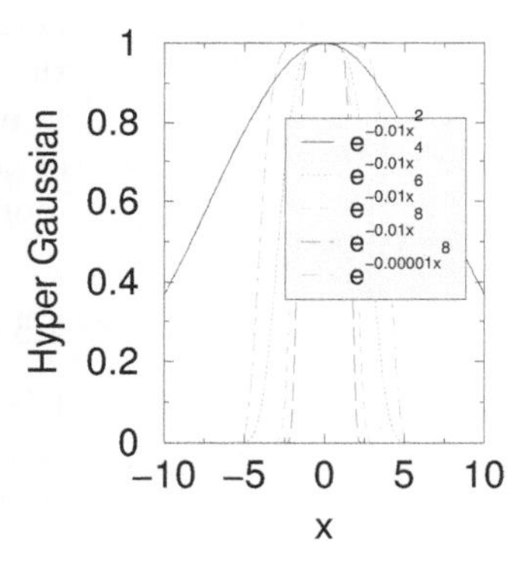

Figure 4.17
This shows how the exponents affect the window function.

There are many hyper Gaussian functions, $\exp(-ax^b)$. But most of them do not work as a window function. The exponent b (here "8") in the function is important. The exponent must be even. The larger it gets, the flatter the peak gets, but gradually the effect becomes smaller. a is the width. This is where one has to watch whether the entire data is covered (see Fig. 4.17).

This method is very effective. We use a hyper or super

[7]Tran Trung Luu (Switzerland, now Hong Kong, China) kindly provided us the details on how the window function is used to Fourier transform a time-dependent signal $\mathbf{P}(t)$ to an actual spectrum.

Gaussian function [Zhang et al. (2018c)],

$$\mathcal{W}(t) = \exp\left[-(at)^8 \times b\right] \tag{4.47}$$

where t is in the unit of fs. a and b are two constants that one has to choose to cover the entire time of $\mathbf{P}(t)$. In our case, $a = 0.035/\text{fs}$ and $b = 5 \times 10^{-9}$ (no units) are found, where the tail of $\mathbf{P}(t)$ drops to zero smoothly at both the leading edge and trailing edge. The following code shows the details, where we have three possible window functions. The reader must change the code to meet her/his needs.

```
1  if(iwindow.eq.1)then
2   window=dtanh((time+ left)*width)
3  &       -dtanh((time+right)*width)
4   window=window/2d0
5  else if(iwindow.eq.2) then
6   window=dexp(-(time*0.035d0)**8*5d-9)
7  else if (iwindow.eq.0) then
8   window=1d0
9  else
10  stop'your window selection incorrect'
11  endif
```

In summary, our theoretical investigation represents a beginning of high-harmonic generation in magnets. One expects more interesting research ahead along this direction. For instance, can one develop HHG as a tool to access some magnetic properties that otherwise are not accessible? Although the photon energy of harmonics is generally lower than those in atoms, would it be possible to explore helicity dependence of the harmonic signals? Any possibilities to resolve them in momentum space? The field is wide open.

4.5.3 Simple picture of HHG in solids

HHG in solids is more complex because it involves the band structure of a solid. To give the reader some insights, we consider an electron in a one-dimensional band dispersion like

$$E_k = E_0[1 - \cos(ka)], \tag{4.48}$$

where a is the lattice constant, k is the crystal momentum, and E_0 is a constant. Its velocity is

$$v_k = \frac{1}{\hbar}\frac{\partial E_k}{\partial k} = \frac{E_0 a}{\hbar}\sin(ka). \tag{4.49}$$

The effect of laser excitation can be written as a shift in the crystal momentum [Zhang et al. (2018a)],

$$k \to k + eA(t)/\hbar\ , \tag{4.50}$$

where $A(t)$ is the vector potential of the laser field.[n] We can substitute Eq. 4.50 into 4.49 and then expand $\sin(ka)$ as

$$a[k + eA(t)/\hbar] + a^3[(k + eA(t)/\hbar]^3/6 + \cdots\ . \tag{4.51}$$

[n] We caution that the vector potential here has no spatial dependence. This means that our magnetic field is zero. See more discussion in Section 3.5.3.

To be more definitive, we suppose $A(t) = A_0 \cos(\omega t)$. The first term in Eq. 4.51 gives the linear response with the same frequency as the incident laser's frequency. The second term contains the term with three times the laser frequency, which corresponds to the third-order harmonic. Caution must be taken that this term also contains $k(eA(t)/\hbar)^2$, which seems to lead to a second-order harmonic. This is incorrect because the system has an inversion symmetry and there is another $-k$ point. When the results from k and $-k$ add up, even order harmonics must be zero. In fact, it can be shown that the average velocity is

$$v_k = \frac{E_0 a}{\hbar} \cos(ka) \sin\left(\frac{eA(t)a}{\hbar}\right), \tag{4.52}$$

whose proof is left as an exercise.

4.6 Exercises

1. Follow the example in Eq. 4.4. Take a hydrogen-like atom as an example, (a) compute the electric fields for $2p$ and $3p$ orbitals, and (b) compare it with the laser field of 2.45 V/Å, which one is larger?
2. Prove that the ponderomotive energy of an electron under a cosine wave field $E = A_0 \cos(\omega t)$ is

$$\bar{E}_{kin} = \frac{1}{4}\frac{e^2 A_0^2}{m_e \omega^2}. \tag{4.53}$$

3. Plot Eq. 4.12 for ωt_i (a) smaller than and (b) larger than 90°. Discuss the physics behind your finding in terms of how the electron precesses under the influence of the laser field.
4. Using the position equation (4.12), find the root for ωt for $\omega t_i = \pi/180$, $11\pi/180$, $13\pi/180$, $15\pi/180$, $17\pi/180$, $19\pi/180$, $21\pi/180$. You can compare your results with the Appendix A.8.
5. Following the previous problem, compute the kinetic energy for each pair $(\omega t, \omega t_i)$, and prove the largest kinetic energy is indeed $3.17U_p$. Hint: Use Eq. 4.10.
6. Explain what happens for ωt_i above 90°.
7. Use the C_{60} code in Appendix A.6 to compute the structure of C_{60}, and use your favorite software to plot it.
8. (a) Prove Eq. 4.52 $v_k = \frac{E_0 a}{\hbar}\cos(ka)\sin(\frac{eA(t)a}{\hbar})$ for the band structure with dispersion $E_k = E_0(1 - \cos(ka))$. (b) Supposing the vector potential is $A(t) = A_0 \cos(\omega t)$, where ω is the laser frequency and t is time, show the harmonic spectrum only exhibits 1st, 3rd, 5th and other odd harmonics. Hint: since harmonics are proportional to the velocity, one can expand the velocity. (c) If the k point is close to the first Brillouin zone boundary ($ka = \pi$), what happens to the harmonic signal? What about $ka = \pi/2$?

5 Femtomagnetism

5.1 History of femtomagnetism
5.2 Magnetic materials
5.3 Time scale of laser-induced demagnetization
5.4 Sample experimental results
5.5 Mechanisms still under debate
5.6 Exercises

After several years of rapid development, femtosecond laser pulses entered many labs around the world in the 1980s. It was an exciting time for optical research and beyond. Such a short pulse allows one to investigate events on several hundred down to six femtosecond time scales [Shank et al. (1990)]. In physics and material science, research focused on atoms, molecules, nanostructure, clusters, insulators, conductors, semiconductors and superconductors. Being able to directly probe electron correlation, scattering and plasma is a big achievement [Kaindl et al. (2003)]. In chemistry, femtochemistry was firmly established, where one can use an ultrafast laser to steer chemical reaction selectively. Spin detection also began [Baumberg et al. (1994)]. However, to that end none of the investigated materials is truly magnetic. This chapter introduces the reader to exciting opportunities at the interface between light and magnetism: How does a laser alter magnetic properties of a magnet on an ultrafast time scale?

Magnetism is traditionally a difficult subject as it involves electron correlation effects, magnetic domain structures, spin-orbit coupling, magnetic anisotropy and spin texture, none of which is covered in an undergraduate physics class. The underlying theory may also involve the Heisenberg exchange spin model, Stoner model and Hubbard model, coupled with advanced first-principles calculation. A complete description of magnetism is desirable, but difficult. This chapter focuses on basic concepts and experimental findings in femtomagnetism, while theoretical treatments that require prior knowledge of advanced quantum physics or chemistry are left to the end of the chapter. Although this greatly limits the scope of this chapter, it is indeed too early to discuss the detailed understanding of femtomagnetism, since the underlying mechanism is still under debate.

5.1 History of femtomagnetism

In 1995, Jean-Yves Bigot made a presentation at the Conference on Lasers and Electric-Optics (CLEO) in Baltimore, USA, where he first discussed the possibility to use a femtosecond laser pulse to manipulate spins. He communicated this information to one of us (GPZ) in an email exchange. Jean-Yves returned to Strasbourg, France, and was joined by

Eric Beaurepaire, J.-C. Merle and A. Daunois to carry out the pioneering experiments in ferromagnetic nickel [Beaurepaire et al. (1996)]. Immediately struck by a talk given by Eric Beaurepaire in Berlin in 1995, Wolfgang Hübner visited the Strasbourg group for six months in 1996/1997 before moving to Halle, Germany. In 1998, Jean-Yves Bigot and Eric Beaurepaire were visiting Wolfgang Hübner and Guoping Zhang in Halle, Germany. The name *femtomagnetism* was invented in a restaurant. At that time, we worried that our field was too small to make any important impact in physics. In particular, at that time, Bigot-Beaurepaire's group was the sole experimental group, while Hübner's group was the sole theory group. The name is modeled after femtochemistry and first appeared in a review article [Zhang et al. (2002b)] in a book edited by Kamel Ounadjela (Strasbourg) and Burkard Hillebrands (Kaiserslautern). This review article was a summary report from the European TMR (Training and Mobility Research) Network in the Europe meeting in Greece, which Eric, Wolfgang and Guoping attended.

To our amazement, attention from the physics community was immediate, with three papers [Hohlfeld et al. (1997); Aeschlimann et al. (1997); Scholl et al. (1997)] published in Physical Review Letters one year after. Now we have an Ultrafast Magnetism Conference every two years. The growth in this field is far beyond our original imagination [Kirilyuk et al. (2010)].

5.2 Magnetic materials

We start with a short introduction to magnetism before we embark on femtomagnetism, as excellent and extensive discussions appear in many books [Kittel (1996); Stöhr and Siegmann (2006); Coey (2010)].

5.2.1 General properties

Magnetism is quantum mechanical. The major component of a magnet is the electron spin. These spins form a spatial pattern in a magnet, so they appear as domains, or magnetic domains, to minimize the total energy. Magnetic domains have sizes, from 10 nm to 1μm or larger. They are the origin of hysteresis loops observed in ferromagnets shown in Fig. 5.1(a). Consider the magnetization $\mathbf{M}$ initially along the $-x$-axis (point a in Fig. 5.1(a)). We apply an external magnetic field $\mathbf{H}$ along the $+x$-axis. As $\mathbf{H}$ increases, domain walls move, and $\mathbf{M}$ becomes less negative and at point b drops to zero where $\mathbf{H}_c$ is commonly called the coercive field. Further increasing $\mathbf{H}$ saturates $\mathbf{M}$ (point c) where the magnetization in the entire sample points in the same direction as $\mathbf{H}$. $\mathbf{M}$ is called the saturated magnetization $\mathbf{M}_s$. If we reduce the magnetic field, $\mathbf{M}$ does not follow the original path because some magnetic energy is still stored in the domains; at $\mathbf{H} = 0$, $\mathbf{M}$ is not zero (point d), which is magnetic remanence. If we increase $\mathbf{H}$ along the $-x$-direction, the situation is similar. There is another way to diminish $\mathbf{M}$. If one heats a sample, $\mathbf{M}$ decreases slowly; once the temperature is above its Curie temperature, its magnetization drops to zero (see Fig. 5.1(b)). Experimental magnetization is given in the unit of kA/m or emu/cc (electromagnetic unit per cubic centimeter). These two units are exactly the same. The magnetic moment (in units of [Am^2]) is defined as the magnetization multiplied by the volume. To convert the magnetic moment from SI units to Bohr magneton (μ_B), one needs to multiply by $10^{-3}/9.274$.

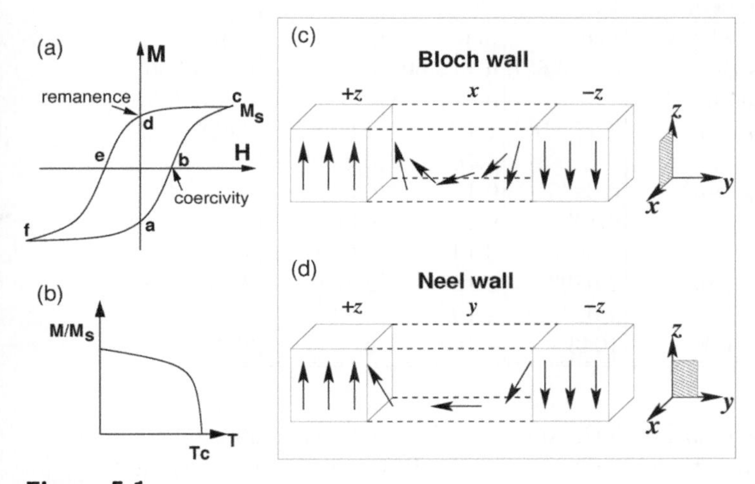

Figure 5.1
(a) Ferromagnetic materials have a characteristic hysteresis loop. Upon increase in an external magnetic field $\mathbf{H}$, the magnetization $\mathbf{M}$ of the sample increases nonlinearly up to a saturated magnetization $\mathbf{M}_s$. (b) Magnetization curve versus temperature. The magnetization is reduced to zero at the Curie temperature T_c. (c) and (d) The origin of the hysteresis loop is the magnetic domain. If we align the magnetic domain wall along the y-axis, there are at least two ways that the spins transition from one orientation $(+z)$ to another orientation $(-z)$. If the spins rotate within the xz-plane (see the shaded region), which are perpendicular to the y-axis and have zero component along the y-axis, this is called a Bloch wall (c). (d) On the other hand, if the spins rotate within the yz-plane, then the wall is called a Neel wall.

Between magnetic domains, there are domain walls. Figures 5.1(c) and (d)) show two common walls: Neél and Bloch walls. Spins inside these walls smoothly transition from one orientation of one domain to another orientation of a neighboring domain. If we arrange the wall normal direction along

Table 5.1
Structure and magnetic properties for four element ferromagnets. bcc: Body-centered cubic with space group $Im3m$ (No. 229). hcp: Hexagonal closed packed with space group $P6_3/mmc$ (No. 194). fcc: Face-centered cubic with space group $Fm3m$ (No. 225). A useful webpage is http://www.cryst.ehu.es/, which lists all 230 space groups. The reader may find Pearson's book [Pearson (1958)] useful. The Curie temperature is taken from Kittel's solid state physics book [Kittel (1996)]. $M_{\rm tot}$: Total moment. M_o: Orbital moment. T_c: Curie temperature. Lattice refers to lattice constant. a: [Wein (1988)]. b: [Frait and Gemperle (1971)]. c: [Kittel (1996)]. d: Ref. [Wijn (1991)]. e: [Farle (1998)]. f: This is the gyroscopic g factor [Ogata et al. (2017)]. g: [Wohlfarth (1980)].

Element	Fe	Co	Ni	Gd
Structure	bcc	hcp	fcc	hcp
Lattice	$a = 2.867$	$a = 2.507$	$a = 3.524$	$a = 3.636$
constant (Å)		$c = 4.070$		$c = 5.783$
$M_{\rm tot}$ ($\mu_{\rm B}$)	$2.22^{a,c,g}$	$1.72^{c,g}$	0.606^c	7.63^c
		1.75^a	$0.62^{a,g}$	
M_o ($\mu_{\rm B}$)	0.09^a	0.16^a	0.05^a	
	0.14^e	0.13^e	0.06^e	
	0.0918^g	0.1472^g	0.0507^g	
g-factor	2.089^b	2.18^d	2.18^a	2.00 ± 0.08^f
	2.09^a	2.19^a	2.21^d	
T_c (K)	1043^c	1388^c	627.2^c	292.5^c

the y-axis, with the spin directions in two neighboring domains along the $+z$- or $-z$-axis, the Bloch wall has the spins that tilt within the xz-plane, but the Neél wall has the spins that tilt within the xy-plane. The width of the wall is determined by the total energy. The total energy in ferromagnets consists of the exchange energy, magnetocrystalline anisotropy energy (MAE), which further consists of volume anisotropy and surface anisotropy energies, and a weaker magnetic dipole-dipole interaction. MAE originates from the spin-orbit coupling. The unquenched orbital moment renders the spins to energetically align them along a few energetically preferred directions, which are called easy axes [Kittel (1996)].

5.2.2 Element ferromagnets and microscopic interactions

Most elements in the periodic table are not magnetic. Magnetic elements start in the third row. Four common ferromagnets are Fe, Co, Ni and Gd, where Fe has a bcc structure, Ni has an fcc structure, and both Co and Gd have a hcp structure. Table 5.1 lists the key properties of these ferromagnets. Ni is the weakest ferromagnet among four, and Gd is the strongest. Due to the crystal field, the orbital moments are universally smaller than the spin moments, which is reflected in the g factor. In solids, due to the translation symmetry, the total angular momentum is not conserved. One cannot use it so as to understand the connection between spin and orbital moments.

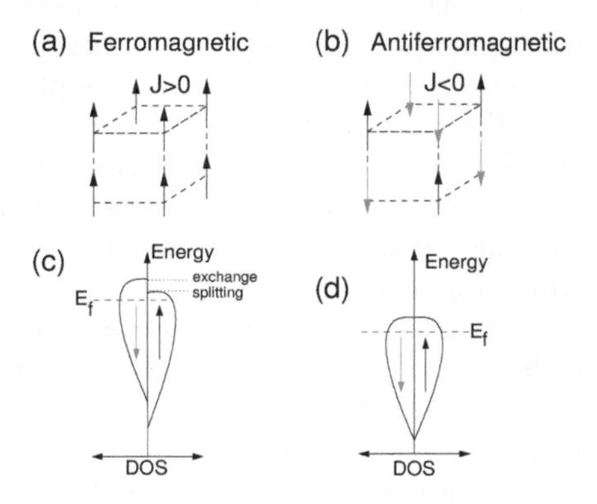

Figure 5.2
(a) In ferromagnets, all the spins point in the same direction, where the exchange interaction J is positive. (b) In antiferromagnets, spins on the neighboring sites point in an opposite direction. Here J is negative. (c) Density of states in a ferromagnet. The number of electrons in spin-up and spin-down channels is different. The one that has more electrons is called majority spin channel, or spin-up channel. The top band energy difference between minority and majority channels is defined as the exchange splitting (see the dotted lines). The long dashed line denotes the Fermi energy E_f. (d) Density of states in antiferromagnets. Both spin channels have the same occupancy.

The reason why a material becomes magnetic is the presence of an exchange interaction between spins. The exchange interaction energetically favors a particular spin configuration. Two commonly used models are: The local spin model (Heisenberg picture), where the spins are considered to be localized at lattice sites, and itinerant spin model (Stoner picture), where spins are allowed to move around. Figures 5.2(a) and 5.2(b) show two different structures with spins at corners of a cube. If the spins in a material at each lattice site point in the same direction, the material is a ferromagnet (see Fig. 5.2(a)). Here the exchange interaction is positive. Such a spin configuration is captured through the Heisenberg model,

$$H = -J \sum_{i>j} \mathbf{S}_i \cdot \mathbf{S}_j, \tag{5.1}$$

where the summation is over lattice sites and only over distinctive pairs $i > j$, J is the exchange interaction, $\mathbf{S}_{i(j)}$ is the spin at site $i(j)$, and $-$ is adopted due to the antisymmetry of the wavefunction (see Problem 1 in the Exercise). One can see that if $J > 0$, in order to minimize the total energy, the spins

S tend to align with each other. This is why the ferromagnets have the spins pointing in the same direction. If we treat the model semiclassically, the energy cost for each tilting of a spin with respect to its neighboring spin can be computed from $-J|S|^2 \cos\theta$, where θ is the angle between two neighboring spins. If neighboring spins point in opposite directions, these magnets are called antiferromagnets (see an example in Fig. 5.2(b)). In this case, J is negative.[1]

In the Stoner itinerant picture, one invokes band structures to explain the magnetism. In a ferromagnet, we say that its spins are polarized. This means that if we plot the density of states (DOS) as a function of energy, the number of electrons in each channel is different. Figure 5.2(c) shows that close to the Fermi energy (see the dashed line),[a] there are more electrons in the spin-up channel than those in the spin-down channel. In the literature, the spin-up channel is also called the majority channel, while the spin-down channel is called the minority channel. Different from ferromagnets, antiferromagnets have a symmetric density of states for both spin channels (see Fig. 5.2(d)).

[a]In solids, at 0 K, if we fill all the electrons in the energy band, the topmost filled band energy is defined as the Fermi energy. The Fermi energy separates the occupied and unoccupied states. Above 0 K, one uses the chemical potential instead. Naturally how we call the spin-up or spin-down is arbitrary, but it is very convenient for the researcher to communicate if we stick to this convention.

To appreciate how the exchange interaction appears, we consider two electrons, 1 and 2, occupying two orbitals $\phi_a(r)$ denoted by $|a\rangle$ and $\phi_b(r)$ denoted by $|b\rangle$. Because electrons are Fermions, their wavefunctions must be made anti-symmetric with respect to particle exchange, i. e., a Slater determinant,

$$\Psi(1,2) = \frac{1}{\sqrt{2}} \begin{vmatrix} \phi_a(1) & \phi_a(2) \\ \phi_b(1) & \phi_b(2) \end{vmatrix}. \tag{5.2}$$

Exchanging electron 1 with 2 results in sign reversal, i.e., $\Psi(1,2) = -\Psi(2,1)$. Since the Coulomb interaction between two electrons separated by a distance r_{12} is $\frac{e^2}{4\pi\epsilon_0}\frac{1}{r_{12}}$, its expectation value is[b]

[b]To simplify our notation, here we adopt the Dirac notation, $|a\rangle \equiv \phi_a$

$$\langle\Psi(1,2)\left|\frac{e^2}{4\pi\epsilon_0}\frac{1}{r_{12}}\right|\Psi(1,2)\rangle = \underbrace{\langle ab\left|\frac{e^2}{4\pi\epsilon_0}\frac{1}{r_{12}}\right|ba\rangle}_{\text{Coulomb}} \tag{5.3}$$

$$- \underbrace{\langle ab\left|\frac{e^2}{4\pi\epsilon_0}\frac{1}{r_{12}}\right|ab\rangle}_{\text{exchange}}, \tag{5.4}$$

where the Coulomb integral is

$$\langle ab\left|\frac{e^2}{4\pi\epsilon_0}\frac{1}{r_{12}}\right|ba\rangle = \frac{e^2}{4\pi\epsilon_0}\int\int d\mathbf{r}_1 d\mathbf{r}_2 |\phi_a(1)|^2 r_{12}^{-1} |\phi_b(2)|^2$$

$$= \frac{e^2}{4\pi\epsilon_0}\int\int d\mathbf{r}_1 d\mathbf{r}_2 \phi_a(1)^* \phi_b(2)^* r_{12}^{-1} \phi_b(2)\phi_a(1), \tag{5.5}$$

[1]This is an overly simplified picture where we use a negative exchange J for antiferromagnets. Just like all the Coulomb integrals, all the exchange integrals are positive. How one can get a negative J is not as simple as one might think. It contains contributions not only from the electron-electron interaction term, but also from the kinetic energy term.

and the exchange integral is

$$\frac{e^2}{4\pi\epsilon_0}\int\int d\mathbf{r}_1 d\mathbf{r}_2 \phi_a(1)^*\phi_b(2)^* r_{12}^{-1}\phi_a(2)\phi_b(1). \tag{5.6}$$

If we compare the Coulomb and exchange integrals, we see that in the exchange integral one electron exchanges one wavefunction with the other electron, taking two different wavefunctions, ϕ_a and ϕ_b. We call this the exchange integral. One can show that this integral is always positive (see the Exercise problem). As explained above, exchange interactions in antiferromagnets are negative. That is why the exchange integral does not capture the entire physics of the exchange interaction.

Another interaction that is relevant to demagnetization is the spin-orbit coupling (SOC). Physically, the electron spin experiences additional interaction from its own orbital motion. In atoms, it can be rewritten as $H_{so} = \lambda\mathbf{L}\cdot\mathbf{S}$, where λ is the spin-orbit coupling, $\mathbf{L}$ is the orbital angular momentum, and $\mathbf{S}$ is the spin angular momentum. In solids, the expression is

$$H_{\rm SO} = \frac{1}{4m^2c^2}(\nabla V \times \mathbf{P})\cdot\mathbf{S}, \tag{5.7}$$

where V is the crystal potential, $\mathbf{P}$ is the momentum, and $\mathbf{S}$ is the spin. The steeper the potential is, the larger the spin-orbit coupling becomes.

5.3 Time scale of laser-induced demagnetization

With several essential interactions explained, we now address the time scale of ultrafast demagnetization. Since the entire demagnetization involves multiple closely connected interactions among electrons, spins and lattices, it is necessary to discuss them separately. Figure 5.3 sketches a commonly accepted picture. We discuss the time scale for electron and spin responses first, followed by the time scale for phonons.

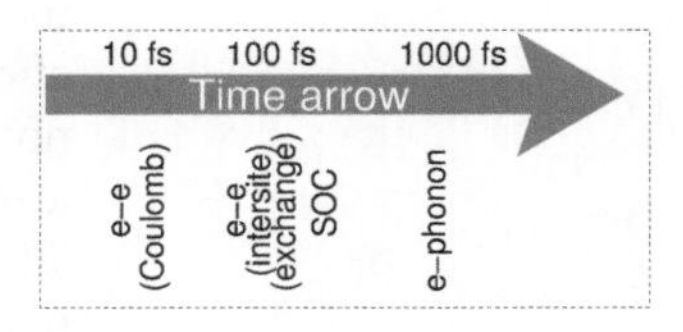

Figure 5.3
Time scale for various interactions. On the shortest time scale, we have electron-electron interactions, followed by the exchange interaction and spin-orbit coupling. At 1 ps, the electron-phonon interaction dominates.

5.3.1 Time scale for electron response

After a sample is excited with a femtosecond laser pulse, electrons respond first. The Coulomb interaction is in control. A useful relation to estimate the time scale is

$$Et = h = 4.14 \text{ eV}\cdot\text{fs}, \tag{5.8}$$

where E is the energy in the unit of eV, t is the time in fs, and h is the Planck constant. $t = 1$ fs corresponds to $E = 4.14$ eV; if $t = 100$ fs, $E = 0.0414$ eV. This means that if the demagnetization occurs within 100 fs, the interaction involved is on the energy order of 0.0414 eV, which is the typical time scale of the exchange interaction and the spin-orbit coupling. In solids, the electron-electron interaction varies a lot. Consider a pair of electrons separated by a distance r, with the potential energy

$$U = \frac{e^2}{4\pi\epsilon_0 r}. \tag{5.9}$$

If $U = 1$ eV, then $r = 14.4$ Å, covering tens of lattice sites. If $U = 10$ eV, the distance is reduced to $r = 1.44$ Å. Therefore, we have another relation,

$$U\,r = 14.4 \text{ eV Å}. \tag{5.10}$$

This equation connects the electron-electron interaction with the spatial separation. If we take fcc Ni's lattice constant of 3.524 Å as an example, we get $U = 4.09$ eV, which is not far off from its actual value. This energy scale corresponds to 1 fs approximately.

During laser excitation, most of electrons do not contribute since their energies are far away from the Fermi energy. Our focus is on those electrons at the Fermi level, whose velocity is defined as Fermi velocity v_f. In fcc Ni, the Fermi velocity is 0.28×10^6 m/s, or 2.8 Å/fs [Petrovykh et al. (1998); Zhang et al. (2018a)]. This means that electrons on the Fermi surface[c] travel 2.8 Å every femtosecond. However, this velocity should not be confused with the collective velocity of electrons, since on the Fermi surface electrons at the $+\mathbf{k}$ and $-\mathbf{k}$ move in opposite directions, so the net velocity is zero and there is no transport of electrons. A laser pulse gives additional velocity to the electrons.

[c]The Fermi surface is an iso-energy surface in the crystal momentum space. One fixes the energy at the Fermi energy and finds the coordinates of the crystal momentum $\mathbf{k}$. Since $\mathbf{k}$ is three-dimensional in general, one has a special shape of the Fermi surface for a particular material.

5.3.2 Time scale for spin response

Light is an electromagnetic wave. A laser pulse has both a magnetic (B) and an electric (E) field. Since $E/B = c$, the effect of the magnetic field is relatively weaker, but this does not mean that it cannot be experimentally observed. Consider a pulse with $E = 0.05$ V/Å, where its magnetic field is 1.67 T. This means that the spins are also affected by the laser magnetic field. Experimentally, it was shown that a terahertz field can affect spins in NiO by means of the Zeeman interaction [Kampfrath et al. (2011)].

In ferromagnetic metals, spins are exchange-coupled with each other (see Fig. 5.4(a)), so the time scale for spin is related to the exchange interaction J. Different materials have different J, so their time scales may differ a lot. It is commonly accepted that J should be weaker than the Coulomb interaction. As an estimate, J is on the order of 0.1 eV, so the time scale is around 40-100 fs (Fig. 5.3). Even though spin response can be fast, this does not necessarily mean that the demagnetization is fast. Figure 5.4(b) shows an example. In the beginning, the spin at site 1 is tilted with respect to its original orientation, and after some time, through exchange coupling the spin at site 2 also becomes tilted (case 1 in Fig. 5.4(b)). The total spin moment does not change. In this case, the demagnetization and spin dynamics are decoupled.[2] Therefore, one cannot directly infer the time scale for demagnetization from the time scale for spin dynamics. On the other hand, if the restoration of spin at site 1 is delayed (case 2 in Fig. 5.4(b)), then the demagnetization is related to the spin dynamics. The first case is a coherent case, a pure spin wave propagation. The second case is an incoherent process related to demagnetization.

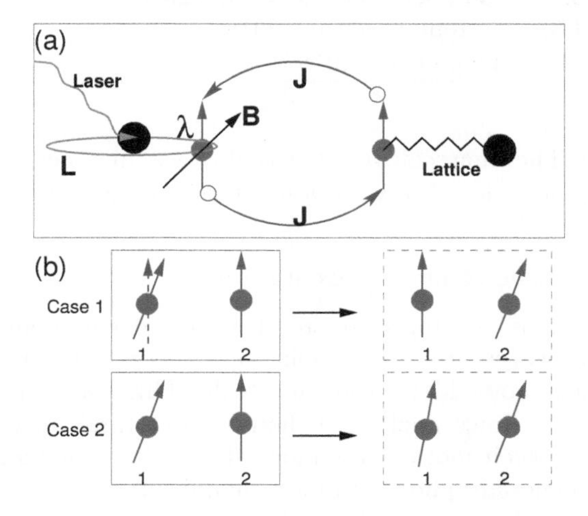

Figure 5.4
(a) Interactions that affect spins (arrow with a filled circle) include the exchange interaction J, spin-orbit coupling λ and electron-phonon interactions. (b) Coherent and incoherent spin wave progression. Case 1 shows a coherent spin wave propagation from site 1 to site 2, with no change to the total spin. Case 2 shows an incoherent spin wave propagation, with demagnetization.

This brings us to the second important interaction, spin-orbit coupling (SOC). Because it is relativistic, λ in $3d$ transition metals is on the order of 0.01 eV, or several hundred fs. On the left of Fig. 5.4(a), we illustrate that each spin is subject to SOC. SOC can retard the spin precession, so spins on each site become incoherent. The second effect of SOC is to allow the spin angular momentum to transfer to the orbital angular momentum.

The third interaction is the magnetic crystalline anisotropy (MCA). MCA destroys the spatial spherical symmetry of the spins. This naturally brings in an incoherent spin precession

[2]This is highly counter-intuitive since a coupling interaction leads to decoupling of observables: The decoupling happens exactly because the spins interact with each other without the necessity of electron transport; thus the spins exhibit slightly different time scales than the charges.

Table 5.2
Element-resolved Hübner times for various compounds. All the numbers are extrapolated to zero fluence. Due to lack of systematic experimental investigations, these times should be used with great care. In particular, we do not have a clear picture of the dependence of the Hübner time on the photon energy dependence and laser pulse duration. For nearly all the cases, only one experiment was performed.

Element	τ_H (fs)
Gd	363 ± 20 ($5d$); 690 ± 20 ($4f$) [Sultan et al. (2011)]
Fe	58.9 ± 30 ($3d$) [Carpene et al. (2008)]
Ni	176 ± 27 [Tengdin et al. (2018)]
CoPt	111 [Zhang et al. (2018d)]

(Fig. 5.4(b)). The time scale of MCA is also system dependent, typically on the order of μeV, or a few picoseconds or longer.

5.3.3 Time scale of phonon excitation

Lattice vibrations are important to thermally driven demagnetization. However, traditional spin wave theory for demagnetization does not address the time scale. This leaves a big time gap to fill. Since nuclei are thousands of times heavier than electrons, their motions are much slower. It is commonly agreed that generally phonons play a significant role only on a longer time scale, on the order of picoseconds. In this regard experimental results are not yet conclusive, and therefore the direct relevance to femtosecond demagnetization is still not fully understood.

5.3.4 Demagnetization time

A particular challenge in femtomagnetism is to accurately determine the demagnetization time τ_M for each element. τ_M depends on the laser fluence. With so many uncertainties in the experimental demagnetization time, we decide to follow Sultan et al. [Sultan et al. (2011)]. We extrapolate all τ_M to the zero fluence limit, in order to eliminate the influence of the laser fluence from all demagnetization times.

This zero-fluence demagnetization time should be unique to each material, and in this book we denote it as the Hübner time, or τ_H. Some elements have more than one Hübner time because more than one orbital participate (Table 5.2). For instance, we notice that the Gd data [Sultan et al. (2011)] have two linear dependences of τ_M on the fluence. The second linear dependence, once it is extrapolated to zero fluence, gives $\tau_H(4f) = 690 \pm 20$ fs [Sultan et al. (2011)] for the $4f$ electron. However, if we extrapolate the first linear dependence to zero fluence, it yields $\tau_H(5d) = 350$ fs, which we believe is from the

5*d* orbitals [Zhang et al. (2017)]. Since not many experiments carry out an intensity-dependent measurement, a complete list of all the investigated materials is not available. Many studies use too strong a laser fluence, so it is not possible to accurately extrapolate to zero fluence. Table 5.2 shows some of the experimental τ_H, which is obtained with good certainty.

5.4 Sample experimental results

Nickel thin films were first investigated, followed by Co, Fe and $CoPt_3$. There are many compounds studied than we can cover, so we focus on two representative experiments. One is on Fe (Fig. 5.5(a)), Ni (Fig. 5.5(b)), permalloy (Fig. 5.5(c)), and permalloy doped with Cu (Fig. 5.5(d)). The other is on the Heusler compounds. Other materials, including experimental techniques used, are summarized at the end of this section.

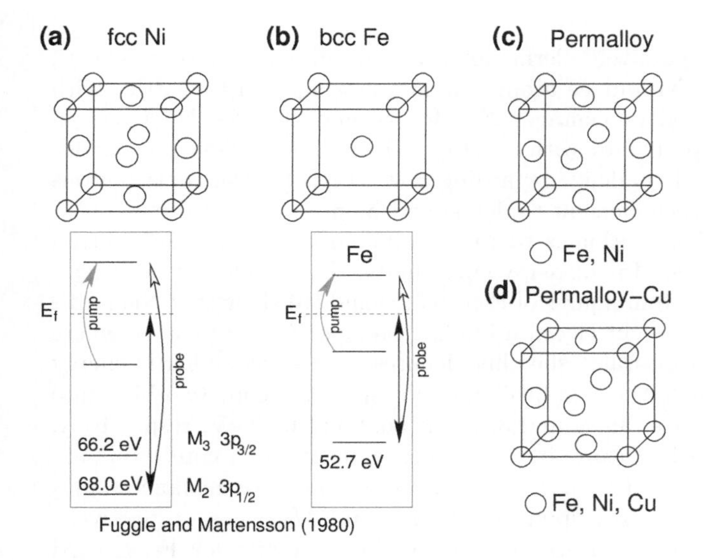

Figure 5.5
(a) fcc Ni structure. Bottom: Core energy levels for Ni at two M edges. (b) bcc Fe structure. The core energies for Fe at two M edges are different from Ni. (c) Permalloy $Fe_{20}Ni_{80}$ structure. Fe and Ni are distributed randomly but with a fixed proportionality, with 20% Fe and 80% Ni. The crystal structure is A1 fcc. (d) Cu-doped permalloy.

5.4.1 Fe, Ni and permalloy

Elemental iron and nickel are of great importance to our understanding of laser-induced ultrafast demagnetization. Table 5.1 already shows the basic parameters of Ni and Fe. Figures 5.5(a) and (b) show the crystal structure of fcc Ni and bcc

Table 5.3
Electron binding energies in electron volts for four element ferromagnets. – denotes the energy is beyond XUV and soft x-ray regions. The splitting of $3p$ is small at the M edge in Fe, but the splitting of $2p$ at the L edge is much larger. All the data are from "X-ray Data Booklet" edited by A. Thompson et al. Courtesy of T. A. Callcott (University of Tennessee).

Element	Fe	Co	Ni	Gd
L_2 $(2p_{1/2})$	719.9	793.2	870	–
L_3 $(2p_{3/2})$	706.8	778.1	852.7	–
M_1 $(3s)$	91.3	101.0	110.8	–
M_2 $(3p_{1/2})$	52.7	58.9	68.0	–
M_3 $(3p_{3/2})$	52.7	59.9	66.2	–
N_5 $(4d_{5/2})$	–	–	–	142.6
N_6 $(4f_{5/2})$	–	–	–	8.6
N_7 $(4f_{7/2})$	–	–	–	8.6

Fe, respectively. Permalloy is an iron-nickel alloy (Fig. 5.5(c)), where Ni and Fe atoms randomly occupy an fcc lattice with the fixed percentage, 20% for Fe and 80% for Ni. One can dope permalloy with Cu (Fig. 5.5(d)). These alloys pose an additional challenge regarding how to resolve the contributions from each element to demagnetizaion.

The solution is to target differences in the core electron energies.[d] In the x-ray community, this is called x-ray edges. The bottom figures of Figs. 5.5(a) and (b) show that Ni and Fe have different electron binding energies for core states, where the horizontal dashed line denotes the respective Fermi energy E_f. In Ni the spin-orbit coupling splits the core $3p$ orbital into two levels, $3p_{1/2}$ and $3p_{3/2}$ at 66.2 and 68.0 eV, respectively. In the literature and x-ray community, same core levels are also labeled with K, L, M, and so on.[e] For instance, $3p_{1/2}$ and $3p_{3/2}$ correspond to the M_2 and M_3 edges, respectively. Table 5.3 provides electron binding energies for Fe, Co, Ni and Gd. These core electron binding energies give us a unique capability to target different elements in a compound, which is called element specificity.

[d]Core energies are insensitive to chemical environment, so that they can be used as a unique identifier.

[e]The labels K, L, M, and so on correspond to the principal quantum numbers $n = 1, 2, 3, \ldots$

The experiment by Mathias and his colleagues [Mathias et al. (2012)] precisely used this capability. Experimentally, a pump pulse with a strong intensity fires at the sample. The pulse has a duration of 25 fs and wavelength of 780 nm. The laser operates at a 2 kHz repetition rate. Each pulse has an energy about 2.2 mJ. Only 10% of the laser power is used as the pump pulse. The remaining 90% is used to generate high harmonics from the neon gas inside a waveguide. This is an excellent example of why high harmonic generation is so useful for the tabletop experiment [Berlasso et al. (2006)]. The generated harmonics have an energy span from the 21st up to the 43rd order, with the energy in the extreme ultraviolet

(XUV) regime (from 35 to 70 eV), and target the M-edge resonances of both Fe and Ni. These harmonic pulses are much shorter, with duration less than 10 fs, and are used as probe pulses in the experiment.

The pump pulse excites the electrons and therefore the absorption of the sample changes. After a short delay, the probe arrives at the sample and is reflected from the permalloy grating sample. Because the probe pulse consists of multiple harmonics with different energies, if the energy matches the absorption edge (i. e. 66.2 and 68.0 eV for Ni, and 52.7 eV for Fe), the light is absorbed, so the reflection spectrum has a smaller amplitude. The bottom panel of Figs. 5.5(a) and (b) helps us understand this process: it shows that the pump pulse first excites electrons from the valence band to the conduction band. However, the probe pulse also samples the same set of the conduction bands, so if the pump excited states are changed, the probe pulse also feels its influence. This is the essential idea of the pump-probe experiment explained in Chapter 2. If we change the delay between the pump and probe pulses, we can investigate the influence of the pump pulse on the electronic system on the appropriate time scale. However, this only gives the charge signal. In order to detect the spin response and the demagnetization, one must move one extra step.

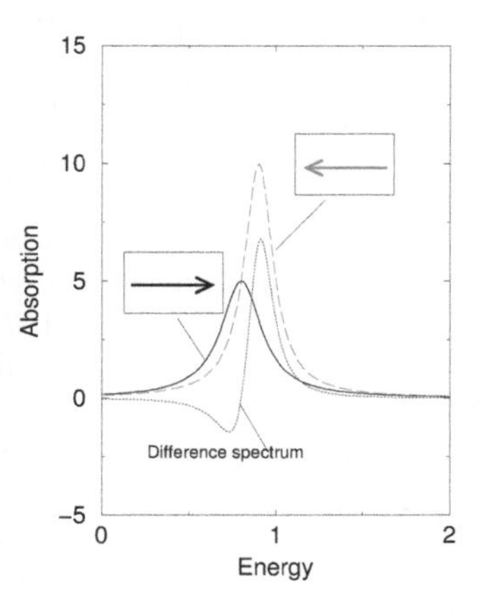

Figure 5.6
X-ray absorption for two different magnetization directions (solid line with magnetization to the right and long-dashed line with magnetization to the left). The dotted line denotes the difference spectrum which is computed by subtracting one absorption spectrum from the other one.

For a magnetic system, the way the system absorbs light depends on the direction of the magnetization. Figure 5.6 shows an example. If we magnetize the sample to the right, the absorption spectrum is given by the solid line. But if we magnetize the sample's magnetization to the left, the absorption spectrum is given by the long-dashed line. The difference between these two spectra is denoted by the dotted line, which is called the difference or asymmetry spectrum. Experimentally, the asymmetry is computed from

$$A = \frac{I_+ - I_-}{I_+ + I_-}, \tag{5.11}$$

where I_+ and I_- are the intensities with the magnetic field along one direction and the opposite direction, respectively. Equation 5.11 computes the difference, $(I_+ - I_-)$, of these two measurements, so the response that does not depend on the external magnetic field is subtracted. What remains is magnetic.[3] This asymmetry is now directly proportional to the magnetization of the sample. It is this asymmetry A that is plotted as a function of time in Fig. 5.7. A in Eq. 5.11 and magnetization are the same thing. Figure 5.7(a) shows the demagnetization as a function of time for elemental Fe and Ni. For this particular fluence of 2 mJ/cm^2, the magnetization is quenched by 19% for Fe, while for Ni it is 45%. One can

[3]This scheme is not perfect since a higher order of magnetic response that is not linear with the external magnetic field may still contribute.

see that the reduction in magnetization is related to the spin moment amplitude, but whether this is generic is still under debate. The demagnetization time τ_M is found through fitting the experimental normalized magnetization $m(t)$, where the normalization is done with respect to the initial value, to an exponential function

$$m(t) = 1 - \Delta m[1 - \exp(-t/\tau_M)] \exp(-t/\tau_r), \qquad (5.12)$$

where Δm is the maximum magnetization reduction, and τ_r is the recovery time. This gives $\tau_M(\text{Fe}) = 98 \pm 26$ fs, $\tau_M(\text{Ni}) = 157 \pm 9$ fs, $\tau_r(\text{Fe}) = 11 \pm 7$ ps and $\tau_r(\text{Ni}) = 9 \pm 1$ ps. There is also another active debate as to whether τ_M is proportional or inversely proportional to the spin moment. We note that the demagnetization time τ_M found here is not the Hübner time τ_H, so τ_M depends on the laser field amplitude. The two lines in 5.7(a) are the fitted functions for Fe and Ni.

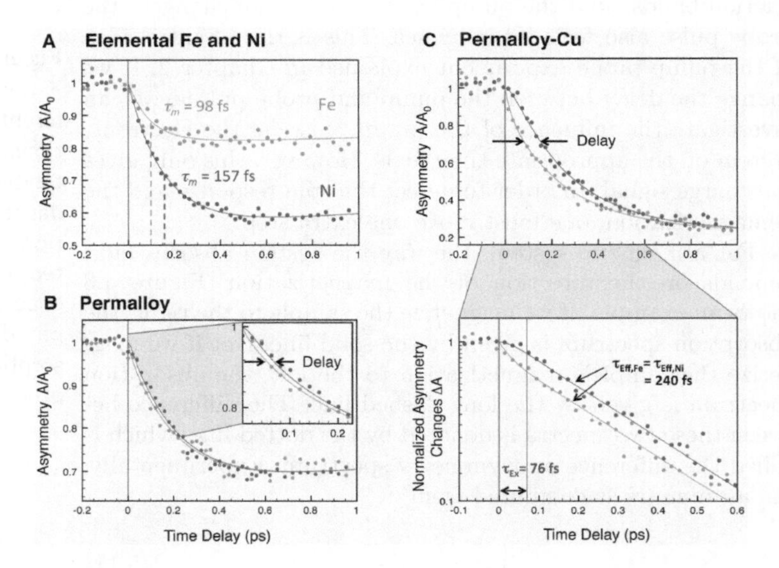

Figure 5.7
Demagnetization in four materials. (a) Under the same laser amplitude, elemental Ni demagnetizes more than Fe. Their demagnetization times are 98 fs for Fe and 157 fs for Ni. All the laser parameters are given in the text. (b) Permalloy demagnetizes about 30% under the same laser conditions. There is a delay of about 10-20 fs between Ni and Fe, but both Ni and Fe demagnetize on the same time scale. (c) The delay between Ni and Fe in permalloy-Cu increases to 76 fs, but their demagnetization time is the same, 240 fs. The figure courtesy of Stefan Mathias. Used with permission from PNAS (USA).

The exchange interaction that was introduced in Section II has an important effect on the demagnetization. One can theoretically prove that the demagnetization time is inversely proportional to the exchange interaction [Zhang et al. (2015c)], but the exact relation is difficult to develop because there are no systematic investigations on the exchange interaction. One has to rely on the experiment to gain the key physics.

Figure 5.7(b) shows demagnetization in permalloy. There are three unusual findings. First, under the same laser conditions, as a whole the permalloy sample demagnetizes by 30%, which is higher than the pure nickel but lower than the pure iron, which is reasonable considering that the Curie temperature of permalloy is lower than iron but higher than nickel. However, it is unexpected that both Fe and Ni in permalloy demagnetize to the same level, if we consider that Fe and Ni demagnetize locally alone in the traditional Heisenberg picture (Eq. 5.1). Therefore, the itinerant picture may account for the amount of demagnetization, but this is not conclusive since we do not know whether the laser wavelength and other parameters may play a role too.

Second, the iron demagnetizes 10-20 fs earlier than the nickel. If the exchange coupling between Ni and Fe is so strong, why is there such a delay? From Eq. 5.8, 10-20 fs corresponds to 0.207-0.414 eV, which is indeed on the order of the exchange interaction. But this delay cannot be easily explained in the itinerant picture, where the majority and minority band shifting lead to demagnetization (see Fig. 5.2). Instead, we believe it is more likely that spin wave propagation in the spin wave picture leads to this delay in demagnetization [Knut et al. (2018); Zhang et al. (2019)]. In diluted permalloy with Cu, or permalloy-Cu (Fig. 5.5(d)), where the sample has 60% permalloy ($Ni_{0.8}Fe_{0.2}$) and 20% Cu and has Curie temperature of 406 K, the delay increases to 76 fs (see Fig. 5.7(d)). With Cu, the exchange coupling becomes weaker, but there are three possible couplings: Fe-Fe, Fe-Ni, and Ni-Ni. If this delay is due to Fe-Ni, this may make sense. These amorphous materials are ideal to fine tune exchange interactions [Jana and et al. (2018)], but are difficult to simulate theoretically without precise structural knowledge.

The demagnetization delay between Fe and Ni appears due to the additive nature of Fe, Ni and Cu, but the demagnetization time is entirely different. In both permalloy and permalloy-Cu, both Fe and Ni demagnetize with the same time constant. Figure 5.7(d) shows that in permalloy-Cu, the effective demagnetization time is 242 ± 12 for Fe and 236 ± 13 fs for Ni. This same demagnetization time is highly unexpected, since the exchange interaction and spin moment at Fe and Ni sites are all different. If demagnetization proceeds only through the magnon generation which is controlled by the exchange interaction, one expects a different demagnetization time for Ni and Fe. Therefore, one way to explain this is the itinerant nature of spins in permalloy. To test this idea, one can use an insulator, but this has not been done experimentally. It seems that to explain laser-induced ultrafast demagnetization one has to invoke both the magnon picture (Heisenberg exchange picture) and the itinerant picture (Stoner's picture), but whether the delay is mainly due to the magnon picture and demagnetization time is due to the itinerant picture is unknown to this end.

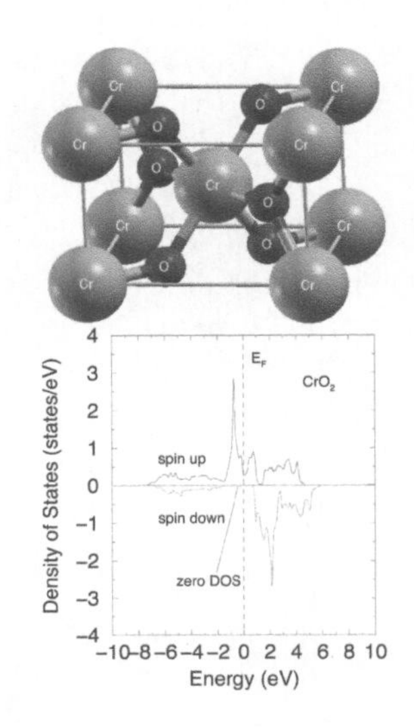

Figure 5.8
(Top) Crystal structure of CrO_2. (Bottom) Density of states for the majority and minority spins. The Cr atom has a spin moment of 1.91 μ_B. The spin moment on O is small, $-0.076\mu_B$.

5.4.2 Half-metallic and Heusler compounds

In ferromagnetic metals, both the majority and minority bands have density of states across the Fermi level (see Fig. 5.2), and both contribute to the laser-induced demagnetization. It is difficult to disentangle whether the demagnetization proceeds through spin-flipping between the majority and minority bands, through spin wave excitation, or through phonons and other mechanisms. If we can shut off some of these processes, maybe we can identify the channels which are essential to demagnetization. In the following, we take CrO_2 and full and half-Heusler compounds as an example. Since full and half-Heusler compounds are important compounds for spintronics, we decide to provide details about their crystal structures.

CrO_2 is half-metallic, where there is no band gap in the majority channel, but there is a gap in the minority channel. Figure 5.8 shows its structure and the density of states (DOS). The minority spin channel has zero DOS at the Fermi level, so it is semiconducting [Schwarz (1986)]. Now, consider a laser pulse exciting electrons out of the majority valence band to the majority conduction band. These excited electrons will quickly relax to the Fermi surface. The minority channel is different. Since the minority channel around the Fermi surface has no states, the majority electrons, unlike in ferromagnetic metals, have no possibility to scatter into the minority states through spin-flip processes. This leads to a very slow demagnetization [Zhang et al. (2006)]. However, Zhang, Nurmikko and coworkers misinterpreted our theory [Zhang and Hübner (2000)], which is based on fcc Ni, and not CrO_2. The slow demagnetization is due to the missing minority channel so the efficient spin flipping is not possible. The system has to proceed through phonons and other high-order terms to dissipate spins.

Heusler compounds are another example of materials that can be half-metallic. They have a face-centered cubic structure (fcc), see the the Wyckoff positions in Table 5.4. Its nominal composition is X_2YZ, where X and Y refer to transition metals, and Z refers to a main group element. If X=Y=Z, there would be four fcc sublattices. Heusler compounds crystallize in the $L2_1$ structure (No. 225, Fm-3m, Cu_2MnAl structure), where $L2_1$ is the Strukturbericht notation and L denotes alloys. Two X atoms are at the eight (8*c*) positions, $(\frac{1}{4},\frac{1}{4},\frac{1}{4})$, $(\frac{3}{4},\frac{3}{4},\frac{3}{4})$, Y is at (4*b*), $(\frac{1}{2},\frac{1}{2},\frac{1}{2})$, and Z is (4*a*), (0,0,0). Note that two X atoms are equivalent in this structure, so one cannot have antiferromagnetic configurations between two X atoms. If we lose one X, XYZ are called half-Heuslers, where one fcc sublattice has no atom.

Table 5.4
Wyckoff positions of Heusler compounds X_2YZ and half-Heusler compounds XYZ. In the Cu_2MnAl structure, where X = Cu, Y = Mn, and Z = Al, two X atoms are equivalent. In the Hg_2CuTi structure, where $X_1 = Hg_1$, $X_2 = Hg_2$, Y = Mn, and Z = Al, the two X atoms are inequivalent. Half-Heusler compounds lose one of two inequivalent X atoms.

Cu_2MnAl structure	$L2_1$, No. 225, Fm-3m
Cu_2 [X_2 (8c)]	$(\frac{1}{4},\frac{1}{4},\frac{1}{4})$,$(\frac{3}{4},\frac{3}{4},\frac{3}{4})$
Mn [Y] (4b)	$(\frac{1}{2},\frac{1}{2},\frac{1}{2})$
Al [Z] (4a)	(0,0,0)
Hg_2CuTi structure	C_{1b}, No. 216, F-43m
Hg_1 [X_1 (4a)]	(0,0,0)
Hg_2 [X_2 (4c)]	$(\frac{1}{4},\frac{1}{4},\frac{1}{4})$
Cu [Y] (4b)	$(\frac{1}{2},\frac{1}{2},\frac{1}{2})$
Ti [Z] (4d)	$(\frac{3}{4},\frac{3}{4},\frac{3}{4})$
MgAgAs structure	C_{1b}, No. 216, F-43m
Mg [X (4d)]	$(\frac{3}{4},\frac{3}{4},\frac{3}{4})$
Ag [Y (4c)]	$(\frac{1}{4},\frac{1}{4},\frac{1}{4})$
As [Z] (4a)	(0,0,0)

What is often confusing is that X_2YZ can also adopt the Hg_2CuTi structure. According to Graf et al. [Graf et al. (2009)], Pearson's Handbook does not uniquely classify X_2YZ. Here the space group is No. 216, F$\bar{4}$3m, or C_{1b}. One X atom takes the 4a position (0,0,0), and the other X atom takes the 4c position $(\frac{1}{4},\frac{1}{4},\frac{1}{4})$. In other words, although we use the same X atom, these two X positions are different structurally. The Y atoms occupy (4b) $(\frac{1}{2},\frac{1}{2},\frac{1}{2})$, and the Z atoms occupy (4d) $(\frac{3}{4},\frac{3}{4},\frac{3}{4})$.

Münzenberg and coworkers pursued this line of research by extending into a group of full and half-Heusler compounds such as Co_2MnSi and Co_2FeAl. Figure 5.9 shows that as the spin polarization increases, the demagnetization time increases as well. This further shows that the spin flipping is very important to demagnetization. For Fe, Co, Ni and permalloy, where the spin polarization is small, both spin channels contribute and spin can flip, so their demagnetization time is shorter. As the spin polarization becomes larger, the spin flipping becomes increasingly difficult, so the demagnetization has to proceed through other means such as phonons.

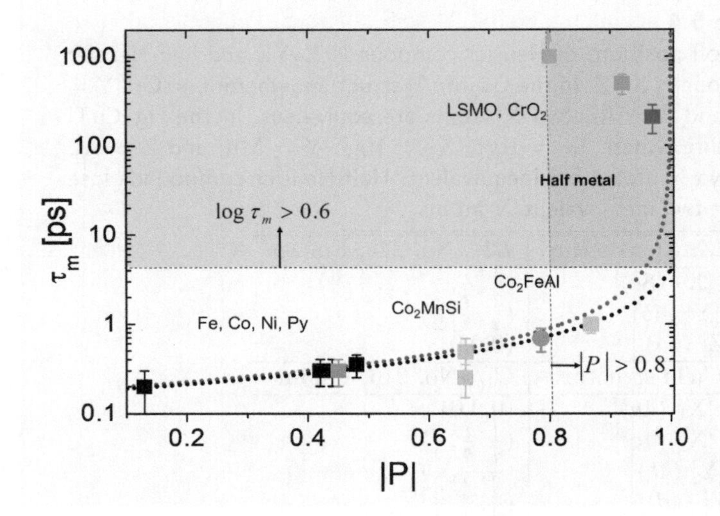

Figure 5.9
Demagnetization time τ_{M} as a function of the spin polarization P for a group of materials. As the spin polarization increases, the demagnetization time increases from hundred femtoseconds up to several hundred picoseconds. Note that they did not investigate the Hübner time. Figure courtesy of M. Münzenberg. Used with permission from Müller et al. [Müller et al. (2009)].

5.4.3 Short experimental summary

It is obvious that decades of investigation have produced a significant amount of experimental results. Experimental techniques include time-resolved magneto-optics, two-photon photoemission, x-ray magnetic circular dichroism, angle-resolved photoemission spectroscopy, and second-harmonic generation.

Table 5.5 shows the experimental demagnetization times for a group of compounds investigated. We only list those which have a clear percentage loss of spin moment. Some of the data is compiled from Radu's 2015 table [Radu et al. (2015)]. Even for the same magnetization loss, there is no agreement among different experiments. In GdFeCo, the disagreement is larger in Gd.

Table 5.6 shows some of the most intensively investigated compounds. There are many more that we cannot cover. The reader may find several review articles helpful [Zhang et al. (2002b); Kirilyuk et al. (2010)]. Further experimental research is necessary. In particular, there is a great need to detect both spin and orbital dynamics on the time scale. The orbital moment change will provide additional insights as to how charge degree of freedom is altered during laser excitation.

Table 5.5
Demagnetization time τ for each element in its pure form or in an alloy. Pulse durations (in units of fs) and photon energy (or wavelength) are different: 25 fs/780 nm (pump), < 10 fs/M-edge for Fe and Ni (35-72 eV) (probe) [Mathias et al. (2012)]; 60 fs/1.55 eV (pump), 100 fs @L_3-edge for Ni, Fe, and Co or M_5 edge for Gd and Dy (probe) [Radu et al. (2015)]; 50 fs/800 nm [Koopmans et al. (2010)]; 40 fs/800 nm [Hennecke et al. (2019)]; 60 fs/780 nm-100 fs @$L_{3,2}$ [Stamm et al. (2010)]; 60±20 fs/790nm-100 fs @$L_{3,2}$ [Boeglin et al. (2010)]; 50 fs @1.5eV-M_5 [Wietstruk et al. (2011)]. References: a [Mathias et al. (2012)], b [Radu et al. (2015)], c [Hennecke et al. (2019)], d [Koopmans et al. (2010)], e [Boeglin et al. (2010)], f [Wietstruk et al. (2011)], g [Stamm et al. (2010)], h [Hofherr et al. (2017)].

Material	τ (fs)	$\Delta M/M_0$	Ref.
$Fe_{20}Ni_{80}$	83 ± 7(Fe)	56%	a
	103 ± 5(Fe)	35%	a
	93 ± 10(Fe)	29%	a
	80 ± 8(Fe)	22%	a
	300 ± 50(Fe), 180 ± 40(Ni)	50%	b
$Fe_{20}Ni_{80}$-Cu	242 ± 12(Fe), 236 ± 13(Ni)	70%	a
$Ni_{50}Fe_{50}$	280±50(Fe), 80±30(Ni)	50%	b
GdFeCo	150±55(Fe), 450±100(Gd)	100%	b
	175±65(Fe), 550±150(Gd)	~40%	b
	201±15(Fe_{L_3})	88%	c
	204±24(Fe_{L_2})	91%	c
	259±42(Gd_{M_5})	79%	c
	270±54(Gd_{M_4})	75%	c
$DyCo_5$	170±30(Co), 1025±150(Dy)	~50%	b
Pure Fe	220±50	~45%	b
	98±26	~19%	a
Pure Ni	130±40	~95%	g
	155±50	~80%	b
	145±30	~65%	b
	157±9	45%	a
	~147	~ 43%	d
	~175	~ 46%	d
	~197	~65%	d
	~211	~69%	d
	~191	~ 93%	d
Ni/Au	42±8	87±1%	h
Pure Co	~156	~14%	d
	~220	~25%	d
	~224	~34%	d
	~240	~48%	d
CoPd	~280±20(Co)	55%	e
Pure Gd	750±250	30%	f

Table 5.6
Selected compounds that have been the focus of research. Experimental tools include TR-MOKE: time-resolved magneto-optical Kerr effect; TPPE: Two-photon photoemission; XMCD: x-ray magnetic circular dichroism; ARPES: Angle-resolved photoemission spectroscopy; SHG: second-harmonic generation; RMOKE: resonant magneto-optical Kerr effect. *a* [Beaurepaire et al. (1996)], *b* [Rhie et al. (2003)], *c* [Weber et al. (2011)], *d* [Andres et al. (2015)], *e* [dalla Longa et al. (2007)], *f* [La-O-Vorakiat et al. (2012)], *g* [Koopmans et al. (2000)], *h* [Krauß et al. (2009)], *i* [Kachel et al. (2009)], *j* [Yamamoto et al. (2014)], *k* [Kruglyak et al. (2005)], *l* [Gong et al. (2012)], *m* [Carpene et al. (2008)], *n* [Vomir et al. (2005)], *o* [Krauß et al. (2009)], *p* [Guidoni et al. (2002)], *q* [Pickel et al. (2008)], *r* [Boeglin et al. (2010)], *s* [Bovensiepen (2006)], *t* [Sultan et al. (2012)], *u* [Carley et al. (2012)], *v* [Teichmann et al. (2015)], *w* [Müller et al. (2009)], *x* [Wüstenberg et al. (2011)], *y* [Kim et al. (2009)], *z* [Mathias et al. (2012)].

Compound	Tool	Ref.
Ni/MgF_2	TR-MOKE	*a*
Ni/W(110)	TPPE, TR-MOKE	*b,c,d*
Ni/SiO	TR-MOKE	*e*
Ni-Fe grating	XUV-TR-MOKE	*f*
Ni/Cu(111),Cu(001)	TR-MOKE	*g*
Ni/Si	TR-MOKE	*h*
Ni/Al	XMCD	*i*
Ni/Ru	RMOKE	*j*
Ni, Pd, Hf	TR-MOKE	*k*
Fe/GaAs	TR-MOKE	*l*
Fe(100)/MgO(100)	TR-MOKE	*m*
Co/Al_2O_3	TR-MOKE	*n*
Co/MgO	TR-MOKE	*o*
$CoPt_3$/Al_2O_3	TR-MOKE	*p*
fcc Co/Cu(001)	TPPE	*q*
CoPd/Si_3N_4	XMCD	*r*
Gd, Tb	SHG, ARPES	*s,t,u,v*
Heusler alloys	TR-MOKE	*w,x*
TbFe	TR-MOKE	*y*
Permalloy	TR-MOKE	*z*

5.5 Mechanisms still under debate

At this time, there is no complete agreement with respect to the underlying mechanism for femtomagnetism. There are at least five different models proposed, which we briefly discuss in their chronological order.

Spin-orbit coupling model In 2000 a theoretical model was proposed to explain ultrafast demagnetization through the cooperation between the laser field and spin-orbit coupling (SOC) [Zhang and Hübner (2000)]. SOC couples spin triplets with singlets, and the population change is initiated by a laser pulse. The idea is that the spin-orbit coupling breaks the spin symmetry, so that the laser field can affect the spin by flipping or shifting relative energies for the majority and minority spins. The state dissipation is important.

Spin-orbit-Coulomb interaction model H. C. Schneider's group proposed a mechanism [Krauß et al. (2009)] that is based on the Elliott-Yafet (EY) formalism but included electron-electron interactions through the Boltzmann equation. They showed that an EY-like mechanism based on electron-electron scattering has the potential to explain ultrafast demagnetization without resorting to a "phononic spin bath".

Spin-phonon interaction model Koopmans and coworkers proposed this model [Koopmans et al. (2010)]. The original idea is based on the Elliott-Yafet theory in semiconductors [Elliott (1954); Yafet (1963)], where the spin flipping is caused by impurities and phonon vibrations. However, the EY theory is not about laser-induced ultrafast demagnetization. Instead, it computes the demagnetization rate with presence of phonon.[f]

[f] One additional drawback of this proposal is that it requires an unphysically large spin-flip-scattering ratio.

Spin superdiffusion model This semiempirical model was proposed by Oppeneer and coworkers [Battiato et al. (2010)]. They suggest that majority spin electrons move faster than the minority electrons, so locally the magnetization is reduced.

Phonon model The ultrafast Einstein—de Haas effect was proposed to allow a rapid transfer of spin angular momentum to the lattice [Dornes et al. (2019)]. This is an interesting proposal, but there is no additional experimental proof. Experimentally, they only measured the x-ray diffraction signal, but not the angular momentum.

From the above discussion, it becomes apparent that a complete theory is still missing. Despite this, we feel that it is necessary to introduce the reader in the simplest terms to what available theories look like. Semiempirical approaches, although useful in many respects, are unlikely to give additional insight into laser-induced ultrafast demagnetization. In the following, we will discuss three commonly used models: the Hubbard model, the spin-orbit coupling model, and the Heisenberg model. Here we will not be covering the details of the superdiffusion theory, which goes beyond the scope of the present book, but the reader may resort to the original paper [Battiato et al. (2010)] for details.

5.5.1 Spin-orbit coupling model

We start with the spin-orbit coupling model. As discussed above, the spin-orbit coupling appears if we take a non-relativistic approximation of the Dirac equation (see Section 3.5.1). Physically, the electron spin sees the electron orbital motion as an effective magnetic field, and therefore experiences an additional interaction. We caution that the spin-orbit coupling preserves the time-inversion symmetry.

If a sample is magnetic, with magnetization along a particular direction, the time inversion symmetry is lost, and the number of symmetry elements is reduced with respect to a non-magnetic sample. We note in passing that although the spin is an axial vector and commonly expressed as a 2×2 matrix, it is much simpler to consider it as a polar three-dimensional vector, but with an additional determinant of the symmetry operations [Birss (1964); Zhang et al. (2018c)]. To be more specific, we can transform spins as

$$\mathbf{S}' = \mathrm{Det}(O)O\mathbf{S}, \tag{5.13}$$

where O is a symmetry operation (see details in Section 3.6).

The key idea of spin-orbit coupling (SOC) as a channel to demagnetization consists of three crucial aspects. (i) It allows angular momentum transfer between spin and orbit. In atoms, with a spherical potential (Eq. 5.7) it can be rewritten as

$$H_{\mathrm{SOC}} = \lambda \mathbf{L} \cdot \mathbf{S}, \tag{5.14}$$

where $\mathbf{L}$ is the orbital angular momentum ($\mathbf{r} \times \mathbf{P}$). This coupling ensures that neither the spin nor the orbital degrees of freedom alone are conserved, but the sum of spin and orbital momenta is only conserved for a system with a spherical potential. The problem with this idea is that in solids the potential is not spherical and obeys translational symmetry, so the orbital angular momentum becomes ill-defined.[g] This very transfer becomes unclear. What makes things more complicated is that experimentally one only measures the z component L_z. In general, we only can talk about the momentum transfer between the electron spin and electron momentum. However, the orbital angular momentum transfer idea is still popular among researchers. (ii) SOC breaks the spin symmetry, so the spin is not conserved and demagnetization is possible. This is manifested in our first theory (see below). (iii) SOC allows spin flipping. This is the main idea of the Elliott-Yafet theory [Elliott (1954); Yafet (1963)]. To see this clearly, we can express Eq. 5.14 as

[g] The reason for this is that in periodic systems the origin of the axes itself is ill-defined.

$$H_{soc} = \lambda \left[L_z S_z + \frac{1}{2}(L^+ S^- + L^- S^+) \right], \tag{5.15}$$

where the last terms reveal an important fact. SOC can switch up-spin to down and also switch down-spin to up at the same rate. The materials that Yafet investigated are nonmagnetic and mostly semiconductors, so he only introduced the spin relaxation time, instead of the demagnetization time. In our case, all materials are magnetic, so spin flipping and flopping rates are different, so demagnetization is possible.[h]

[h]Flipping switches the up-spin to down, while flopping switches the down-spin to up, or the other way around.

5.5.2 Hubbard model

Traditional $3d$ transition metals are dominated by the electron correlation effect. A standard model to describe the ground state is the Hubbard model [Hubbard (1963)], although Gutzwiller [Gutzwiller (1963)] and Kanamori [Kanamori (1963)] proposed a similar model at the same time. The Hubbard model is a simplification of an otherwise complicated model. It considers a narrow band structure. In solid state physics, a narrow band means that the band width is smaller than the electron-electron interaction, so that electrons are more localized in nature, with a smaller kinetic energy. Quantitatively the kinetic energy is measured by a hopping integral t. Hopping means that electrons jump from one site to another. In the second quantization notation, one writes $H_0 = -t\sum_{i,j,\sigma}(c^\dagger_{i,\sigma}c_{j,\sigma} + c^\dagger_{j,\sigma}c_{i,\sigma})$, where $c^\dagger_{i,\sigma}$ is called the electron creation operator that creates an electron at site i with spin σ, and $c_{j,\sigma}$ is called the electron annihilation operator that destroys an electron at site j with spin σ. The hopping integral t is an integral over the kinetic energy operator

$$t = -\int d\mathbf{r}\phi^*(\mathbf{r})\left(\frac{\hbar^2\nabla^2}{2m_e}\right)\phi(\mathbf{r}), \tag{5.16}$$

where $\phi(\mathbf{r})$ is the single-particle wavefunction[i] of a particular state of interest, i.e., $3d$ orbitals. The second quantization notation is completely equivalent to the matrix form. For instance, if we have two lattice sites, $i, j = 1, 2$, H_0 is

$$H_0 = \begin{bmatrix} 0 & -t \\ -t & 0 \end{bmatrix}.$$

[i]These wavefunctions should not be confused with the basis functions above. In solid states, they are Wannier functions. Wannier functions are real-space representations of Bloch functions, and can be computed through a Fourier-transform.

If we diagonalize it, we get two eigenvalues, $\pm t$. The band width is $t - (-t) = 2t$. So when we compare our bandwidth with the electron-electron interaction, we use $2t$ to compare with. Models of this kind are often called tight-binding models, because one only includes the overlap of wavefunctions between the nearest neighboring atoms. This is analogous to the Hückel model in chemistry, which also only keeps resonance integrals between connected atoms.

The second crucial element is the electron-electron interaction U, which is at the center of the Hubbard model. Without counting the spin indices, the U term is a tensor with four indices because there are two electrons, and each electron is described by two wavefunctions which can be centered at different lattice sites,

$$U_{ijkl} = \frac{e^2}{4\pi\epsilon_0} \int d\mathbf{r}_1 \int d\mathbf{r}_2 \, \phi_i^*(\mathbf{r}_1)\phi_j^*(\mathbf{r}_2) r_{12}^{-1} \phi_k(\mathbf{r}_2)\phi_l(\mathbf{r}_1), \tag{5.17}$$

where ϵ_0 is the vacuum permittivity, i, j, k, l are the lattice site indices, and r_{12} is the distance between two electrons. This equation is a generalization of Eqs. 5.5 and 5.6. Note that we call them interactions, instead of integrals, if the wavefunctions used in the above equations are eigenstates of a single-particle Hamiltonian such as Wannier wavefunctions. This avoids the unnecessary confusion that may arise if one uses a regular basis function to replace $\phi(\mathbf{r})$, where the value of U_{ijkl} varies a lot. In the Hubbard model, where the Wannier wavefunction is used, the multi-centered U_{ijkl} is the largest if $i = j = k = l$, which is retained in the Hubbard model. All other terms are left out. We denote $U_{iiii} \equiv U$. So far, we have only taken care of the spatial part of the wavefunction. In addition, electrons have spin, so U also carries spin indices. The Hubbard model only retains the terms where two electrons have opposite spins, or $U_{\sigma\bar{\sigma}}$, but we can put those indices on the electron occupation number operator n. This leads to the following standard Hubbard model,

$$H = -\sum_{ij,\sigma} t \left(c_{i,\sigma}^\dagger c_{j,\sigma} + c_{j,\sigma}^\dagger c_{i,\sigma} \right) + U \sum_{i\sigma} n_{i\sigma} n_{i\bar{\sigma}} \tag{5.18}$$

where $n_{i\sigma}$ is the electron number operator and σ takes two values, one up and one down. If $n_{i\sigma}$ applies to a state with one electron $|1\rangle_\sigma$, $n_{i\sigma}|1\rangle_\sigma = 1|1\rangle_\sigma$. $|1\rangle_\sigma$ is the Dirac notation, meaning that an electron exists at lattice site i, with spin σ. To be more specific, if we have an electron with spin up (down), we write the state as $|\uparrow(\downarrow)\rangle$.

It is worthwhile to dig into the Hubbard model a little further. Let us consider a system with two electrons and two lattice sites, with only one type of orbital, such as d orbitals. If both electrons have spin up, the only eigenstate is $|\uparrow\rangle_1|\uparrow\rangle_2$, with spin $S = 1$ $(=\frac{1}{2}+\frac{1}{2})$. However, the eigenvalue is 0, because the application of the hopping term in Eq. 5.18 to $|\uparrow\rangle_1|\uparrow\rangle_2$ gives zero. To see this, we write

$$c_{1\uparrow}^\dagger c_{2\uparrow}|\uparrow\rangle_1|\uparrow\rangle_2 = c_{1\uparrow}^\dagger|\uparrow\rangle_1 c_{2\uparrow}|\uparrow\rangle_2 = |\uparrow\uparrow\rangle_1 \times |0\rangle_2 = 0. \tag{5.19}$$

Here $|\uparrow\uparrow\rangle_1$ is zero due to the Pauli exclusion principle that forbids the electron to occupy the same orbital with the same spin twice. And the second term in Eq. 5.18 also gives zero since this Hubbard term only includes the double occupation at the same site, not single occupation at two different sites:

$$U n_{1\uparrow} n_{2\downarrow}|\uparrow\rangle_1|\uparrow\rangle_2 = 1|\uparrow\rangle_1 \times 0|\uparrow\rangle_2. \tag{5.20}$$

Since there is no electron at site 2 with spin down, we have the 0 in the second term. The situation is similar for the spin down case $|\downarrow\rangle_1|\downarrow\rangle_2$. To demagnetize the system, one has to flip the spin, but because the total spin permutes with the Hamiltonian [Zhang and Hübner (2000)], $[S, H] = 0$, the spin is conserved, so a direct on-site demagnetization is not possible. The only possible channel is nonlocal such as spin transport, where electrons flow out of the system [Zhang et al. (2018a)] (see Fig. 5.10(a)). In this case, the majority spins have a larger velocity than the minority electron spins, so they move out of the sample quickly. What remains is a comparable number of electrons in both majority and minority channels. Because the spin moment is proportional to the population difference $(\langle n_\uparrow - n_\downarrow \rangle)$, the spin moment decreases. If this mechanism were completely correct, one would expect that sample 2 (commonly a substrate) should see the same amount of spin moment increase, so the global sum of the spin moment over system 1 and system 2 remains constant. However, at this time, there are not enough experimental data to conclude this. Another difficulty with the spin transport picture is that both samples must be conducting. If a metallic sample is grown on the top of an insulating substrate, then the spin transport is not possible. This is in fact the case for the first experiment [Beaurepaire et al. (1996)].

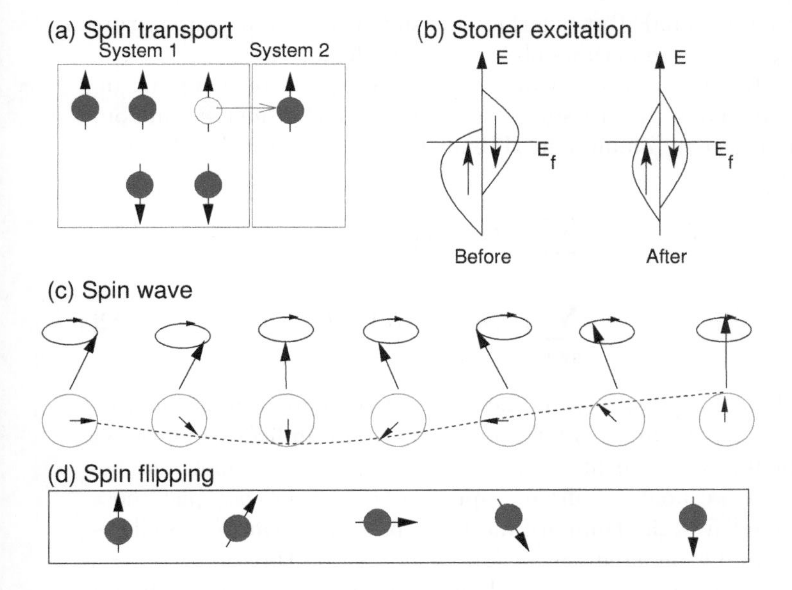

Figure 5.10
(a) Demagnetization through spin transport, where electrons move out of the system. (b) In the Stoner picture, the majority band moves up with respect to the Fermi energy E_f, while the minority moves down. The electrons, which carry spins, are itinerant and mobile. (c) Traditional spin wave excitation, with spin tilting away from the z-axis. In the xy-plane, the spin vector rotates spatially with spin wave wavelength of λ_s. (d) Spin flipping across many lattice sites.

The reader may wonder how the traditional Hubbard model, and its mean-field version, the Stoner model, can get away with demagnetization. The Stoner model employs the Hartree-Fock mean field approximation to the electron-electron interaction, so the Hamiltonian becomes.

$$H_{\text{Stoner}} = -\sum_{ij,\sigma} t\left(c^{\dagger}_{i,\sigma}c_{j,\sigma} + c^{\dagger}_{j,\sigma}c_{i,\sigma}\right) + U\sum_{i\sigma}\langle n_{i\sigma}\rangle n_{i\bar{\sigma}}, \quad (5.21)$$

where $\langle n_{i\sigma}\rangle$ is the expectation value and is no longer an operator. This becomes a single-body problem. In the Stoner model, one introduces a shift in the band structure in the crystal momentum space (see Fig. 5.10(b)). The majority spin band moves up so it loses electrons to the minority band which moves down. How this band shift is induced is not specified in the theory. Typically, one introduces a temperature which is often attributed to lattice vibrations (phonons). Phonons in the majority spin channel of the electron potential are different from those in the minority spin channel, so the electron-phonon coupling becomes spin dependent. But the spin-dependent electron-phonon coupling still does not allow mixing of two channels of spins, because it acts on two separate spins. Under laser excitation, the band structure change is indeed detected. But whether this shift is due to a temperature increase or something else is not yet clear.

In summary, one cannot avoid the spin-orbit coupling introduced in the last subsection. This is the key idea behind our first theory [Zhang and Hübner (2000)]. We use the Hamiltonian

$$\begin{aligned} H &= \sum_{\substack{i,j,k,l \\ \sigma,\sigma',\sigma'',\sigma'''}} U_{i\sigma,j\sigma',l\sigma''',k\sigma''} c^{\dagger}_{i\sigma} c^{\dagger}_{k\sigma'} c_{k\sigma'',l\sigma'''} \\ &\quad + \sum_{\nu,\sigma,K} \mathcal{E}_{\nu}(K) n_{\nu\sigma}(K) + H_{\text{SO}}, \end{aligned} \quad (5.22)$$

where $U_{i\sigma,j\sigma',l\sigma''',k\sigma''}$ is the on-site Coulomb electron interaction. Note that this term has many more indices, because we include both $3d$ and $4sp$ states in our Hamiltonian and because we include different spin configurations, not just those included in the Hubbard model. This way we can very well describe ferromagnetism in full generality by three parameters: Coulomb repulsion U, exchange interaction J, and exchange anisotropy ΔJ. The values for ferromagnetic Ni are $U_0 = 12$ eV, $J_0 = 0.99$ eV, and $(\Delta J)_0 = 0.12$ eV. $\mathcal{E}_{\nu}(K)$ is the spin-independent band structure of a Ni monolayer, $n_{\nu\sigma}(K)$ the particle number operator of band ν in K space, and H_{SO} the spin-orbit coupling.

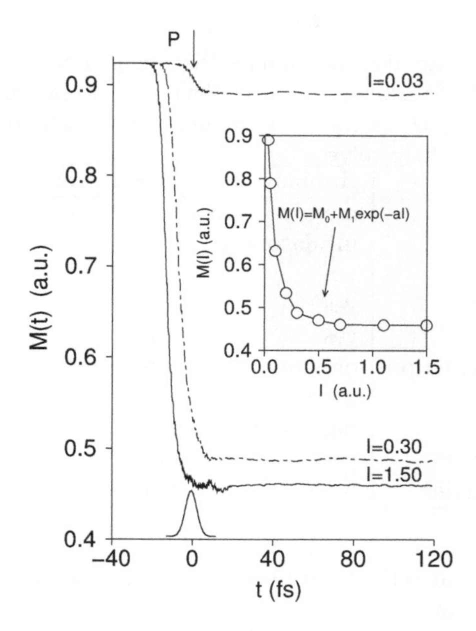

Figure 5.11
A femtosecond laser pulse can effectively demagnetize fcc Ni. The intensity I (a. u.) is 0.03 (long-dashed line), 0.3 (dot-dashed line), 1.50 (solid line). Inset: the exponential dependence of $M(I)$ on the laser intensity I. Both the figure and caption are adopted from Ref. [Zhang and Hübner (2000)], with permission from the American Physical Society.

It turns out that this Hamiltonian is enough to address most of the features of the spin dynamics on the ultrafast time scale as it contains all the necessary ingredients. For the subsequent results, 66 states[j] for each individual Ni atom are solved analytically, and the solution is embedded in a crystal field given by the band structure including the translational invariance. Then one can diagonalize the resulting Hamiltonian [Hübner and Falicov (1993)]. For the more advanced readers, this treatment very much resembles a k-dependent self-energy correction as typically performed in the dynamical mean-field theory.

To investigate laser-induced demagnetization, we add an interaction between the laser and system,

$$H_I = -e\mathbf{E}(t) \cdot \mathbf{D}, \tag{5.23}$$

where $\mathbf{E}(t)$ is the laser field and $\mathbf{D}$ is the dipole transition operator. We then solve the time-dependent Schrödinger equation numerically,

$$i\hbar\frac{\partial}{\partial t}|\Psi(t)\rangle = H|\Psi(t)\rangle \tag{5.24}$$

where $|\Psi(t)\rangle$ is the many-body wavefunction. The spin moment can be computed as $M(t) = \langle\Psi(t)|S_z|\Psi(t)\rangle$. Figure 5.11

[j] 66 states comes from $12!/10!2!$, where 10 is the number of electrons of Ni atom and 12 is the total number of available states: there are 10 d orbitals and 2 sp orbitals, so in total we have 12. These 66 states contain two sets of triplets with spin up and spin down, but we only need one set since fcc Ni is ferromagnetic, plus another set with total spin $s_z = 0$ along the z axis.

Table 5.7
Comparison between the traditional Heisenberg model and the Hubbard model. The table only lists their native features. The combined model such as the $t - J$ model is more powerful, but is also more difficult to solve.

Property	Hubbard	Heisenberg
Material type	metals and insulators	insulators
Kinetic energy	yes	no
Band structure	yes	no
Fermi energy	yes	no
Exchange interaction	on-site	inter-site
Difficulty level	hard	easy
Excitation	band shift	spin wave
Laser excitation	yes	no
Excitation barrier	high	low

shows the first theoretical results. One sees that indeed a 10-fs laser pulse can demagnetize fcc Ni very rapidly. When we look into the details of how demagnetization occurs, we find that lots of lower spin states high in energy are occupied during laser excitation, so the spin moment is reduced.

5.5.3 Heisenberg model

The Heisenberg model (see Eq. 5.1) is a different model for magnetic systems. This model has no electrons, only spins, so inherently, direct laser field excitation is not possible if we ignore the magnetic field of the laser.[k] Table 5.7 compares the Heisenberg model with the Hubbard model. One can see they complement each other very well. In general, the Heisenberg model is much easier to solve since it only focuses on the potential energy term. This model has its own demagnetization mechanism. Upon increase of temperature, a tilting of spin at one site will propagate through the exchange interaction across the space, so the spins at the neighboring lattice sites are affected (see Fig. 5.10(c)). The spin precesses around the z axis. The bottom part of Fig. 5.10(c) shows a view from the top. Spins at each lattice site have a phase difference from one site to next. However, the module of the spin at each site remains constant.

[k] Additionally, since it neglects spatial degrees of freedom it inevitably overestimates the importance of the Pauli exclusion principle, yielding harder magnetic systems.

Consider a one-dimensional Heisenberg spin chain. Although such a chain does not allow long-range magnetic ordering, it provides some useful insight into spin wave excitation. We rewrite the Heisenberg model as

$$H = -J \sum_i \mathbf{S}_i \cdot \mathbf{S}_{i\pm1}, \tag{5.25}$$

where the summation runs over the two nearest neighbors of

i, $i-1$ and $i+1$, except at the two ends where only one neighbor remains (if we consider a finite spin chain without using periodic boundary conditions). One should be cautious about the units in the Hamiltonian. Since the spin angular momentum has units of $\hbar$, to have a correct energy unit for the Hamiltonian, the exchange interaction J must be in J$/\hbar^2$ or eV$/\hbar^2$. We adopt the Heisenberg equation of motion, $i\hbar\dot{A} = [A, H]$, to find the dynamic equation for the spins. Here A is an operator. The spin at site i ($\mathbf{S}$) precesses according to

$$\frac{d\mathbf{S}_i}{dt} = J\mathbf{S}_i \times (\mathbf{S}_{i-1} + \mathbf{S}_{i+1}). \tag{5.26}$$

Physically, the right site represents a torque on $\mathbf{S}_i$, and $(\mathbf{S}_{i-1} + \mathbf{S}_{i+1})$ is an effective field. The detailed derivation is left as a homework assignment. Although the above equations are simple, there is no analytic solution to them. When the spin precession amplitude is small, we can consider the z component of the spin constant, or $dS_{iz}/dt = 0$. For the other two components, we have[l]

$$\frac{dS_{ix}}{dt} = -JS_{iz}(S_{i-1,y} + S_{i+1,y} - 2S_{iy}) \tag{5.27}$$

$$\frac{dS_{iy}}{dt} = -JS_{iz}(S_{i-1,x} + S_{i+1,x} - 2S_{ix}). \tag{5.28}$$

[l]In the continuum limit, the right side represents a second-order derivative, $J(\partial^2 S_{x(y)}/\partial x^2)(\delta x)^2$. This is the origin of the effective magnetic field in a Landau-Lifshitz-Gilbert simulation.

A stationary state solution [Kittel (1996)] is $S_{ix} = u\cos(ika - \omega t)$ and $S_{iy} = u\sin(ika - \omega t)$, where u is the amplitude of the transverse spin components, k is the spin wavevector[m] (see Fig. 5.10(c)), a is the lattice constant, ω is the spin wave frequency, and t is time. If we substitute these two solutions into the above two equations and replace S_{iz} by S, we obtain the dispersion of the spin wave (magnon),

$$\hbar\omega_k = 2\hbar JS(1 - \cos(ka)). \tag{5.29}$$

[m]The spin wave normally has a much longer wavelength λ_s, or a much smaller k. This means that if we want to simulate a low-energy spin wave excitation using first-principles methods, the unit cell has to be extremely large, which is very challenging.

We note again that J is in eV$/\hbar^2$ and S is in $\hbar$.

With the energy spectrum, we can now investigate demagnetization. The quantized spin waves, called magnons, are bosons and have energies $(n + \frac{1}{2})\hbar\omega_k$, so their number distribution follows the Bose-Einstein distribution for each k,

$$\langle n_k \rangle = \frac{1}{\exp(\beta\hbar\omega_k) - 1}, \tag{5.30}$$

where $\beta = 1/k_B T$ is the temperature factor, k_B is the Boltzmann constant and T is the temperature. The magnetization change $(M_s - M)/M_s = -\Delta M/M_s$, where M_s is the saturated magnetization, is equal to $\sum n_k/NS$ [Kittel (1996)], where N is the number of lattice sites. If the temperature increases, $\sum n_k/NS$ increases, and $\Delta M/M_s$ becomes more negative, or larger demagnetization.

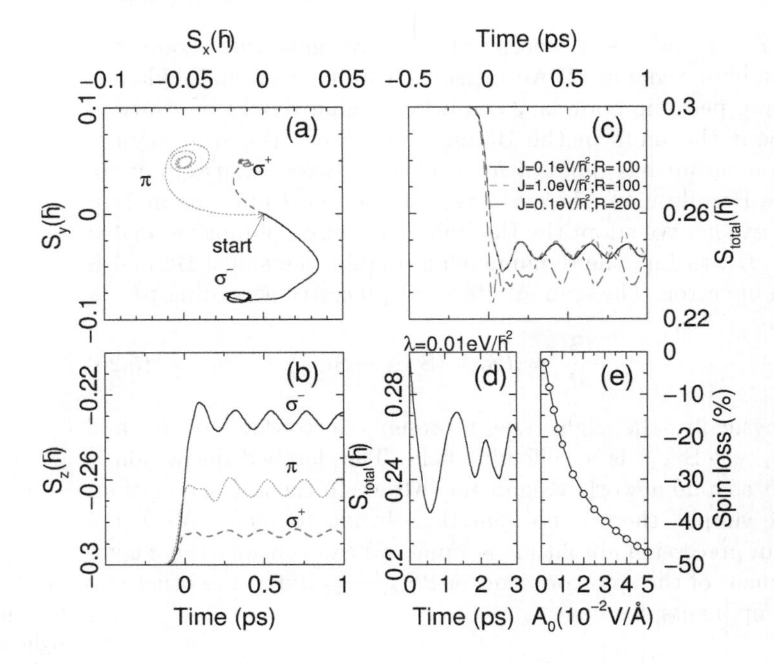

Figure 5.12
(a) Phase diagram of the transverse components of the total spin for σ^+, σ^- and π light. The initial point is denoted by "start". (b) z component of the spin for σ^+, σ^- and π light. σ^- has the strongest influence because of the spin configuration. (c) Dependence of the total spin as a function of time for two exchange interactions J and two beam radii R. Here a linearly polarized pulse (π) is employed. (d) Spin reduction with a reduced spin-orbit coupling. (e) Percentage loss of the spin as a function of the laser field amplitude A_0. Figure and caption from [Zhang et al. (2019)] used with permission from the American Institute of Physics.

In femtomagnetism, we are not interested in how demagnetization occurs with temperature; instead we have a different question: how the spin system increases its temperature (if the concept of temperature is still valid). Upon laser excitation, electrons first receive the energy. In ferromagnets, electrons in the majority and the minority spin channels absorb different amounts of energy.[n] However, if there is no energy exchange between majority and minority spins, electrons in different channels have different temperatures. As far as there is no electron exchange between majority and minority channels, the spin moment of the sample remains unchanged, independent of temperature. If electron spins at different lattices are collinear, they remain collinear. The only way that the system can demagnetize is to drive some electrons out of a system, or transport. Other mechanisms must invoke spin-orbit coupling, where the entire process is not necessarily about the transfer of angular momentum between spin and orbital degrees of free-

[n] Here we use the itinerant electron picture, rather than the localized spin picture. The reason that we have to discuss electrons under laser excitation, instead of spin under laser excitation, is that the spins do not couple directly to the laser. Note that we ignore the interaction of the laser's magnetic field with the system.

dom. For instance, through electron-phonon interactions, each spin channel transfers energy to the lattice, but the number of electrons in each spin channel is unchanged, so there is no demagnetization.

It seems that the Heisenberg model is unsuitable for laser-induced ultrafast demagnetization. Spins at each site have a fixed amplitude, $|\mathbf{S}_i| = \sqrt{S^2 + u^2}$, where S is the spin amplitude along the z-axis, and u is the transverse amplitude in the xy-plane. However, if we have a way to reorient spins, we still can have demagnetization. Consider the following vector sum:

$$\sum_i \mathbf{S}_i = \left(\sum_i S_{ix}, \sum_i S_{iy}, \sum_i S_{iz} \right). \qquad (5.31)$$

Take the stationary state solution as an example, $S_{ix} = u\cos(ika - \omega t)$ and $S_{iy} = u\sin(ika - \omega t)$. Now, if we sum over a period of spin wavelength $\lambda_s = 2\pi/k$, both $\sum_i S_{ix}$ and $\sum_i S_{iy}$ are zero. So the spin wave picture is extremely efficient for demagnetization, as far as one can find a way to tilt spins.

We have recently found a way to introduce this spin tilting through spin-orbit coupling, and we noticed an important neglected fact about the Heisenberg model [Zhang et al. (2019)], namely that its demagnetization barrier is much lower than that in both the Stoner model (see Table 5.7) and time-dependent density functional theory calculations (see below). Our idea is to combine the Heisenberg model with the spin-orbit coupling model. Figure 5.12 shows an example. Our model system has a slab geometry, with 501 spins along the x- and y-directions and 4 layers along the z-axis. The entire system has over 1 million spins ($501 \times 501 \times 4$). Figure 5.12(a) shows the x- and y-components of the total spin under an ultrashort laser pulse, with three different helicities (linear π, right circularly σ^+, and left circularly σ^-). Each helicity leaves a hallmark on the spins.[o] If the spin is initialized pointing down, only σ^- can effectively flip it. Other helicities require a stronger laser pulse. This is consistent with traditional magneto-optics. In our case, this effect is larger because we are not in a perturbative limit. Figure 5.12(b) reveals that the z component of the total spin also depends on the laser helicity.[p] If we compute the magnitude of the total spin, we find a strong demagnetization, which has a weak dependence on the exchange interaction J (Fig. 5.12(c)). If the laser beam radius is smaller, the spin reduction is smaller, as expected. However, the demagnetization strongly depends on the spin-orbit coupling λ. Under the same laser parameter, a small λ leads to spin oscillations with longer periods (Fig. 5.12(d)), reminiscent of all-optical spin switching in rare-earth compounds.

Naturally, the most critical parameter is the amount of demagnetization. Except for a semiempirical treatment such as the spin-diffusion model, none of the existing theories is capable of reproducing the same amount of demagnetization

[o] The spatial quantization axis is set by the spin direction. Once that direction is determined, π, σ^+ and σ^- are no longer equivalent.

[p] In $3d$ metals, the helicity dependence is weak for two reasons. Electrons are mobile, so the orbital angular momentum is small. The second is the relatively weak spin-orbit coupling. Our model adopts the harmonic potential. We use the fcc Ni's spin moment as an example.

under the experimental laser parameters. Figure 5.12(e) represents a clear departure, where the percentage demagnetization reaches 50% with a laser field amplitude of 0.05V/Å. The key message here is that although the traditional treatment of the Heisenberg model does not include spin flipping, once the Heisenberg model includes spin-orbit coupling, it can describe spin flipping, thus the laser-induced demagnetization.

5.5.4 Time-dependent Liouville density functional theory

Time-dependent density functional theory (TDDFT) is arguably the most accurate method for femtomagnetism. It is a first-principles method, without any fitting parameters and experimental input, except for the types of atoms and their positions. It has great potential to help us understand how at the earlier time scale the laser pulse excites electrons and demagnetizes a sample, without relying on models or semiempirical procedures.

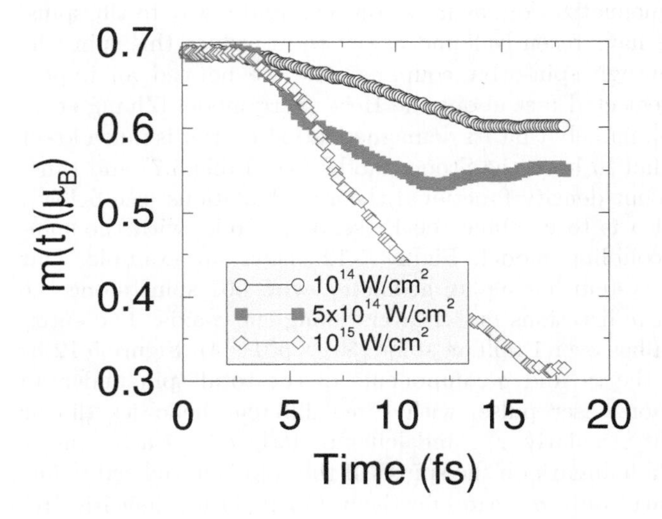

Figure 5.13
TDDFT results for fcc Ni under a 6-fs pulse [Krieger et al. (2015)]. The field peak intensity increases from 10^{14} W/cm^2 to 10^{15} W/cm^2. These intensities are more than 100 times stronger than those used in experiments [Beaurepaire et al. (1996); Mathias et al. (2012)]. If we compare it with Fig. 5.7(a), the demagnetization time here is too short, even if we consider the 4-fs difference in laser pulse duration between the experiment and theory.

In Chapter 3, we extensively review this technique. However, its application to laser-induced demagnetization has a mixed record of success [Krieger et al. (2015)]. Figure 5.13 shows that to reach 50% magnetic moment reduction, one needs 10^{15}W/cm^2, or a fluence of 1J/cm^2.[q] Experimentally, Beaurepaire et al. used 7 mJ/cm^2 to reach a similar amount of demagnetization [Beaurepaire et al. (1996)]. Mathias et al. used 2 mJ/cm^2 to obtain 45% demagnetization [Mathias et al. (2012)]. Therefore, there is a discrepancy of 1000 times between theory and experiment.

[q] We can convert the field intensity to fluence by multiplying the intensity by the pulse duration. For a 6-fs pulse, we have 6×10^{-15}s $\times$ 10^{15} W/cm^2 = 1 J/cm^2.

The time-dependent Liouville density functional theory (TDLDFT) [Zhang et al. (2014)] has the same issue as TDDFT. If one uses the same laser fluence as the experiment, the maximum spin moment reduction only reaches 10%. Using a monolayer of Ni helps to some extent but does not completely resolve this discrepancy. This is an open question at this time. However, there are several issues found with TDDFT.

Since TDDFT works with Kohn-Sham wavefunctions, the electron occupations are fixed from the beginning. Consider that we have a system with two electrons: Electron 1 is in an s orbital and electron 2 is in a p orbital. Without the second electron, electron 1 may be excited from the s orbital to p orbital during laser excitation. But because electron 2 is already in the p orbital, electron 1 cannot be transitioned into the p orbital. In TDDFT, if we evolve these two electrons independently, electron 1 is still allowed to transition into the p orbital, regardless of whether the p is occupied or not. This clearly violates the Pauli exclusion principle. The problem becomes severe in materials such as metals where many electrons are excited. In our TDLDFT calculation, density matrices are used,

$$i\hbar\frac{\partial\rho}{\partial t} = [H_0 + H_I, \rho], \tag{5.32}$$

where ρ is the density matrix, H_0 is the field-free Hamiltonian, and H_I is the interaction Hamiltonian between the system and laser field. In the density matrix formalism, the Pauli exclusion principle is rigorously enforced.[r]

[r] Density matrices ρ have three indices: one for the crystal momentum k and two for the energy bands i and j. The permutation relation zeroes out the terms if the original state is already occupied. The proof is left as an exercise.

In TDLDFT, we only calculate the density matrices for a small time step. The new density matrices — to be more precise, the diagonal elements of the density matrices — are fed into the self-consistent Kohn-Sham equation to find a new potential and density, so we compute both spin matrices and optical transition matrix elements for the next time step [Zhang et al. (2014)]. The calculations are very accurate, but also time consuming.

Another challenge is that the unit cell size used in TDDFT and TDLDFT is too small to accommodate spin waves over hundred lattice sites. It misses the low-energy excitation spectrum. However, there is no easy way to include many lattice sites in a supercell.

5.5.5 Time-dependent magneto-optics theory

Magneto-optics Kerr effect (MOKE) is a standard experimental technique to measure magnetic responses of a sample (see

Chapter 2). It uses the optics to probe magnetism. When a beam of light with a well-defined polarization gets reflected from the surface of a nonmagnet, the light polarization remains the same with respect to the original polarization.[s] If a sample is magnetic, the polarization of light changes after reflection from the surface. Linearly polarized light may become elliptically polarized [Zvezdin and Kotov (1997)]. Figure 5.14 shows that the x- and y-components of the light electric field, E_x and E_y, trace an ellipse as time evolves. The magnetic properties are correlated with the Kerr rotation angle θ and Kerr ellipticity $\epsilon = E_y/E_x$ or b/a in Fig. 5.14, which are measured experimentally.

[s]Here we assume that the material is completely isotropic. A nonmagnetic but anisotropic sample may also induce a polarization change. This is not the case which we consider here.

In general, θ and ϵ are both frequency dependent, and under continuous wave (CW) excitation, they can be computed from

$$\theta(\omega) + i\epsilon(\omega) = -\frac{\sigma_{xy}(\omega)}{\sigma_{xx}(\omega)}\sqrt{\left(1 + \frac{4\pi i}{\omega}\sigma_{xx}(\omega)\right)^{-1}}, \quad (5.33)$$

where σ_{xx} and σ_{xy} are the diagonal and the off-diagonal conductivities, respectively. Although θ and ϵ depend on the conductivities in such a complicated form, it is well established that θ and ϵ follow each other well. However, this may be no longer the case in the time domain.

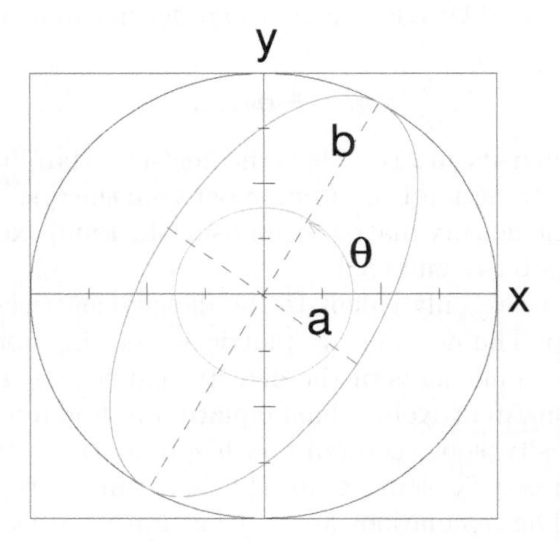

Figure 5.14
Elliptically polarized light. The Kerr rotational angle is measured from the x-axis. The ellipticity ϵ is the ratio of the major (b) and minor (a) axis length, $\epsilon = b/a$. There are two different conventions to name left or right elliptically polarized light. One is along the light propagation direction, and the other is against the propagation direction.

The emitted light field $\mathbf{E}$ is determined by the polarization $\mathbf{P}(t)$ through the Maxwell equation. For an ellipse centered at the origin (0,0), the generic equation (in the xy-plane) is

$$\left(\frac{\cos^2\theta}{a^2}+\frac{\sin^2\theta}{b^2}\right)P_x^2(t)+\sin 2\theta\left(\frac{1}{a^2}-\frac{1}{b^2}\right)P_x(t)P_y(t)$$
$$+\left(\frac{\sin^2\theta}{a^2}+\frac{\cos^2\theta}{b^2}\right)P_y^2(t)=1. \quad (5.34)$$

In order to extract θ and ϵ, the trajectory of $P_x(t)$ and $P_y(t)$ must form an ellipse. However, there is no guarantee that this is true at all the times. Our practical calculations show that the trace formed by $P_x(t)$ and $P_y(t)$ does always not look like an ellipse [Zhang et al. (2018c)] at any five time instants.[t]

[t] At least five points are needed to uniquely determine an ellipse.

If we limit ourselves to a linear regime to first-order susceptibility, the situation is relatively simple. This means that we compute the first-order polarization $P_{xy}^{(1)}$ which in turn means that the laser is polarized along the x-axis, while the y-component of the polarization is computed. In the density matrix (ρ) notation,

$$P_{xy}^{(1)}=\mathrm{Tr}\left[\rho^{(1)}(E_x)D_y\right]. \quad (5.35)$$

Although in general density matrices have no spatial index, once we choose one particular polarization direction of the laser field, the density matrix has a "direction", which is why we write the density as $\rho^{(1)}(E_x)$ if the laser field is polarized along the x-axis. If a crystal has a symmetry, the symmetry operation must apply to both the density matrices $\rho^{(1)}$ and the dipole moment $\mathbf{D}$. If there are N symmetry operations, one carries out N separate calculations. Naturally, some symmetry operations may produce equivalent matrix elements for a particular laser configuration, in which case one simple calculation is enough, and we can simply multiply the final results by the degeneracy.

A useful test of the symmetry operation for transition matrix elements is to use the following summation: Take three photon transitions among four states ($i-l$) as an example. We compute the following summation over all the symmetry operations,

$$\sum_O \langle i|O\mathbf{D}|j\rangle\langle j|O\mathbf{D}|k\rangle\langle k|O\mathbf{D}|l\rangle\langle l|O\mathbf{D}|i\rangle, \quad (5.36)$$

where O is the symmetry operations. If the system has a well-defined parity and inversion symmetry, the above summation is nonzero. But if we have three states from i to k (this corresponds to two-photon transitions), the following sum is zero:

$$\sum_O \langle i|O\mathbf{D}|j\rangle\langle j|O\mathbf{D}|k\rangle\langle k|O\mathbf{D}|i\rangle. \quad (5.37)$$

We have verified this relation up to 7 eigenstates at different k points computed with the Wien2k code [Blaha et al. (2018)], and we find that the symmetry of the matrix elements is correct.

$\rho^{(1)}$ in Eq. 5.35 is the first-order density matrix where the system interacts with the laser field once. We have to solve the following equation numerically at each k point:

$$i\hbar\frac{\partial\rho^{(1)}}{\partial t} = [H_I, \rho^{(0)}]. \tag{5.38}$$

Here $\rho^{(0)}$ is the zero-order density matrix of the ground state, and H_I is the interaction between the laser and system and only appears once. In general, if we have two pulses, pump and probe, it is customary to write the density matrix as $\rho^{(n|m)}$ [Zhang and George (2007a)],[u]

[u] $n(m)$ refers to the number of times that the system interacts with the pump(probe) field. If n is positive, this means that the pump propagates along the $+\mathbf{k}$ direction with the phase $e^{i\mathbf{k}\cdot\mathbf{r}}$. If negative, $e^{-i\mathbf{k}\cdot\mathbf{r}}$. See Section 2.6.2.

$$\begin{aligned} i\hbar\dot{\rho}_{ij}^{(n|m)} &= -\hbar(n\omega_1 + m\omega_2)\rho_{ij}^{(n|m)} + \sum_l (t_{il}\rho_{lj}^{(n|m)} - \rho_{il}^{(n|m)}t_{lj}) \\ &+\vec{E}_1\cdot(\vec{r}_i - \vec{r}_j)\rho_{ij}^{(n-1|m)} + \vec{E}_1^*\cdot(\vec{r}_i - \vec{r}_j)\rho_{ij}^{(n+1|m)} \\ &+\vec{E}_2\cdot(\vec{r}_i - \vec{r}_j)\rho_{ij}^{(n|m-1)} + \vec{E}_2^*\cdot(\vec{r}_i - \vec{r}_j)\rho_{ij}^{(n|m+1)} \\ &+U\sum_{\alpha,\beta}[\rho_{ii}^{(n-\alpha|m-\beta)} - \rho_{jj}^{(n-\alpha|m-\beta)}]\rho_{ij}^{(\alpha|\beta)}, \end{aligned} \tag{5.39}$$

where we use a simple tight-binding model to illustrate the key idea (see Section 4.3.2.1)). For three pulses, one has to use $\rho^{(n|m|l)}$. The first superscript identifies the first pulse interacting with the system, while the second superscript identifies the second pulse, and so on.

We provide the code below to show the reader how to implement the symmetry operations. Note that we only compute the z component of the orbital moment, so there is no symmetry operation applied to it. *nsym* is the number of symmetry operations. For a spin-polarized calculation, the first half of the symmetry operators keeps the spin invariant, but the second half reverses it, so one has to treat it separately.

```
1     do i=1,nm                    !over state i
2       do j=1,nm                  !over state j
3         k=(i-1)*nm+j             !compute the state index
4         do is=1,nsym             !over symmetry operation
5           do ii=1,3              !transform spin, dipole matrix
6         os(ii)=(0d0,0d0) !symmetry operation on spin
7          ol(ii)=(0d0,0d0)!symmetry operation on angular
8          od(ii)=(0d0,0d0)!symmetry operation on dipole
9            do jj=1,3
10             os(ii)=os(ii)+sym(jj,ii,is)*spin(i,j,jj)
11             if(ii.eq.jj.and.jj.eq.3)then
12                ol(ii)=ol(ii)+sym(jj,ii,is)*lz(i,j)
13             endif
14             ctmp=dcmplx(or(i,j,jj),oi(i,j,jj))
15             od(ii)=od(ii)+sym(jj,ii,is)*ctmp
16           enddo
17          enddo
18          !there are six directions, +/- X, +/- Y, +/- Z
```

```
19           do ii=1,6         !transform the density matrix
20              op(ii)=(0d0,0d0)
21              do jj=1,6
22                 id=(jj-1)*nc
23              ! y is density matrix (real and imag)
24                 ctmp=dcmplx(y(k+ns11+id),y(k+ns11+id+ns/2))
25                 op(ii)=op(ii)+o6(jj,ii,is)*ctmp
26              enddo
27           enddo
28
29 !      expectation values
30               do ii=1,3
31                  do jj=1,6
32                     if(is.le.nsym/2)then
33                        sp1(ii,jj)=sp1(ii,jj)+dconjg(os(ii))*op(jj)
34                        if(ii.eq.3)then
35                           lp1(ii,jj)=lp1(ii,jj)+dconjg(ol(ii))*op(jj)
36                        endif
37                     else
38                        sp1(ii,jj)=sp1(ii,jj)+os(ii)*dconjg(op(jj))
39                        if(ii.eq.3)then
40                           lp1(ii,jj)=lp1(ii,jj)+ol(ii)*dconjg(op(jj))
41                        endif
42                     endif
43 !                 check(ii,jj,is)=check(ii,jj,is)+dconjg(os(ii))*op(jj)
44                     if(is.le.nsym/2)then
45                        dp1(ii,jj)=dp1(ii,jj)+dconjg(od(ii))*op(jj)
46                     else
47                        dp1(ii,jj)=dp1(ii,jj)+od(ii)*dconjg(op(jj))
48                     endif
49                  enddo
50               enddo
51            enddo                     !over symmetry operation
52
53         enddo                        !over state j
54      enddo                           !over state i
```

We have carried out a massively parallel first-principles calculation using an extremely dense k mesh ($104 \times 104 \times 104$) for fcc Ni. Our laser pulse duration is 12 fs, and laser field amplitude is 0.05 V/Å. Figure 5.15(a) shows the normal optical response $\mathrm{Im}[P_{xy}^{(1)}](t)$ as a function of time in fs with the photon energy changing from 1.5 to 2.6 eV. We clearly see that the signal strongly depends on the photon energy. In the top inset, we show the laser pulse. The signal still changes even after the laser pulse is gone. Figure 5.15(b) shows the magnetic counterpart. Its change shows a stronger dependence on photon energy. Below 2 eV, there is good correspondence between the magnetic and optical signals, and above 2 eV, there is a large discrepancy. These two signals are quite different, where the magnetic response can increase with time, instead of a monotonic decrease.

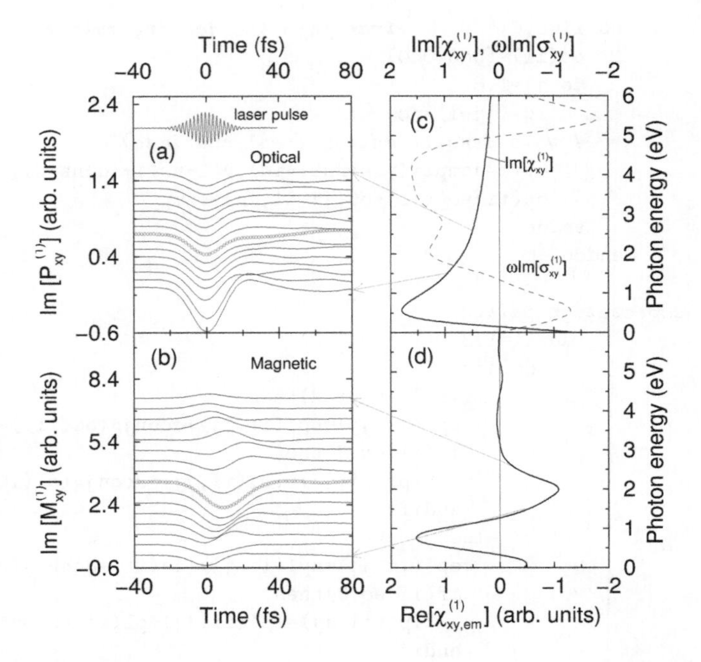

Figure 5.15
(a) Time evolution of the first-order off-diagonal polarization as a function of incident photon energy. From bottom to top, the photon energies are 1.5, 1.7, 1.8, 1.9, 1.925, 1.95, 1.975, 2.0, 2.025, 2.075, 2.1, 2.15, 2.3, 2.4, 2.5 and 2.6 eV. The empty-circle line denotes the result at $\hbar\omega = 2.0$ eV. Inset: Laser pulse shape. The laser pulse duration is 12 fs and laser field strength is 0.05 $\mathrm{V/\AA}$. (b) Off-diagonal magnetization change as a function of time. The change with incident energy forms a crescent shape. (c) Off-diagonal susceptibility (solid line) and conductivity (dashed line) as a function of photon energy. Note the x-axis is the amplitude of the spectrum, and the y-axis is the photon energy. Two arrows identify where the excitation energy is used in (a) and (b). (d) Magneto-optical susceptibility as a function of photon energy. Used with permission from Nature Physics [Zhang et al. (2009)].

These differences can be understood from the frequency information. Figure 5.15(c) shows the first-order susceptibility and conductivity. For fcc Ni, there are characteristic features that oscillate with the pump energy, where the signal does not change sign (see the two arrows). But this is not the case if we compute the magnetic counterpart. We notice a clear sign change with different photon energies. This reflects the intrinsic difference when one uses a shorter laser pulse. This does not occur if we use a longer pulse [Zhang et al. (2009)].

5.6 Exercises

1. Starting from the wavefunctions from the particle-in-a-box problem [Zhang et al. (2014)],

$$\phi_n = \sqrt{\frac{2}{L}} \sin\left(\frac{n\pi x}{L}\right), \qquad (5.40)$$

where L is the length of the box, n is the main quantum number, and x is the position of the particle, show that the exchange integral is positive. Hints: you can check Ref. [Zhang et al. (2014)].

2. (a) Use Eq. 5.9 to plot a curve with the energy (in units of eV) as the vertical axis and the distance two electrons as the horizontal axis. (b) On the curve, identify each distance that one can get for an electron-electron interaction of 10 eV.

3. Sultan et al. [Sultan et al. (2011)] measured the characteristic time as a function of relative fluence. The following table is created by directly reading off from Fig. 3 of [Sultan et al. (2011)] using https://automeris.io/WebPlotDigitizer/ by Tyler Jenkins.

Relative Fluence	τ_M(ps)
0.2022082242	0.5498407531
0.2748106731	0.6169509519
0.3059947625	0.7202349777
0.4029301437	0.6823271286
0.4653903319	0.7430356005
0.5174746974	0.7368320476
0.603595442	0.7122655531
0.7156628211	0.7552976148
0.807013943	0.7160980961
0.917127893	0.7712258475
0.9570599476	0.7681683063
1.006539741	0.7447590063

From this table, find the Hübner times for the $4f$ and $5d$ orbitals.

4. If we start with the Ni M_2 edge, what conduction bands do we mostly likely probe? To be specific, is there any difference if we start with M_3 edge?

5. The core electron has a certain lifetime. Imagine that one knocks an electron out of the $3p$ shell. Can we use it as a way to clock the dynamic process?

6. According to Chapter 2, if one directly tunes the laser cavity length to get 2 KHz repetition rate, how long is the laser cavity length?

7. The absorption spectrum in Fig. 5.6 is plotted using

the function

$$\alpha(\omega) = \frac{A}{(\omega - E)^2 + \Gamma} \quad (5.41)$$

for two magnetization directions, where A is fixed at 0.1, but E takes two values 0.8 and 0.9 and Γ takes 0.02 and 0.01. Use your favorite software to plot this function and find the difference absorption between them. Next, try different parameters to understand why in some edges the difference absorption spectrum becomes negative while others are positive.

8. Derive the equation of motion for spin at site i (see Eq. 5.26)
9. (a) Consider a two-level system with two states $|n\rangle$ and $|m\rangle$ occupied by two electrons each. The density matrices $\langle n|\rho|n\rangle = \langle m|\rho|m\rangle = 1$. Using Eq. 5.32, show that if $\langle n|\rho|m\rangle = 0$ initially, $\langle n|\rho|m\rangle(t)$ at any time step remains zero. (b) If $\langle n|\rho|n\rangle = 0$ and $\langle m|\rho|m\rangle = 1$, under the external field, show $\langle n|\rho|m\rangle(t) \neq 0$.

6 All-optical spin switching

6.1 Basic optics
6.2 Background
6.3 Key ingredients of all-optical spin switching
6.4 Theory
6.5 Exercises

Using light to switch spins is technologically important for the magnetic storage industry. Magneto-optical recording is just an example [Gau (1989)] and employs a group of special ferrimagnets such as $(\mathrm{Tb},\mathrm{Gd})_\mathrm{x}(\mathrm{Fe},\mathrm{Co})_{1-\mathrm{x}}$ thin films, with a perpendicular magnetic anisotropy [Chaudhari et al. (1973)]. The recording speed is slow by modern standards and is mainly given thermally by a laser field [Kryder (1985); Shieh and Kryder (1986)]. By the end of the first decade after the pioneering discovery of femtomagnetism [Beaurepaire et al. (1996)], intensive research efforts have borne fruit, such as the discovery of all-optical spin switching (AOS) [Stanciu et al. (2007)]. Since an ultrafast laser pulse is used, the switching is much faster. This chapter introduces the basic principles and surveys the basic ideas and experimental findings of all-optical spin switching.

6.1 Basic optics

It is instructive to review how light propagates in a sample before we present the key experimental results. There are several excellent books on this topic. Here we follow the study by Si and Zhang [Si and Zhang (2010)].[a]

[a] During the investigation of laser-induced ultrafast demagnetization, there was a debate as to whether there are enough photons for each atom. This idea came from a suggestion that photons could instill their angular momentum directly into the spin systems. If one follows this logic, to completely quench a ferromagnet, one would need lots of photons. This idea is no longer popular.

Suppose that we have a beam of light propagating along the positive z-axis (see Fig. 6.1), with the light polarization linearly along the x-axis. Here we have to use a complex form of the laser field, since the light wave propagates with a specific direction. A real form is possible but contains less information about the propagation direction. Its generic form is

$$E_x(\omega,\tau,z;t) = E_0 \exp(-t^2/\tau^2) \exp\left[i\omega(\frac{z}{v} - t)\right], \qquad (6.1)$$

where ω is the laser carrier frequency, τ is the laser pulse duration, z is the coordinate, t is the time, E_0 is the amplitude of the laser electric field, and v is the phase velocity. The phase velocity can be rewritten using the speed of light, c, with the index of the complex refraction $\tilde{n}$ as $v = c/\tilde{n}$. Note that $\tilde{n}(\omega) = n(\omega)+ik(\omega)$, where n and k are the real and imaginary parts of the index of refraction. We rewrite the above equation as

$$\mathcal{E}_x(\omega, \tau, z; t) = E_0 \exp(-t^2/\tau^2) \exp\left[i\omega\left(\frac{n(\omega)z}{c} - t\right)\right] \exp(-\frac{\omega k}{c} z). \quad (6.2)$$

The magnitude of the Poynting vector $\mathbf{S} = \mathbf{E} \times \mathbf{B}/\mu_0$, the energy flux density, is the transient intensity ι of light,[b]

$$\begin{aligned} \iota(\omega, \tau, z; t) &= n(\omega)\epsilon_0 c |\mathcal{E}_x(\omega, \tau; t)|^2 \quad (6.3) \\ &= n(\omega)\epsilon_0 c E_0^2 \exp(-2t^2/\tau^2) \exp(-\frac{2\omega k}{c} z) \\ &\equiv I_0(\omega, \tau; t) \exp(-\frac{z}{d}), \quad (6.4) \end{aligned}$$

[b]It should be noted that if circularly polarized light is used, with the x- and y-components $[\cos(\omega t), \sin(\omega t)]$, Eq. 6.3 is still valid, as is Eq. 6.5. For linearly polarized light, Eq. 6.5 has a coefficient of 1/2 before $n(\omega)$, due to the time average over $\cos^2(\omega t)$. In Eq. 6.3, some books have a factor of 2 [Boyd (1991)].

where ϵ_0 is the permittivity of free space, and $I_0(\omega, \tau; t)$ is the laser intensity before penetrating the sample. This is the well-known Beer-Lambert law. Here d is the penetration depth, defined as $d = c/2\omega k = \lambda/4\pi k$, which describes how the light intensity falls off starting from the surface of a sample. It is clear that d itself is intensity independent. Since we are interested in the pulse energy fluence $F(\omega, \tau, z)$, we integrate the above equation over time to find,

$$F(\omega, \tau, z) = n(\omega) c \epsilon_0 \int_{-\infty}^{\infty} |\mathcal{E}_x(\omega, \tau, z; t)|^2 dt, \quad (6.5)$$

which can be simplified as[c]

$$F(\omega, \tau, z) = n(\omega) c \epsilon_0 E_0^2 \sqrt{\frac{\pi}{2}} \tau \exp(-\frac{z}{d}) \equiv F_{\max} \exp(-\frac{z}{d}), \quad (6.6)$$

[c]For linearly polarized light, we have an extra factor of 1/2. See the side note b. So for the same fluence, the electric field of the linearly polarized light is stronger by $\sqrt{2}$ than that of circularly polarized light.

where $F_{\max}$ is the initial laser fluence given in experiments. This is the exact expression for the laser energy fluence at depth z, if the pulse is a Gaussian function.

If a laser pulse of fluence $F(\omega, \tau, z)$ shines on a spot with area A, the total number of photons within A at depth z is

$$N_{\text{photon}}(\omega, z) = \frac{F_{\max} A}{\hbar\omega} \exp(-\frac{z}{d}), \quad (6.7)$$

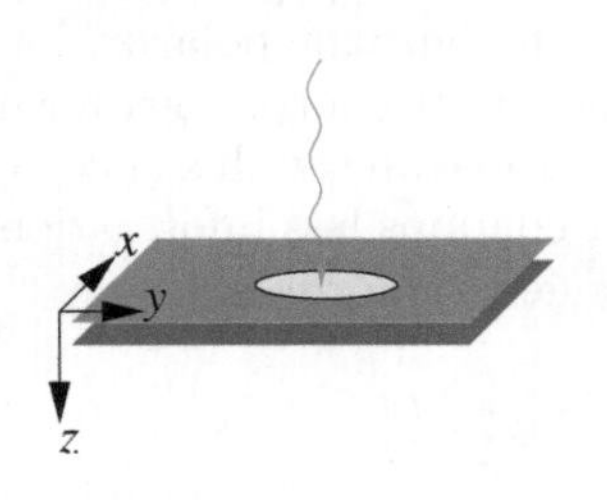

Figure 6.1
When a light beam impinges on a surface of a material, it penetrates with a small depth over an area of finite radius. The electric field is perpendicular to the propagation direction.

where $\hbar\omega$ is the photon energy, and $\hbar$ is the reduced Planck's constant. A couple of things are interesting to note. Each layer of a magnetic film receives different amounts of light. Equation 6.6 is extremely useful to convert the laser fluence to the field amplitude E_0. This conversion is not simple since it involves the index of refraction of the sample and the laser pulse duration. For the same fluence, a longer pulse has a smaller E_0. From Ref. [Si and Zhang (2010)], for fcc Ni, the experimental fluence ranges from 0.6 to 35 mJ/cm^2. This imposes a limit on the magnitude of the laser field that a realistic theory can use in the simulation. We will address this critical issue in later sections, since not all theories use actual experimental fluences.

In order to compute the number of photons per atom, we can divide the number of photons by the number of atoms on the surface. If there is more than one layer of atoms, one has to divide the number again by the number of layers. However, caution must be taken, since not all atoms in each layer receive the same intensity of light due to the penetration depth. It is incorrect to divide the total number of photons by the total number of atoms in the sample to get the number of photons per atom. Since the atoms deep inside the sample receive nearly zero photons, doing so would greatly exaggerate the limited number of photons per atom.

6.2 Background

6.2.1 Ferrimagnets and magneto-optical recording

Magneto-optical recording is effectively thermomagnetic recording [Furlani (1997)]. A laser beam is employed to heat the sample, commonly a ferrimagnet. This increases the temperature of the sample. From the last chapter, we know that if the temperature is above the Curie temperature, the magnet loses its magnetic moment. If we apply a magnetic field during the sample cooling, we can reorient the spin to a different direction, in order to write a spin bit into the disk. In ferrimagnets, there is another critical temperature besides the Curie temperature.

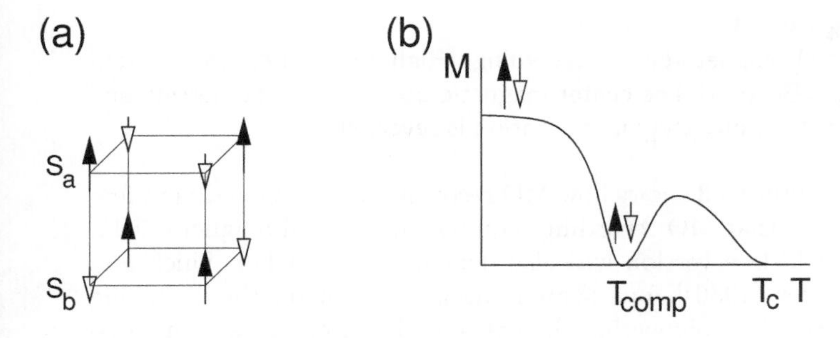

Figure 6.2
(a) Magnetic structure of a ferrimagnet. There are two sublattices with spins pointing in two opposite directions: One sublattice has a stronger spin moment, and the other has a weaker spin moment. (b) Magnetization curve versus temperature. There are two critical temperatures at which the magnetization is zero, where the first one is the compensation temperature and the second the Curie temperature.

Figure 6.2(a) shows that ferrimagnets have two spin sublattices, a and b, with spins pointing in the opposite directions, just like antiferromagnets. But different from antiferromagnets, the magnitudes of the sublattice spins are different. As temperature increases, both sublattice spins reduce, but the rate is different, so there is a temperature at which both spins have the same magnitude and the total spin moment cancels out, $S = S_a - S_b = 0$. This temperature is called the compensation temperature $T_{\rm comp}$ (see Fig. 6.2(b)) and is normally lower than the Curie temperature. This is important, as we see in the last chapter, that Curie temperatures in element ferromagnets are extremely high – higher than the room temperature. To raise the sample temperature to such a high temperature poses technological challenges. Since $T_{\rm comp}$ can be tuned gradually toward room temperature by changing composition, ferrimagnets represent a big advantage. In addition, most magneto-optical (MO) materials are amorphous, so their composition can be easily changed.

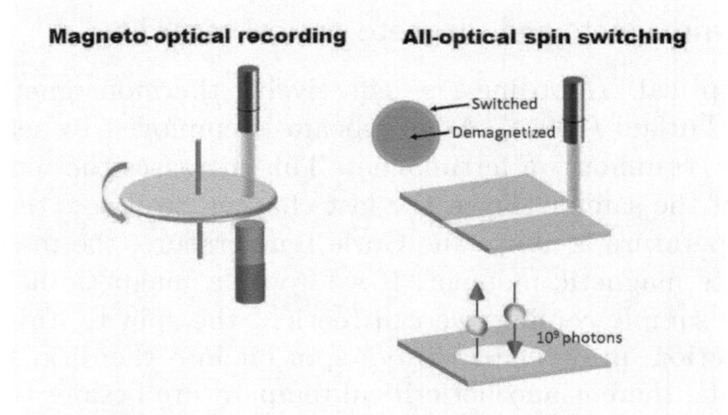

Figure 6.3
(Left) Magneto-optical recording. (Right) All-optical spin switching. (Bottom) The center magnetic domain is demagnetized and in the vicinity the magnetization is reversed.

Figure 6.3 shows how MO recording works. One salient feature is that MO recording requires an external magnetic field. A field-free version was also demonstrated earlier [Shieh and Kryder (1986)], but the emphasis was not on the switching speed. The pioneering discovery of Beaurepaire and his colleagues [Beaurepaire et al. (1996)] inspired new investigations using ultrafast laser pulses.

6.2.2 Experimental discovery

In 2007, Stanciu and coworkers [Stanciu et al. (2007)] carried out an experiment with $Gd_{22}Fe_{74.6}Co_{3.4}$ in the magnetic field free settings. The right part of Fig. 6.3 schematically shows how all-optical spin switching works. The sample is 20 nm thick and has a saturation magnetization of 1000 G. The small amount of Co is added to ensure that the sample has a strong magnetic perpendicular anisotropy. Because $Gd_{22}Fe_{74.6}Co_{3.4}$

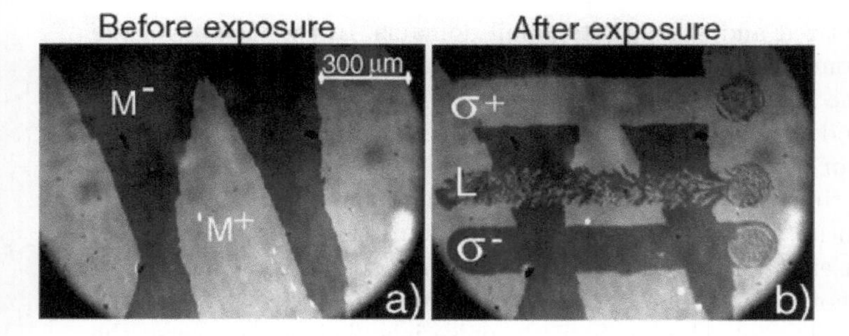

Figure 6.4
All-optical helicity-dependent spin switching (AO-HDS) in $Gd_{22}Fe_{74.6}Co_{3.4}$ [Stanciu et al. (2007)] and the effect of ultrashort polarized laser pulses on magnetic domains in $Gd_{22}Fe_{74.6}Co_{3.4}$. (a) Magneto-optical image of the initial magnetic state of the sample before laser exposure. White and black areas refer to up (M^+) and down (M^-) magnetic domains, respectively. (b) Domain pattern obtained by sweeping linear (π), right-handed (σ^+), and left-handed (σ^-) circularly polarized beams across the surface of the sample, with a laser fluence of about 11.4 $\mathrm{mJ/cm^2}$. The central area of the remaining spots at the end of each scan line consists of small magnetic domains, where the ratio of up to down magnetic domains is close to 1. This represents a demagnetized state. Used with permission from American Physical Society.

has a large Faraday rotation, the idea is to see whether the generated effective magnetic field is strong enough to reverse spins upon laser excitation. A "single" 40-fs pulse with wavelength 800 nm and laser repetition rate 1 kHz is used. Figure 6.4(a) shows that the sample has initial magnetization up (white) or down (black). The image is obtained with the magneto-optical Faraday effect, not the magneto-optical Kerr effect,[d] so the sample has to be very thin. Three different laser helicities, left-circularly polarized (σ^-), right-circularly polarized (σ^+), and linearly polarized (π), are employed. If a σ^+ pulse of 11.4 $\mathrm{mJ/cm^2}$ is scanned across the sample at 30 μm/s, it only switches the magnetic down-domains up, while leaving those up-domains intact. If a σ^- pulse is scanned across the sample, it only switches the magnetic up-domains down, while leaving those down domains intact. The sample has a clear preference to the helicity of the light. If a π pulse is scanned across the sample, it only creates small domains with spins randomly pointing up or down. Switching of this type is called all-optical helicity-dependent spin switching (AO-HDS).

[d] Recall from Section 2.3 that the Faraday effect refers to the phenomenon where light polarization changes after it passes through a magnetic sample, while the Kerr effect refers to the reflected light polarization change.

Since the laser pulse sweeps with a speed of 30 μm/s, within 1 kHz a single pulse covers 30 nm. Because the domain size is normally larger than 30 nm, a single domain experiences more than one pulse. The laser beam has a spot size of 100 μm, and at the center of excitation the entire domain is demagnetized and broken into small domains. In the vicinity of the domain, the magnetization is switched over (see Fig. 6.3). If the laser intensity is weakened, one can avoid the overheating and demagnetization. For instance, at 2.9 mJ/cm^2, only σ^- can reverse up-spin down, and σ^+ has no effect. One interesting aspect of all-optical spin switching (AOS) is that the entire switched area is very small, only 20 μm at such a weak laser intensity [Stanciu et al. (2007)], so the total number of photons which hit the sample is over 10^9.

To test whether a single pulse is sufficient to reverse the magnetization, one can move the laser beam across the sample surface very quickly, so that a single domain only receives a single pulse. It is demonstrated that a 40-fs circularly polarized laser pulse (single pulse) of 2.9 mJ/cm^2 can permanently switch magnetization.

6.3 Key ingredients of all-optical spin switching

The pioneering experiment by Stanciu and coworkers has generated huge interest worldwide. How is the angular momentum information instilled into the electronic system via electric dipole interaction within 40 fs (their laser pulse duration), and how can the electronic system keep this angular momentum information so long without getting destroyed by phonons? What are the key ingredients of all-optical spin switching?

Soon after the discovery of AOS in GdFeCo, it was found that the laser intensity has an important effect on how the laser helicity affects spin switching. At low laser fluences, the switching, if it happens, depends on the helicity of the light. One type of helicity switches one type of spins. But if the fluence is increased above a threshold, the helicity does not matter any more [Ostler et al. (2012)]. In this case the switching is called all-optical helicity-independent spin switching or AO-HIDS.

A complete review of the existing research activities would be too long for one section. In 2016 and 2018, we published two review articles [Zhang et al. (2016b, 2018b)] that detail the controversies and challenges facing AOS. The reader may refer to these two articles for further details. There are several proposed mechanisms, including inverse Faraday effects, different absorption abilities for different light helicities in magnetic circular dichroism, or simply pure heating. Nearly all theoretical investigations are phenomenological ones, where the laser field is replaced by an effective magnetic field or a thermal field. In the following, we restrict ourselves to two closely-related lines of research to examine both the ongoing research and possible new directions. The one is to investigate the origin of AOS, and the other is to explore new materials. Because GdFeCo is a ferrimagnet, its composition and the compensation temperature is a focus in the beginning. Since the composition influences the compensation temperature, we present them together.

6.3.1 Composition and compensation temperature

The composition of $Gd_xFe_yCo_{100-x-y}$ is determined by x and y. A small amount of Co atoms ensures a perpendicular magnetic anisotropy as mentioned before. In the ground state, the element Gd has a spin moment of 7.63 μ_B and iron has 2.2 μ_B. To have a compensated spin moment $M_{\rm tot} = 0$, x and y must scale proportionally to their respective spin moments,[e]

$$M_{\rm tot} = xM_{\rm Gd} + yM_{\rm Fe} = 0, \tag{6.8}$$

so we have $x/y = 2.2/7.63 = 22/76.3$. This is the reason why GdFeCo has a composition of $x = 22$, and $y = 76$. We can also understand a similar trend in TbCo and TbFe. Naturally, by changing x and y, one can tune the compensation temperature above or below room temperature, and thus investigate whether the compensation temperature affects AOS. Experimentally, it is found that not all x and y combinations lead to AOS. There is a limited range in which AOS occurs. Figure 6.5(a) shows four different materials, where the boxes represent the range of x where AOS is observed. Beyond these limits one observes either demagnetization or no switching. We see that the widest range of x is in TbFe. This feature is closely connected to the effect of the magnitude of the spin moments on AOS (which will be explained below). Different research groups report slightly different results, but the main conclusion is the same: only a narrow region of x allows AOS.

[e]A similar equation is used to compute spin moments for each element. We also provide a Fortran code in [Zhang et al. (2018b)]

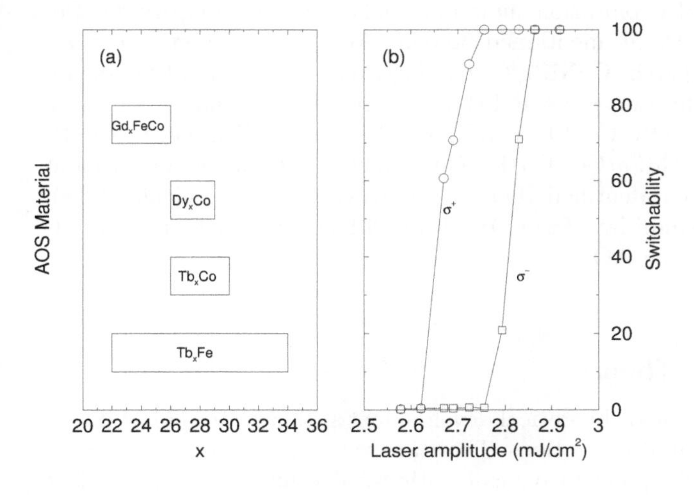

Figure 6.5
(a) Composition for a group of AOS ferrimagnets. The all-optical spin switching only occurs when a specific composition is met. This does not apply to other types of materials. (b) Switchabil-

ity of spins as a function of laser intensity in $Gd_{22}Fe_{68.2}Co_{9.8}$ for right circularly polarized light (σ^+) and left circularly polarized light (σ^-) [Vahaplar et al. (2009)]. σ^+ appears to be more powerful than σ^-. Data are read from their paper.

6.3.2 Laser parameters

When a laser pulse impinges on a material, electrons are first excited, followed by spin excitation. The energy absorbed into the system is used to switch spins. It is expected that the laser parameters must have an important effect [Ostler et al. (2012)]. Similar to the composition dependence, the laser has a narrow region of fluence to switch magnetization. If the fluence is too high, it normally breaks the magnetic domains into small random domains and the structure is entirely demagnetized. Vahaplar et al. [Vahaplar et al. (2009)] showed that in $Gd_{22}Fe_{68.2}Co_{9.8}$ the right circularly polarized light (σ^+) is more powerful to switch spins than the left circularly polarized light (σ^-). In Fig. 6.5(b), we reproduce their data, where one can see the onset of σ^+ is at 2.62 mJ/cm^2, but that of σ^- is at 2.75 mJ/cm^2. This 0.13 mJ/cm^2 difference is not due to the different absorption ability as in magnetic circularly dichroism. Instead, it is due to the magnetization axis difference in ferrimagnets. Further increase in the laser intensity allows both helicities of light to switch the magnetization.

6.3.3 New materials

For a long time, it was believed that GdFeCo is the only material for AOS. After several years of intensive experimental exploration, new materials were also discovered. Besides those mentioned above, AOS ferrimagnets now include Co/Ir/CoNiPt/Co/Ir, Tb/Co multilayer and Pt/Co/Gd; AOS ferromagnets and those mixed with ferrimagnets include Co/Pt, Pt/Co/Pt, Cu/Ni/Co/Ni/Cu, Co/Ni, FePt, Co/Pt/Co, GdFe/Co, Co/Pt/Cu/GdFeCo. The latest experiment research identified Heusler Mn_2RuGa [Banerjee et al. (2019)] as a candidate for AOS. Table 6.1 shows a list of materials exhibiting AOS.

6.4 Theory

In this field theoretical development still lags far behind experimental investigations. There are several proposed mechanisms without quantitative calculations. The inverse Faraday effect has long been thought as a possible origin of AOS, but it does not produce a strong enough magnetic field. The magnetic circular dichroism and the angular momentum transfer between

Table 6.1
Chronicle of AOS materials whose magnetic domain images are taken. Ordering refers to magnetic ordering. IFE: inverse Faraday effect. SF-SRS: spin-flip stimulated Raman scattering. FIM: ferrimagnetic. AFM: antiferromagnetic. PM: paramagnetic. Under Mechanism, only a selected few are listed. Linear: linear reversal; HD-AOS: helicity-dependent all-optical switching; HID-AOS: helicity-independent all-optical switching; T_{comp}: compensation temperature dependent; MCD: magnetic circular dichroism; SDC: superdiffusive current; LR: low remanence; DM: magnetic domain size; ST: stochastic. The underlined compounds are the only ferromagnets that show a single-shot switching. The slanted lines denote those that are disapproved by a referenced paper. The table is adopted from [Zhang et al. (2018b)] and used with permission from the World Scientific Publishing Company.

Compound	Ordering	Mechanism	AOS/Non-AOS	Ref.
$TmFeO_3$	AFM		Non-AOS	[Kimel et al. (2004)]
$DyFeO_3$	AFM	IFE	Non-AOS	[Kimel et al. (2005)]
$HoFeO_3$	AFM	IFE	Non-AOS	[Kimel et al. (2009)]
$NaTb(WO_4)_2$	PM	IFE	Non-AOS	[Jin et al. (2010)]
$Gd_{22}Fe_{74.6}Co_{3.4}$	FIM	IFE	HD-AOS	[Stanciu et al. (2007)]
$Gd_{22}Fe_{74.6}Co_{3.4}$	FIM	~~Thermal~~	AOS	[Hohlfeld et al. (2009)]
$Gd_{24}Fe_{66.5}Co_{9.5}$	FIM	Linear	HD-AOS	[Vahaplar et al. (2009)]
$Gd_{26}Fe_{64.7}Co_{9.3}$	FIM	~~IFE/SF-SRS~~	HD-AOS	[Steil et al. (2011)]
$Gd_{23}Fe_{68}Co_9$	FIM	IFE	HD-AOS	[Ohkochi et al. (2012)]
$Gd_{x=20\leftrightarrow28}Fe_{90-x}Co_{10}$	FIM	IFE/Linear	H(I)D-AOS	[Vahaplar et al. (2012)]
$Gd_{26}Fe_{65}Co_9$	FIM	MCD	HD-AOS	[Khorsand et al. (2012)]
$Gd_{24,25}Fe_{66.5}Co_{9.5}$	FIM	Thermal	HID-AOS	[Ostler et al. (2012)]
$Gd_{24,25}Fe_{65.6}Co_{9.4}$	FIM	Thermal	HID-AOS	[Ostler et al. (2012)]
$Gd_{24}Fe_{66.5}Co_{9.5}$	FIM		H(I)D-AOS	[Alebrand et al. (2012b)]
$Tb_{x=0.12\leftrightarrow0.34}Co_{1-x}$	FIM	T_{comp}	H(I)D-AOS	[Alebrand et al. (2012a)]
$Tb_{x=19\leftrightarrow38.5}Fe_{100-x}$	FIM	~~T_{comp}~~	HD-AOS	[Hassdenteufel et al. (2013)]
Co/Ir/CoNiPtCo/Ir,$Tb_{26}Co_{74}$	FIM	T_{comp}	HD-AOS	[Mangin et al. (2014)]
Tb/Co multilayer	FIM	T_{comp}	HD-AOS	[Mangin et al. (2014)]
$Tb_{36}Fe_{64}/Tb_{19}Fe_{81}$	FIM	LR	HD-AOS	[Schubert et al. (2014a)]
$Tb_{29}Fe_{71}$,$Tb_{34}Fe_{66}$	FIM	LR	HD-AOS	[Hassdenteufel et al. (2015)]
$Tb_{30}Fe_{70}$	FIM	conductivity	HD-AOS	[Hassdenteufel et al. (2014)]
$Tb_{22}Fe_{69}Co_9$	FIM	IFE,~~MCD~~	HD-AOS	[Gierster et al. (2015)]
$Tb_{x=8\to14.5}Co_{100-x}$(< 6.5nm)	FIM	DM, ~~LR~~	HD-AOS	[Hadri et al. (2016)]
$Tb_{x=16.5\to30.5}Co_{100-x}$(< 15nm)	FIM	DM, ~~LR~~	HD-AOS	[Hadri et al. (2016)]
$Tb_{x=22\to34}Fe_{100-x}$(5-85nm)	FIM	LR	HD-AOS	[Hebler et al. (2016)]
Pt/Co/Gd	FIM	Thermal	HID-AOS	[Lalieu et al. (2017)]
$[Co(4Å)/Pt(7Å)]_{2\to3}$	FM		HD-AOS	[Lambert et al. (2014)]
Pt/Co(6Å $\leftrightarrow$ 15Å)/Pt	FM		HD-AOS	[Lambert et al. (2014)]
$[Pt/Co_{1-x}Ni_x(6Å)]_{2\to4}$	FM		HD-AOS	[Lambert et al. (2014)]
$Cu/[Ni(5Å)/Co(1Å)]_2/Ni/Cu$	FM	~~SDC~~	HD-AOS	[Lambert et al. (2014)]
$[Co(2Å)/Ni(6Å)]_2$	FM	DM	HD-AOS	[Hadri et al. (2016)]
$[Pt(7Å)/Co(6Å)]_{1-2}$	FM	DM	HD-AOS	[Hadri et al. (2016)]
FePt	FM	ST	HD-AOS	[John et al. (2017)]
Co/Pt/Co/GdFeCo	FIM/FM	~~transport~~	HID-AOS	[Gorchon et al. (2017)]
[Co/Pt]/Cu/GdFeCo	FIM/FM	transport	HID-AOS	[Iihama et al. (2018)]
Pt/Co/Pt	FM	~~IFE~~	HID-AOS	[Vomir et al. (2017)]

spin sublattices are also proposed. Since nearly all AOS materials are amorphous, a first-principles calculation is difficult, if not impossible. Most of the theoretical investigations are highly phenomenological, where the laser field is replaced by an effective magnetic field so that the initial excitation is not taken into account.

A model based on the Heisenberg exchange model uses the following Hamiltonian [Ostler et al. (2011)]:

$$H = -\frac{1}{2}\sum_{ij} J_{ij}\mathbf{S}_i \cdot \mathbf{S}_j - \sum_i D_i(\mathbf{S}_i \cdot \mathbf{n}_i)^2 - \sum_i \mu_i \mathbf{B} \cdot \mathbf{S}_i. \quad (6.9)$$

Here the first term is the familiar Heisenberg exchange interaction between two spins at site i and j, J_{ij} is the exchange integral, and $\mathbf{S}_i$ is the normalized vector $|\mathbf{S}_i| = 1$. D_i is the uniaxial anisotropy vector and allows for the spins pointing along a particular direction space in the ground state, and $\mathbf{n}_i$ is the direction of the anisotropy vector. μ_i is the magnetic moment of the site i, and $\mathbf{B}$ is the applied field.[f] In the actual calculation, one often treats the model classically, so millions of spins can be computed. While this model is capable of describing many experimental observations, it intrinsically builds in an effective magnetic field. However, for a long time, there has been no alternative.

[f] $\mathbf{B}$ is given in a later paper [Vahaplar et al. (2012)] as,

$$\mathbf{B}(t,r) = \sigma\frac{2\beta F}{c\tau} f(t) e^{-\frac{r^2}{2r_0^2}} \hat{k},$$

where σ denotes the laser helicity (not to be confused with the dielectric function above). $\sigma = \pm 1$ or 0 for right and left circularly polarized light and linearly polarized light. $\sigma = 0$ represents a big deficiency in the theory since linearly polarized light does switch spins [Ostler et al. (2012)]. τ is the pulse duration. F is the fluence. From Eq. 6.6, one can see F/τ is related to the electric field squared, but in their calculation they used $F/\tau = c\epsilon_0 E_0^2/2$, without $\sqrt{\pi/2}$. $\hat{k}$ is the unit vector in the direction of the wave vector of light, so their magnetic field is an effective field, not related to the light's own magnetic field. For the same reason, the envelope function $f(t)$ has different time dependence. The last term is the spatial profile. β is the susceptibility 2×10^{-6}m/A, estimated using the element Fe and Gd magnetizations.

6.4.1 Birth of the first single spin switching model

All-optical spin switching only involves a laser pulse, [Si and Zhang (2010); Zhang (2011); Zhang and George (2013); Zhang et al. (2014, 2015a,b)]. Using an effective magnetic field is not satisfactory. In November 2014, one of us (GPZ) attended the 59th Annual Conference on Magnetism and Magnetic Materials in Honolulu, Hawaii. Discussions with colleagues from the University of Colorado convinced us to introduce a model that is based on the oscillator model often used in magneto-optics. J.-Y. Bigot was the first one who suggested this to us in 1999. So in the following subsections, we present the details of our model that works reasonably well. We provide a MatLab code, so the reader can test our results and adopt them for his/her own need.

Before we introduce the spin switching model, it is instructive to review the traditional oscillator model. The oscillator model is often used as an introduction for nonlinear optics [Shen (1984); Boyd (1991)]. Consider that an electron is confined in a harmonic potential, with its fundamental frequency Ω. Without external fields, it oscillates with Ω. If we subject the electron to an electric field $E(t)$, where t is time, the electron's response (the displacement x) contains linear and quadratic terms and higher orders in $E(t)$:

$$\frac{d^2x}{dt^2} + \gamma\frac{dx}{dt} + \Omega^2 x + \underbrace{ax^2 + bx^3 + \cdots}_{\text{nonlinear}} = \frac{-eE(t)}{m_e}. \quad (6.10)$$

Here the first term is the acceleration, the last terms on the left side are nonlinear, γ is a damping, and a and b are nonlinear coefficients.

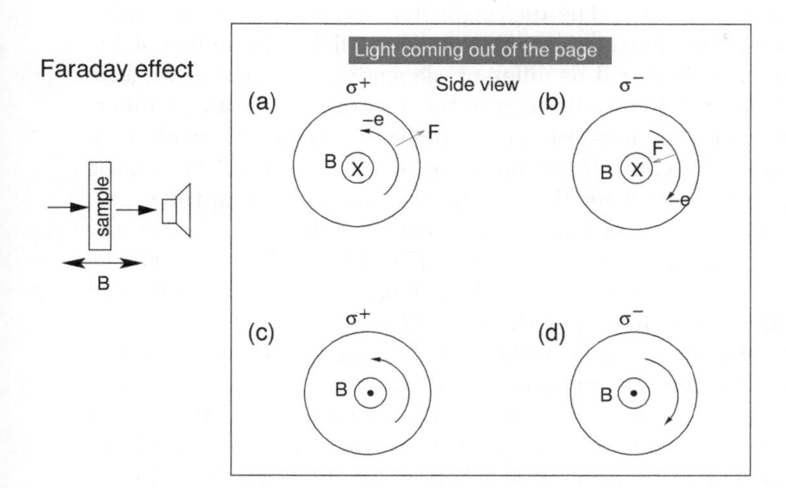

Figure 6.6
Classical interpretation of the Faraday effect. (Left) Schematic shows that light passing through the sample is detected, with a static external magnetic field applied in the same direction as the light propagation direction. (Right) (a)-(d) are viewed from the camera, where light comes out of the page. Classically, the electron orbital motion depends on the light helicity and the static magnetic field. (a) The electron orbits counterclockwise (since it carries negative charge) under right circularly polarized light σ^+ alone. With the presence of the magnetic field $\mathbf{B}$, which points into the page, the Lorentz force from $\mathbf{B}$ pulls the electron away from the center. (b) Left circularly polarized light σ^- with the magnetic field B into the page. (c) Right circularly polarized light σ^+ with the magnetic field B out of the page. (d) Left circularly polarized light σ^- with the magnetic field B out of the page.

In the classical linear magneto-optics, these nonlinear terms are removed, and the above equation is modified (in a three-dimensional system) to [Qiu and Bader (1998)],

$$\frac{d^2\mathbf{r}}{dt^2} + \Omega^2\mathbf{r} + \gamma\frac{d\mathbf{r}}{dt} = \left(-e\mathbf{E}(t) - e\frac{d\mathbf{r}}{dt} \times \mathbf{B}\right)/m_e, \qquad (6.11)$$

so that an extra term is added on the right side.[1] This term is the contribution from an external magnetic field $\mathbf{B}$ and is not associated with the light's magnetic field. Let us compare two cases: One with a magnetic field, and the other without a magnetic field. If the magnetic field is zero, then the left (σ^-)

[1]Although not in the same form, the inclusion of the magnetic field in the conductivity tensor for the Faraday effect dates much earlier back [Roth (1964); Bennett and Stern (1965)].

and right (σ^+) circularly polarized light produce no difference in the orbital motion of the electron, with the same radius, except that the orbital angular momentum direction is opposite for σ^- and σ^+. The dielectric function is exactly the same. Now, if we have an external magnetic field, the radius of the electron orbit will be different, depending on whether we have σ^- or σ^+. Figure 6.6 shows the Faraday effect under different configurations. Figure 6.6(a) has σ^+ light, so the electron moves clockwise. If the magnetic field points into the page, the Lorentz force from **B** pulls the electron away from the center (see the force in Fig. 6.6(a)). For σ^-, the force pushes the electron toward the center (see Fig. 6.6(b)). If we flip **B**, the opposite effects appear (see Figs. 6.6(c) and (d)). This classical picture also applies to the Kerr effect.

When we were investigating the magneto-optical effect, we realized that the main problem with Eq. 6.11 is the magnetic field. One has to replace it with something else that has a magnetic origin, because without a magnetic field there is no magneto-optic effect (see Chapter 2). Since we had already worked on the ultrafast demagnetization for a long time, we asked ourselves whether it is possible to replace it with spin-orbit coupling. We proposed the following model Hamiltonian [Zhang et al. (2015a)] with great success:[g]

[g] The Hamiltonian does not have an external magnetic field.

$$H = \frac{\mathbf{p}^2}{2m_e} + \frac{1}{2} m_e \Omega^2 \mathbf{r}^2 + \lambda \mathbf{L} \cdot \mathbf{S} - e\mathbf{E}(t) \cdot \mathbf{r}, \qquad (6.12)$$

where m_e is the electron mass. One can see that there is no major difference between this Hamiltonian and the harmonic oscillator model. Here, the first term is the kinetic energy operator of the electron. We do not use the second quantization format since using it would restrict our working space.[h] The second term is the harmonic potential energy operator with system frequency Ω. This shows that the form of the potential really matters. λ is the spin-orbit coupling in units of $\mathrm{eV}/\hbar^2$. **L** and **S** are the orbital and spin angular momenta in units of $\hbar$, respectively. One crucial element here is that we use a real space orbital angular momentum, so we can compute the orbital angular momentum under any laser excitation conditions. **p** and **r** are the momentum and position operators of the electron, respectively. The last term in the equation is the laser-system interaction. Here we use the dipole approximation. The laser field direction can be specified through $\mathbf{E}(t)$ (see below for details). The real space implementation has a huge advantage.

[h] The second quantization requires us to choose the basis function beforehand. This is not possible in general since we do not have the information as to how strong the laser excitation can be. We also have tried to use other potentials, but they do not work well for AOS.

To appreciate the physical insight of our Hamiltonian, we set the spin along the $+z$-axis, and consider a cw optical field polarized along the x direction, $E_x(t) = A_x e^{i\omega t} + cc.$, where A_x is the amplitude and ω is the laser frequency. Under cw excitation, we can derive the linear susceptibility[i] [Zhang et al. (2015a)] (the details of the derivation are left as a homework

[i] In SI units, the linear susceptibility $\chi^{(1)}$ is dimensionless. In general, for the nth order $\chi^{(n)}$, the units are $(\mathrm{m/V})^{n-1}$ [Butcher and Cotter (1990)].

assignment),

$$\chi_{xx}^{(1)}(\omega) = -\frac{Ne^2}{\epsilon_0 m}\frac{\Omega^2-\omega^2-\lambda^2S_z^2}{(\Omega^2-\omega^2-\lambda^2S_z^2)^2-(2\lambda S_z\omega)^2} \quad (6.13)$$

$$\chi_{xy}^{(1)}(\omega) = -i\frac{Ne^2}{\epsilon_0 m}\frac{2\lambda S_z\omega}{(\Omega^2-\omega^2-\lambda^2S_z^2)^2-(2\lambda S_z\omega)^2}, \quad (6.14)$$

where N is the number density in units of m^{-3} and ϵ_0 is the permittivity in vacuum in units of $\mathrm{C}^2/(\mathrm{Nm}^2)$. These two equations beautifully contain all magneto-optics ingredients. The spin-orbit coupling λ and the spin angular momentum S_z always appear together. The diagonal susceptibility $\chi_{xx}^{(1)}(\omega)$ is dominated by the charge response, (see $\Omega^2-\omega^2$ in the numerator). Since λS_z[j] is always small, the charge dynamics is determined by $\Omega^2-\omega^2$. Magneto-optics always considers $\chi_{xx}^{(1)}(\omega)$ as a quantity carrying less information about the spin.[k] The off-diagonal susceptibility $\chi_{xy}^{(1)}(\omega)$ is quite different, and reflects the spin response. One can see from Eq. 6.14 that $\chi_{xy}^{(1)}(\omega)$ is proportional to $\lambda S_z\omega$. When the spin-orbit coupling or the spin angular momentum is zero, $\chi_{xy}^{(1)}(\omega)$ is zero. This is the basis for magneto-optics as a tool to detect spin changes. $\chi_{xy}^{(1)}(\omega)$ goes to zero if ω goes to zero, but $\chi_{xx}^{(1)}(\omega)$ does not.[l]

[j] λS_z unit is frequency, same as Ω.

[k] But one has to be careful since the diagonal term still contains the spin contribution.

[l] This is because a DC field can still drive electrons.

6.4.2 Numerical solutions and MatLab codes

To compute the spin reversal, we use the Heisenberg equation of motion, rather than the Schrödinger equation. The former has the advantage that we can compute the spin expectation directly, without worrying about the convergence of the basis functions. Mathematically, the Heisenberg and the Schrödinger pictures are equivalent. The Heisenberg equation of motion works with operators,[m]

$$i\hbar\frac{dA}{dt} = [A, H], \quad (6.15)$$

[m] Note that there is a sign change with respect to the time evolution of the density matrix ρ since the latter is in the Schrödinger picture (Eq. 5.32).

where A is an operator such as spin, orbital or position, and H is the total Hamiltonian that includes the system and the laser field. After a lengthy but straightforward calculation, we find[2]

$$\frac{d\mathbf{r}}{dt} = \frac{\mathbf{p}}{m} - \lambda(\mathbf{r}\times\mathbf{S}), \quad (6.16)$$

$$\frac{d\mathbf{p}}{dt} = -m\Omega^2\mathbf{r} + e\mathbf{E}(t) - \lambda\mathbf{p}\times\mathbf{S}, \quad (6.17)$$

$$\frac{d\mathbf{S}}{dt} = \lambda(\mathbf{L}\times\mathbf{S}), \quad (6.18)$$

$$\frac{d\mathbf{L}}{dt} = -e\mathbf{E}(t)\times\mathbf{r} - \lambda(\mathbf{L}\times\mathbf{S}). \quad (6.19)$$

[2] The term $-\lambda\mathbf{p}\times\mathbf{S}$ in Eq. 6.17 is of relativistic origin (see [Hübner and Bennemann (1989)], although historically this is not the first time this term was mentioned in the literature).

These four sets of differential equations can be solved numerically using Matlab.

Here are two codes: `Liouville.m` is the main code, and `spinswitching.m` shows the above differential equation. They use eV as the energy unit, fs as the time unit and Å as the length unit. Spin and orbital angular momenta are in the unit of $\hbar$. The initial spin is along the $-z$-axis, $(0, 0, -2.2)\hbar$, and the initial velocity is $(0, 0, 1)$Å/fs. The initial position is at (0,0,0). We find that the MatLab differentiation solver is not very accurate and only `ode45` works.

```
%%%%%%%%%%%%%%%%%%%%%%%%%%%%%%%%%%%%%%%%%%%%%%%%%%%%%%%%%%%%%%%%
% Spinswitching code                                           %
% Authors: Guoping Zhang & Yihua Bai at                        %
% Indiana State University, USA, May 4, 2015,                  %
% supported by U. S. Department of Energy grant,               %
% Contract No. DE-FG02-06ER46304.                              %
%                                                              %
%This code is based on the paper G. P. Zhang, Y. H. Bai%
%and Thomas F. George, "A␣new␣and␣simple␣model␣for␣␣␣␣␣%
%magneto-optics␣uncovers␣an␣unexpected␣spin␣switching",%
%EPL, 112 (2015) 27001, and  can be extended to            %
%include exchange interactions, G. P. Zhang, Y. H. Bai %
%and Thomas F. George,"Switching␣ferromagnetic␣spins␣by%
%an␣ultrafast␣laser␣pulse:␣Emergence␣of␣giant␣optical␣␣%
%spin-orbit␣torque", EPL, 115 (2016) 57003.                %
%                                                              %
%This code is free of charge to anyone who is                 %
%interested in %all-optical spin switching. The authors%
% have no responsibility for any consequence of this    %
% code. It is copyrighted by                                  %
%Guoping Zhang and Yihua Bai, (2020).                         %
%%%%%%%%%%%%%%%%%%%%%%%%%%%%%%%%%%%%%%%%%%%%%%%%%%%%%%%%%%%%%%%%

% liouville.m.

dy0=[0;0;0;0;0;1;0;0;-2.2]; options =
 odeset('RelTol',1e-6,'AbsTol',1e-8); ...
 [tv,Yv]=ode45('spinswitching',
 [-1000 1000], dy0, options); plot (tv,Yv(:,7)); ...
 hold on plot
 (tv,Yv(:,8)); plot (tv,Yv(:,9)); hold off

%The following is the differentiation equation,
%where the parameters are given as well.

function Fv=funcl(t,Y)
%constants
echarge=1.60217653;
emass=9.1093826;
h=6.62606957;
%This is used to ensure the unit correct.
constant=(2*pi/h)^2*echarge*emass*0.01;
eh=2*pi*echarge/h;
em=echarge/emass*100.0;
```

```
45 % position of the electron
46 xx=Y(1);
47 yy=Y(2);
48 zz=Y(3);
49 % spin of the electron
50 sx=Y(7);
51 sy=Y(8);
52 sz=Y(9);
53 % system parameters
54 soc=0.06;
55 %damping
56 gamma=0;
57 %laser parameters
58 tau=60;
59 amplitude=0.035;
60 photonenergy=1.6;
61 electronenergy=photonenergy;
62 omega0=2*pi*echarge*electronenergy/h;
63 w=2*pi*echarge*photonenergy/h;
64 %Left circularly polarized light
65 asin=amplitude*exp(-t^2/tau^2)*sin(w*t);
66 acos=amplitude*exp(-t^2/tau^2)*cos(w*t);
67 ex=-asin;
68 ey=acos;
69 ez=0;
70 lx=Y(2)*Y(6)-Y(3)*Y(5);
71 ly=Y(3)*Y(4)-Y(1)*Y(6);
72 lz=Y(1)*Y(5)-Y(2)*Y(4);
73 Fv(1,1) = Y(4)+soc*eh*(zz*sy-yy*sz);
74 Fv(2,1) = Y(5)+soc*eh*(xx*sz-zz*sx);
75 Fv(3,1) = Y(6)+soc*eh*(yy*sx-xx*sy);
76 Fv(4,1) = -2*gamma*Y(4)-omega0^2*Y(1)
77           +em*ex+soc*eh*(Y(8)*Y(6)-Y(9)*Y(5));
78 Fv(5,1) = -2*gamma*Y(5)-omega0^2*Y(2)
79           +em*ey+soc*eh*(Y(9)*Y(4)-Y(7)*Y(6));
80 Fv(6,1) = -2*gamma*Y(6)-omega0^2*Y(3)
81           +em*ez+soc*eh*(Y(7)*Y(5)-Y(8)*Y(4));
82 Fv(7,1) = soc*constant*(ly*Y(9)-lz*Y(8));
83 Fv(8,1) = soc*constant*(lz*Y(7)-lx*Y(9));
84 Fv(9,1) = soc*constant*(lx*Y(8)-ly*Y(7));
```

The reader can compare his/her own results with Fig. 6.7. Here we initialize the spins along the $-z$-axis, and the magnitude of the spin is 2.2 $\hbar$. The starting time is -1000 fs and the ending time is 1000 fs. The laser pulse duration is 60 fs, the photon energy is 1.6 eV, and the field amplitude is optimized to 0.035 V/Å.

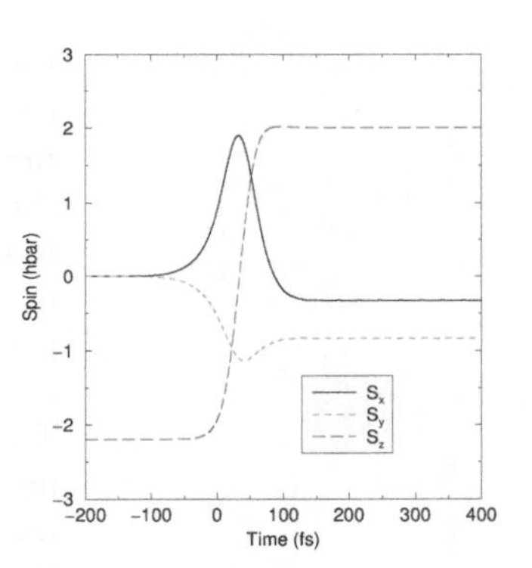

Figure 6.7
Example figure. The reader can use our Matlab code to reproduce this figure.

Experimentally, Vahaplar et al. [Vahaplar et al. (2009)] proposed a linear reversal.[n] This code itself is capable of testing whether these assumptions are true [Zhang (2011)].

[n] In the linear reversal picture, the spin is first reduced to zero, and then after some time, it regrows in the other direction. However, to this end, there is no theoretical proof [Zhang (2011)].

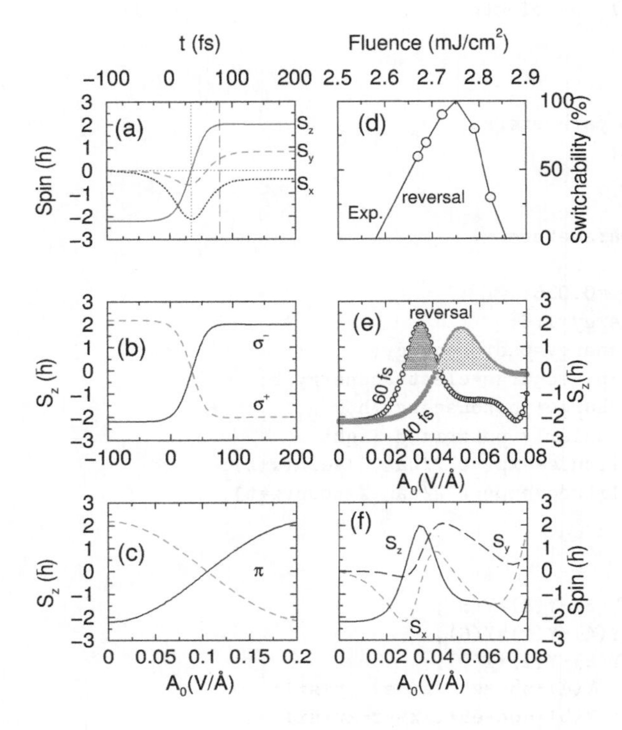

Figure 6.8
(a) All-optical spin reversal for S_x, S_y and S_z as a function of time t. The vertical dotted line denotes the time when S_z passes through zero, and the vertical dashed line denotes the time when the spin reversal starts. Here σ^- is used, the laser pulse duration is $\tau = 60$ fs, and the field amplitude is 0.035 V/Å. (b) The σ^- pulse (solid line) only switches spin from down to up, while the σ^+ pulse (dashed line) only switches spin from up to down. (c) The π pulse can switch spin from up to down or from down to up, but at a much higher field amplitude. (d) Experimental spin reversal window from Vahaplar et al. [Vahaplar et al. (2012)]. (e) Final spin angular momentum S_z as a function of the laser field amplitude for laser durations $\tau = 60$ fs (empty circles) and 40 fs (filled circles). The shaded regions are the spin reversal window. (f) As the field amplitude increases, the spin angular momentum changes from non-switching, canting along the $-x$-axis, switching, and canting along the $+y$-axis [Zhang et al. (2015a)]. Used with permission from EPL.

Figure 6.8(a) shows that upon laser excitation, the spin first precesses on the xy-plane and then switches toward the $+z$-axis [Zhang et al. (2015a)], not linear reversal as proposed. In fact, such a linear reversal is forbidden because the spin operators must satisfy the permutation relation $[S_x, S_y] = i\hbar S_z$ [Zhang (2011)]. Figure 6.8(b) reveals a strong helicity dependence: If the spin points along the $-z$ axis, σ^- light can effectively switch it to the $+z$-axis, while the σ^+ does the opposite. The linear polarized light π works as well (see Fig. 6.8(c)). The most striking experimental result is that the spin reversal only occurs in a narrow region of the laser fluence (see Fig. 6.8(d)) [Vahaplar et al. (2012)]. Note that the experimental peak is asymmetric with respect to the maximum switchability: the switchability[o] slowly increases with the fluence, and after the maximum, it drops quickly. This feature is not reproduced in the phenomenological model based on Eq. 6.9. However, our model reproduces this effect to a large extent. Figure 6.8(e) nicely reproduces the key feature. If we use a shorter laser pulse, the peak is pushed toward a strong laser field. We can also understand why this asymmetric peak appears. Figure 6.8(f) shows the entire process where three components of the spin have a convoluted dependence on the laser field amplitude. If the laser field is too weak, the spin only precesses on the xy-plane. If the laser field is too strong, it also falls onto the xy-plane. AOS only occurs over a narrow region.

[o] Switchability is defined as the ratio of the final spin angular momentum over the initial spin angular momentum along the same direction, or $\eta = \frac{S_z^f}{S_z(0)} \times 100\%$, where S_z^f is the final spin angular momentum. We only use S_z since other components are quite small. The experimental switchability is defined as the probability difference between right and left circularly polarized light excitation.

6.4.3 Reversing millions of spins

The success of switching a single spin is very encouraging. To move one step forward, we include the exchange interaction terms, so we can describe a true magnetic system [Zhang et al. (2016a)]:

$$\begin{aligned} H &= \sum_i \left[\frac{\mathbf{p}_i^2}{2m} + V(\mathbf{r}_i) + \lambda \mathbf{L}_i \cdot \mathbf{S}_i - e\mathbf{E}(\mathbf{r}, t) \cdot \mathbf{r}_i \right] \\ & -\sum_{ij} J_{ex} \mathbf{S}_i \cdot \mathbf{S}_j. \end{aligned} \tag{6.20}$$

Here the summation is over all the lattice sites in the system, the first term is the kinetic energy operator of the electron, and the second term is the potential energy operator. We choose a spherical harmonic potential $V(\mathbf{r}_i) = \frac{1}{2} m\Omega^2 \mathbf{r}_i^2$ with system frequency Ω. As discussed above, other forms of potentials were also tested, but they do not yield proper results. λ is the spin-orbit coupling in units of eV$/\hbar^2$, $\mathbf{L}_i$ and $\mathbf{S}_i$ are the orbital and spin angular momenta at site i in unit of $\hbar$, respectively, and $\mathbf{p}$ and $\mathbf{r}$ are the momentum and position operators of the electron, respectively. The last term in Eq. 6.20 represents the spin-spin interaction between lattice sites. We can initialize the spins along one particular direction, so we can have different spin configurations. This includes various domain walls of different sizes.

The beauty of our model is that we can realistically include the laser pulse shape and time-profiles. For instance, we can

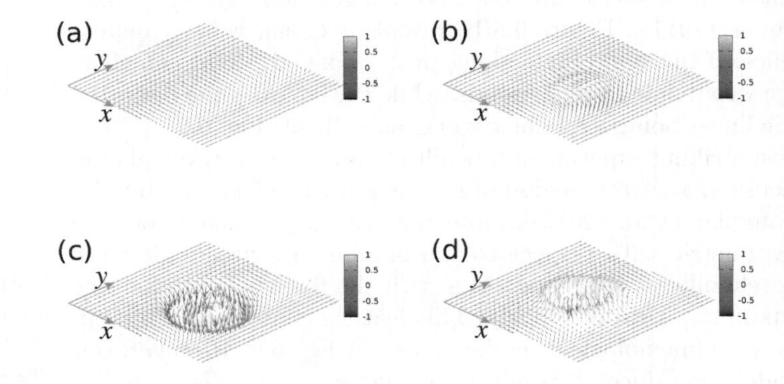

Figure 6.9
Spin reversal across a Néel wall along the y-axis. (a) Initial spin configuration. (b), (c) and (d) Snapshot of spins at 123 fs under linearly polarized light, right- and left-circularly polarized light, respectively. σ^+ tends to switch spin down, while σ^- tends to switch spin up, which creates a basin (c) and mound (d) of spins, respectively. Around the vicinity of the basin and mound, spins are clearly reversed. Figures and captions are taken from Ref. [Zhang and Murakami (2019)], with permission from IOP Publishing Ltd.

center the laser field on magnetic domains, so the laser pulse propagates vertically down along the $-z$-axis, with the electric field

$$\mathbf{E}(\mathbf{r},t) = \mathbf{A}(t)\exp\left[-\frac{(x-x_c)^2+(y-y_c)^2}{R^2}-\frac{z}{d}\right], \quad (6.21)$$

where x and y are the coordinates in the unit of the site number, x_c and y_c are the center of the laser spot, and R is the radius of the laser spot. The lattice site information can be converted to the real distance if the lattice constant of a sample is known. Because we work in real space, the laser helicity can be very easily taken care of. For left-(right)-circularly polarized field $[\sigma^-(\sigma^+)]$ in the xy-plane, $\mathbf{A}(t)$ is

$$\mathbf{A}(t) = A_0 e^{-t^2/T^2}\left[\mp\sin(\omega t)\hat{x} + \cos(\omega t)\hat{y}\right], \quad (6.22)$$

where ω is the laser carrier frequency, T is the laser pulse duration, A_0 is the laser field amplitude, t is time, and $\hat{x}$ and $\hat{y}$ are unit vectors. For a linearly polarized field (π) along the x-axis, $\mathbf{A}(t)$ is

$$\mathbf{A}(t) = A_0 e^{-t^2/T^2}\cos(\omega t)\hat{x}. \quad (6.23)$$

With these fields defined, we can solve the Heisenberg equation of motion in the same way as for the single site model. Here the sample has $501 \times 501 \times 4$ (over a million) spins. The

Néel wall is constructed by two functions, $S_i^x = S_0 \cos(\xi_i)$ and $S_i^y = S_0 \sin(\xi_i)$, where $S_0 = 1\hbar$, and i is the lattice site index running from W_L to W_R. $\xi_i = i\pi/(W_L - W_R) - \pi(W_L + W_R)/(W_L - W_R)/2$, where W_L and W_R are the left and right limits of the wall, respectively. W_L and W_R determine the width of the wall. In our case, the width is 200 sites. Figure 6.9(a) shows the initial stage of the Néel wall, where the spins align along the $+y$-axis and then gradually rotate toward the $-y$-axis. Due to the huge file size, we only show one out of ten spins along the x- and y-directions. We choose the pulse duration $T = 60$ fs, field amplitude $A_0 = 0.05$V/Å, $R = 50$ and photon energy $\hbar\omega = 1.6$ eV. Figures 6.9(b)-(d) show the instantaneous spin configurations for three helicities, at 123 fs after laser excitation with the laser pulse centered above the Néel wall. Figure 6.9(b) is the image of spins after linearly polarized laser pulse radiation. The laser spot size is 50 lattice sites. We see that the change mainly occurs in the vicinity of the laser spot, where spins slightly turn to the negative $-z$. The change is rather small with this field amplitude. The situation is quite different for right-circularly polarized light σ^+. Figure 6.9(c) shows that the spins around the perimeter of the pattern are switched down, similar to the experimental observation [Stanciu et al. (2007)]. Within the center of laser excitation, spins are rather chaotic and strongly misaligned, which is also similar to the demagnetization state observed in experiments. If we switch to σ^- light, the spins are switched up, with strong helicity dependence observed (Fig. 6.9(d)).

6.4.4 Importance of spin moments

From the above discussion, we see that there are varieties of laser and system parameters that affect whether and how laser pulses can switch spins. The benefit of a workable model is that we can test quantitatively whether there is an agreement between experiment and theory. Table 6.1 reveals something quite unusual. Most of the compounds have a strong spin moment. Gd and Tb are known to have a big spin moment (see Table 5.1). While the majority of experimental research focuses on other parameters, it is quite instructive if we examine how the spin angular momentum itself affects AOS.

Figure 6.10(a) plots the switchability η as a function of the initial spin angular momentum $S_z(0)$ for two different radii[3] of the laser beam. We start with the beam radius of 100. The laser pulse vertically impinges on the sample [Zhang et al. (2016a)], so the laser polarization is in the xy-plane (see Fig. 6.1). We notice that if $S_z(0)$ is very small, the switchability is small. For

[3]Since in the Heisenberg model we only have lattice sites and, rigorously speaking, there is no spatial information, we use the number of lattice sites as a measure of the size of the laser beam. The radius of the laser spot is in units of lattice size. One can of course convert those site sizes to an actual dimension if we know a particular material.

instance, for Ni, the spin angular momentum is about $0.3\hbar$, so we only have 30% chance to be switched. This explains why in all earlier studies Ni is not a favorite system. If we increase $S_z(0)$ up to $0.8\ \hbar$, or the threshold spin angular momentum S_z^c, which is about the same size as hcp Co (Table 5.1), η increases above 60%, and it gets better with a stronger $S_z(0)$. Since Fe, Gd and Tb have a much higher $S_z(0)$, one can now appreciate the fact that Table 6.1 is dominated by these three elements. The fundamental reason why spin angular momentum matters is because the spin-orbit torque $\lambda\mathbf{L}\times\mathbf{S}$ requires a larger spin.[p]

[p]The orbital angular momentum is mainly provided by the laser pulse.

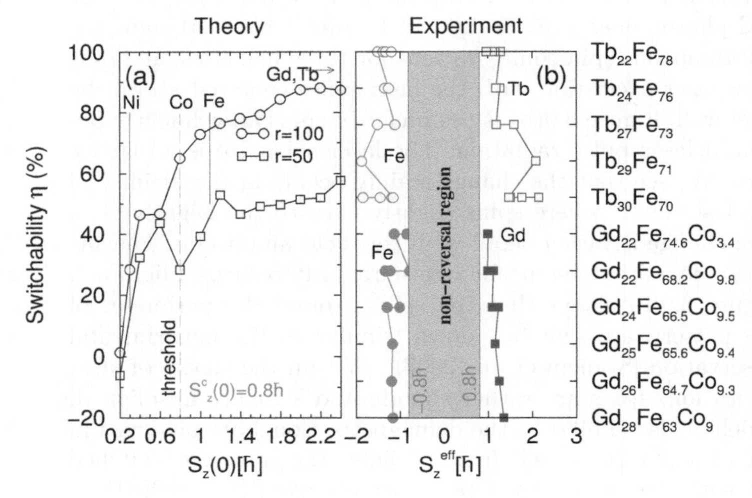

Figure 6.10
(a) Spin switchability versus the initial spin angular momentum $S_z(0)$ at the respective optimal laser field amplitudes.[Zhang et al. (2016a)] The empty circles and boxes refer to the results with the laser beam radii $r = 100$ and $r = 50$, respectively. The long-dashed line denotes the critical spin S_z^c. Two thin vertical lines represent the spins for Ni and Fe. Co is on the border line, while Gd and Tb are way above S_z^c. The arrow on the top right refers to the fact that Gd and Tb have a much higher spin angular momentum. (b) Computed experimental effective spin angular momentum for each element in 11 GdFeCo and TbFe alloys [Hassdenteufel et al. (2015)]. Without exception, all elements have spin larger than S_z^c. Used with permission from EPL [Zhang et al. (2016a)]

If we reduce the size of the laser beam to 50 so there are fewer lattice sites illuminated, η gets smaller as expected (see Fig. 6.10). However, the general trend is the same: The stronger the spin angular momentum is, the larger the switchability becomes. The next challenge is to compare with the experimental results, but no experiment gives the spin angular momentum in unit of $\hbar$. And what is even more problematic is that the experiments give a sum of magnetization, which requires the volume of the sample to convert to magnetic moment. We have figured out a way how to do this [Zhang et al. (2018b)], where we provide a pertinent Fortran code. The key is to use the unit cell of each element as an estimate. Then we use Eq. 6.8 to compute the spin angular momentum of each constituent. This works very well [Zhang et al. (2016a)]. Figure 6.10(b) compares the experimental data with our predicted threshold of S_z^c. We see all the materials meet this threshold.

6.5 Exercises

1. In Fig. 6.4(a), the magnetic image is mapped by the magneto-optical Faraday effect. Explain why the magneto-optical Kerr effect cannot image the magnetic domain in this case.
2. Under the continuous wave approximation, derive the diagonal and off-diagonal susceptibilities in Eqs. 6.13 and 6.14.
3. (a) Compute the number of photons for an area of 20 μm with the laser intensity of 2.9 mJ/cm^2. Suppose the laser wavelength is 800 nm. (b) Under the same area of bcc Fe, how many Fe atoms are there?
4. Derive the equation of motion in Eqs. 6.16-6.19.
5. Use the Matlab code to investigate the spin reversal for the single spin. (a) Use the default parameters that are given in the code. (b) Use different laser parameters. (c) Use different system parameters, in particular the spin-orbit coupling.
6. If we scan the laser beam 50 mm/s [Stanciu et al. (2007)] with repetition 1kHz, find the distance between two spots where two separate laser pulses land.
7. Show that the linear susceptibilities in Eqs. 6.13 and 6.14 are dimensionless.

7 Spin manipulations in magnetic nanostructures

7.1 Computer memory and magnetic storage
7.2 Experimental discovery
7.3 Spin precession
7.4 Rabi oscillation
7.5 Spin-orbit coupling in an atom
7.6 Magnetic resonance in NiO clusters
7.7 Exercises

The surge of technological advances in the late 20th century associated with personal computers and the World Wide Web are now recognized as the third industrial revolution.[1] Just like the first two industrial revolutions, the third revolution has made the world more productive by way of effective information processing. Computer systems store information in digital format called a binary system, which is in the form of zeros and ones. The potential impact of ultrafast technology on modern technology can be understood through some examples. For a comprehensive review (which includes both theoretical methods and experimental techniques), see Refs. [Zhang et al. (2002b); Kirilyuk et al. (2010)]. In this chapter, we introduce key concepts that are fundamental to the mechanisms of magnetization switching, namely, spin precession, Rabi oscillation, and spin-orbit coupling. Then, the magnetization reversal of an antiferromagnetic NiO cluster is discussed in detail, based on the numerical solution of the time-dependent Schrödinger equation.

7.1 Computer memory and magnetic storage

A *computer memory* is any physical device capable of storing the binary information. There are two types of computer memory: volatile and non-volatile. A volatile memory requires electric power to maintain stored information. For example, an operating system of your computer uses random-access memory (RAM) to process your commands. Most RAM available in the market uses semiconductor technology, i.e., it is electric-current dependent and therefore inevitably volatile. Each bit of information in semiconductor RAM is stored in a memory cell which consists of a tiny capacitor and a transistor called MOSFET (metal-oxide-semiconductor field-effect transistor). Billions of such memory cells are mounted on an integrated circuit (IC) or so-called microchip. A typical speed of RAM in a personal computer is a few GHz, which means that it can

[1] The first industrial revolution refers to the invention of the steam engine and the economic growth it brought to the mechanical industry in the late 18th century. The second one was driven by American engineers such as Thomas Edison and Nikola Tesla in the late 19th century, whose work led to the establishment of electric-power stations and telecommunication networks.

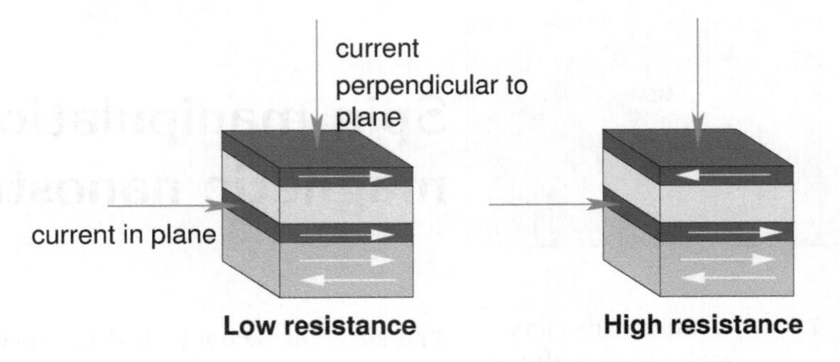

Figure 7.1
Giant magneto-resistance. A nonmagnetic layer is between two magnetic layers. If the magnetization in two ferromagnetic layers is in the same direction, the resistance is small. Otherwise, it is large. There are two different configurations associated with the relative orientation between the thin film plane and the current direction. If the current is perpendicular to the plane, it is CPP. If the current is in the plane of the film, it is called CIP.

read or write a few billion bytes per second. Roughly speaking, it is said that the number of transistors in an IC doubles about every two years, which is called Moore's Law.[a] (There is of course a limit as to what extent in time this law is valid.)

[a] Gordon Moore is the co-founder of Intel Corporation.

While volatile memories are lost upon power-down of a computer, non-volatile memories are semi-permanent and thus typically used for mass storage. A flash drive (or commonly called a USB drive, where USB stands for universal serial bus), for example, uses polarization of a dielectric material (called a floating gate) to store digital data semi-permanently without external power. A hard disk drive (HDD) in your computer, on the other hand, uses magnetic polarities to store data on a circular disk coated by a ferromagnetic thin film. The film is subdivided into small (< micrometer) magnetic regions, each of which has a specific magnetization direction (north or south) and can be used to represent a single binary unit of information (1 or 0). The magnetization direction can be reversed by using an electromagnet placed at the tip of an actuator arm (read-and-write head). Once the data are recorded on a HDD, they will be retained unless an external magnetic field affects the HDD or excessive heat demagnetizes it. Moreover, magnetic storage is read-and-write, meaning that the storage can be reused over and over again by overwriting older data.

7.1.1 Giant magneto-resistance

Another revolution is the discovery of giant magneto-resistance. A current passing through the ferromagnetic layers experiences high and low resistance, depending on the relative orientation of magnetizations between two layers. Figure

Table 7.1
Various computer memories. [i]FeRAM (ferroelectric RAM) or Fe-FET (ferroelectric field-effect transistor) uses ferroelectric materials (which have permanent electric polarization) for a memory cell to achieve non-volatility. [ii]EPROM (erasable programmable read-only memory) or EEPROM (electrical EPROM) is organized as arrays of floating-gate transistors on a microchip. A flash drive is a type of EPROM designed for higher speed and storage capacity.

Types	Volatile	Examples
Semiconductor RAM	Yes	Dynamic RAM, Static RAM
Ferroelectric RAM	No	FeRAM,[i] FeFET
Magnetoresistive RAM	No	Toggle MRAM, STT-MRAM SOT-MRAM
Electrical	No	Flash drive, EPROM[ii] EEPROM
Magnetic	No	VHS tape, HDD, Floppy disk Zip drive
Optical	No	CD, DVD, Blu-ray
Magneto-Optical	No	MiniDisk (MD)
Mechanical	No	punched tape, punched card

7.1 schematically shows a commonly employed structure, with one nonmagnetic layer sandwiched between two ferromagnetic layers. If the current goes through the structure perpendicular to the film plane, this configuration is called CPP (current perpendicular to plane). If the current flows within the plane of the film, the configuration is called CIP (current in plane). In the late 1980s, Albert Fert and Peter Grünberg discovered independently that the magnetic resistance is enhanced when non-magnetic film is inserted between two ferromagnetic layers (giant magneto resistance, or GMR). The application of GMR has led to a dramatic improvement of HDD, with Fert and Grünberg receiving the Nobel Prize in Physics in 2007.

The micrometer scale of magnetic domains allows one to store large amounts of data into a very small area. In the late 1980s, less than 1% of the world's technologically stored information was in digital format, while it was 94% in 2007 and more than 99% by 2014 [Hilbert and López (2011)]. This means that magnetic storages had increased their capacity 1,000-fold within a decade and a half; the pace was in fact much faster than the two-year doubling time of semiconductor chip density posited by Moore's law [Kryder and Kim (2009)]. The major limitation of magnetic storage, however, is that accessing the data can be quite slow. The typical read-and-write speed for HDD is about 100 megabytes per second, more than 10 times slower than the semiconductor RAM. This is the reason why most computer systems use semiconductor RAM as a main memory for their operation system.

Table 7.1 summarizes various types of computer memory. Of particular interest is a magneto-resistive RAM (MRAM), which is a kind of non-volatile RAM. Commercially available MRAM uses magnetic tunnel junctions (MTJ), which consist of two ferromagnets separated by a thin insulator. The first-generation MRAM used a magnetic field to control the direction of spins in the MTJ (toggle MRAM), whereas newer generations of MRAM use spin-transfer torque (STT)[2] [Razdolski et al. (2017)] or spin-orbit torque (SOT) [Garello et al. (2013, 2014); Wadley et al. (2016)]. In some magnetic systems lacking inversion symmetry, an in-plane charge current exerts SOT on spins and subsequently causes oscillations of the spin magnetic moment, similarly to STT. In SOT-based MRAM, the read and the write heads run on separate layers, which improves the read-and-write reliability and saves power consumption.

7.1.2 Magneto-optical recording technology

Although it was not successful commercially, MiniDisk[3] deserves special mention because of its innovative usage of the laser as an actuator for magnetization control as follows: (i) First, a laser heats one side of a ferromagnetic disk above its Curie point, demagnetizing the disk. (ii) As it cools down to the Curie point, an electromagnetic head on the other side of the disk alters the polarity of the disk into desired directions. (iii) Playback is accomplished with the laser alone, taking advantage of the Faraday effect; i.e., as a laser beam transmits through the disk, its polarization rotates depending on the magnetization of the disk. The resulting rotation angle can be interpreted as 1 or 0.

The magneto-optical recording technique used in MiniDisk was heat-assisted, whose speed is limited by the thermal relaxation time [Hübner and Bennemann (1996)]. If a femtosecond laser pulse is used to actuate the demagnetization, magnetic recording can be achieved more quickly by utilizing the precession of electron spins. We will discuss its mechanism in Section 7.3, but in short it works as follows: A laser pulse induces an alternating electric current along its polarization direction. The electric current induces an accompanying magnetic field $\mathbf{B}$ (Ampere's Law). If the initial direction of electron spin $\mathbf{S}$ is not in parallel with $\mathbf{B}$, then the spin magnetic moment $\boldsymbol{\mu}_B$ ($\propto \mathbf{S}$) precesses around $\mathbf{B}$ at a frequency proportional to $|\mathbf{B}|$

[2]When an electron with misaligned spin passes into a ferromagnetic layer, the mismatch gives rise to a small twisting force, i.e., a torque, between the electron and the magnet. In certain situations, this spin-transfer torque (STT) can cause the magnetization of a ferromagnetic layer to precess at a frequency controlled by the current.

[3]The MiniDisk is a magneto-optical drive developed in 1990's by Sony, primarily for audio storage. Its production was ceased in 2013, as flash-memory based audio players, such as iPod Shuffle, dominated the market.

(Larmor precession). After the laser pulse turns off (and so does **B**), the precession relaxes to the equilibrium where $\boldsymbol{\mu}_B$ aligns with **B** (magnetization damping).[4] In particular, when $\mathbf{S} \perp \mathbf{B}$, then $\boldsymbol{\mu}_B$ rotates on a plane perpendicular to **B**, so that the cycle-averaged magnetization becomes zero.

7.1.3 Emergence of ultrafast demagnetization

The ultrafast demagnetization using femtosecond (fs) laser pulses was first demonstrated experimentally by Beaurepaire et al. in 1996 [Beaurepaire et al. (1996)]. It was found that an intense, short (60 fs), 620-nm laser pulse could demagnetize a ferromagnetic nickel (Ni) film in less than one picosecond (ps), much shorter than the heat-assisted process that is on the order of a few hundred ps [Scholl et al. (1997)]. In the following year, Hohlfeld et al. reported the demagnetization of Ni in 280 ($\pm$30) fs [Hohlfeld et al. (1997)], induced by a 150-fs, 800-nm laser pulse.

The next milestone experiment came in 2007 by Stanciu et al. [Stanciu et al. (2007)]. They discovered that a single circularly-polarized laser pulse can reverse the magnetization of ferrimagnetic,[b] rare-earth transition metal alloy films (GdFeCo). One possible mechanism of their helicity-dependent magnetization switching is the inverse Faraday effect (IFE) [Pitaevskii (1961); Pershan et al. (1966)], which is a process where a circularly-polarized laser field **E** induces an effective magnetic field $\mathbf{H} \propto \mathbf{E} \times \mathbf{E}^*$ in a material via spin-orbit coupling (Section 7.5). The right (σ^+) and left (σ^-) circularly-polarized driving laser pulses induce magnetic fields of opposite signs and selectively reverse only those spins that are parallel to them (and not the antiparallel ones) [Kimel et al. (2005)].

[b] In a ferrimagnetic material, neighboring magnetic moments are antiparallel, but their magnitudes are unequal. In contrast to antiferromagnetic materials, in ferrimagnets the opposite moments do not cancel out, so there is a net magnetic moment.

7.2 Experimental discovery

Research in ultrafast magnetization switching using femtosecond laser pulses continues to this day. Spin manipulations in magnetic nanostructures are an active research field. This section discusses some of the experiments that are representative of the current research.

7.2.1 Experiments in permalloy

In 2002, Gerrits et al. reported the precessional magnetization reversal of a ferromagnetic permalloy (NiFe) [Gerrits et al. (2002)]. They found that the precession induced by a laser pulse could be terminated by sending another laser pulse with

[4] The magnetization damping occurs because of the energy transfer from spin to orbital motion, lattice vibrations (phonons), spin waves (magnons), etc.

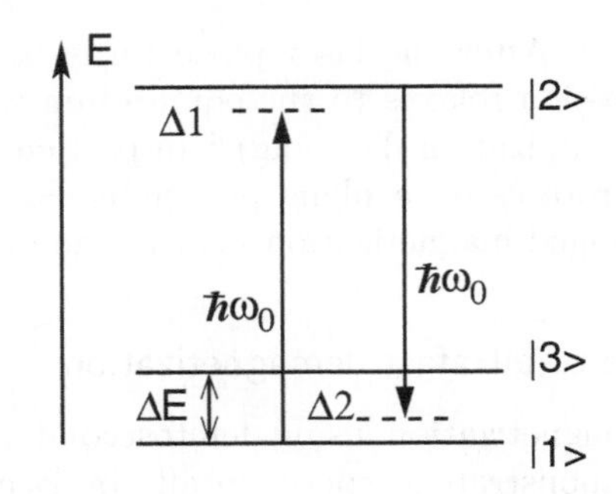

Figure 7.2
Energy levels of the spin states coupled by a laser field of frequency ω_o, with an intermediate state in between. The twofold degeneracy of the stationary spin states is removed by the spin-orbit interaction inside an atom, so they have an energy difference of ΔE. The difference between the laser-field energy $\hbar\omega_o$ and the exact transition energy is called detuning (Δ_1 or Δ_2 in the figure). For dipole transitions to take place, the intermediate state $|2\rangle$ needs to have different parity from the spin-up $|1\rangle$ and the spin-down $|3\rangle$ states.

a sufficient time delay in between (> 400 ps in their experiment). The magnetization reversal time they measured (200 ps), however, turned out to be on the same order as the thermal relaxation time. In principle, faster magnetization reversals are possible based on spin magnetic resonance [Gómez-Abal et al. (2004)].

The fundamental limit for this type of magnetization reversal is half the Rabi oscillation period. As we will discuss in Section 7.4, the Rabi oscillation period of a two-level system is inversely proportional to the energy difference between the two quasistationary states that are coupled by a driving laser field. The magnetization reversal on the sub-picosecond time scale therefore requires a quasistationary energy difference of a few meV and greater. This corresponds to a THz laser frequency. In the limit of the electric dipole approximation, a direct transition between the spin-up and spin-down states by emission or absorption of a single photon is forbidden. To alleviate this problem, we could include an additional transition to an excited state of opposite parity in between the spin-up and the spin-down states, as shown in Figure 7.2. The spin reversal using such a 3-level scheme is discussed in Section 7.6.

7.2.2 Coherent spin manipulation in NiO

NiO is a prototype antiferromagnet insulator, with a fcc NaCl structure (see Fig. 7.3). There is an energy gap of above 4 eV. NiO has the Neel temperature of 523 K. Its lattice constant is 4.17-4.18 Å. Ni spins in the (111) planes (see side note *c* below) are ferromagnetically coupled, and antiferromagnetically stacked along the [111] direction [Hutchings and Samuelsen (1972)].

The first demonstration of the 3-level magnetization switching in NiO was reported in 2006 [Satoh et al. (2006)]. It was controlled by a pair of 100-fs, 1.55-eV laser pulses. These time-resolved measurements revealed oscillations of second-harmonic intensity with a period of about 20 ps.[5] The corresponding frequency of this oscillation, however, does not correspond to the Rabi frequency associated with the 3-level magnetic resonance of NiO, but to the magnetic anisotropy energy of 0.11 meV, i.e., the energy required to collectively reorient spins from the easy axis $[11\bar{2}]$ to the hard axis [111] direction of NiO [Duong et al. (2004)]. When a time lag between the two laser pulses is set to Δt=9.5 ps, an oscillation in the second-harmonic intensity induced by the first pulse is terminated by the second pulse, which signifies the change in magnetic anisotropy by 90°. The magnetization reversal based on the magnetic resonance of NiO can, in principle, be as fast as a few hundred femtoseconds (Section 7.6). The 9.5-ps time lapse needed for the magnetization switching of NiO [Satoh et al. (2006)] is nevertheless an order of magnitude shorter than in Gerrits's experiments on ferromagnetic permalloy (NiFe) [Gerrits et al. (2002)].

A quite different scheme was demonstrated in 2011. As discussed above, a laser pulse has both electric and magnetic fields. The electric field is much stronger than the magnetic counterpart, so the latter can be safely ignored. However, the situation changes if we have an insulator such as NiO with a 4-eV gap, where the contribution of the electric field is significantly attenuated. This is particularly true if the laser photon energy is off resonance, with little electronic and phononic excitation. Consider a THz pulse with frequency from 0.1 to 3 THz and electric field strength of $E = 0.4$ MV/cm [Kampfrath et al. (2011)]. Its magnetic field (0.13 T) can be computed from $B = E/c$, where c is the speed of light in vacuum. The g-factor for NiO is 2.19, so we have the magnetic energy of 1.65×10^{-5} eV, which is insignificant if we compare it with the electronic energy scale. However, the spin wave excitation is efficient with such a low energy. Figure 7.4 shows the Faraday rotation angle as a function of the time delay between the strong THz pump pulse and a weak near-infrared probe pulse with duration of 8 fs, photon energy of 1 eV, and repetition rate of 1 kHz. The polarization change of the probe pulse is measured. One sees that a highly coherent spin wave excitation lasts up to 20 ps.

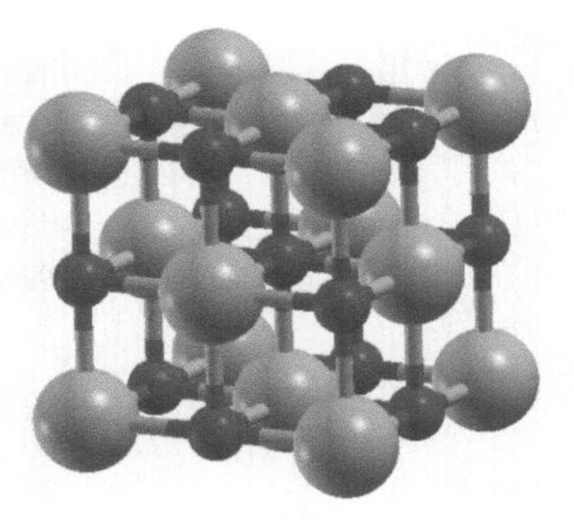

Figure 7.3
Crystal structure of NiO. The large balls refer to Ni atoms, while the smaller ones are O atoms. The lattice constant is 4.17-4.18 Å.

[5]Second-harmonic (SH) radiation is a coherent light of frequency $2\omega_o$, following absorption of photons of frequency ω_o. NiO is a centrosymmetric crystal, and therefore a SH radiation is forbidden by an electric dipole transition but allowed by a magnetic dipole transition between the spin-up and spin-down states. The SH from NiO thus signals a change in magnetization.

From 0 to 10 ps, there are ten oscillations, 1 THz, overlapping the magnon resonance [Hutchings and Samuelsen (1972)]. Because the optical phonon frequency is above 12 THz, the signal is not strongly affected by the phonons.

These oscillations can be reproduced by the following Hamiltonian [Hutchings and Samuelsen (1972)] with an additional time-dependent magnetic field $\mathbf{B}(t)$,[c]

[c] This Hamiltonian is slightly modified for clarity. We adopt the plane and direction notations from the solid state physics [Kittel (1996)]. (i, j, k), with three integers i, j and k, denotes planes, and $[i, j, k]$ denotes directions. In the original reference, they use both () and {} to denote planes and both ⟨⟩ and [] to denote directions.

$$H = H_{[111]} + H_{(111)} + D_1 \sum_i (\mathbf{S}_i^x)^2 + D_2 \sum_i (\mathbf{S}_i^y)^2 + g\mu_B \mathbf{B}(t) \cdot \sum \mathbf{S}_i, \tag{7.1}$$

where all the spins $\mathbf{S}$ are made dimensionless, the spin direction is defined as the z-axis (the [111] direction), and the x-axis is within the (111) plane. Here $H_{[111]}$ is the Heisenberg exchange term between spins along the [111] direction, i.e. between spins in the adjacent (111) planes,

$$H_{[111]} = -J_1 \sum_i \sum_{\delta_i}^{[111]} \mathbf{S}_i \cdot \mathbf{S}_{i+\delta_i}, \tag{7.2}$$

where the first summation is over all the lattice sites, and the second summation is over distinctive pairs of neighboring spins on the adjacent (111) planes. We do not use terms like nearest neighbors because there is an oxygen atom between two Ni atoms. The spin ordering is antiferromagnetic, so J_1 is negative. Note in the original paper by Hutchings and Samuelsen they used $+J_1$, so our convention differs from theirs by a negative sign. Our convention is consistent with the modern day practice.

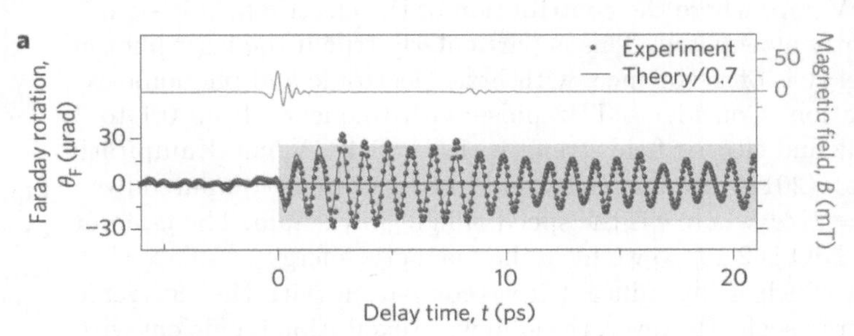

Figure 7.4
Faraday rotation of the probe polarization as a function of time delay between the strong THz pump pulse and the weak near-infrared probe pulse. The probe polarization changes at 0 ps, and the change lasts up to 20 ps. On the top is shown the laser pulse. The first one is the pump pulse and the second is the probe pulse. The time zero is defined as the delay time zero. Used with permission from [Kampfrath et al. (2011)].

$H_{(111)}$ is the Heisenberg exchange term between spins in

Table 7.2
Parameters used in the model Hamiltonian (Eq. 7.1) [Hutchings and Samuelsen (1972); Kampfrath et al. (2011)]. Hutchings and Samuelsen used the unit of Kelvin as the unit of exchange and anisotropy constants, which are all converted to meV or μeV for the present book. The spin S is dimensionless in this table because its unit is absorbed into the J or D.

J_1	-19.01 meV
J_2	1.37 meV
D_1	92.22 μeV
D_2	48.97 μeV
g	2.19
S_z	0.9242±0.0003 (Theory), 0.82±0.04[a](Exp.)

[a] The experimental value includes covalency reduction of 11% [Hutchings and Samuelsen (1972)].

the (111) planes,

$$H_{(111)} = -J_2 \sum_i \sum_{\delta_i}^{(111)} \mathbf{S}_i \cdot \mathbf{S}_{i+\delta_i}, \tag{7.3}$$

where the first summation is over all the lattice sites and the second summation is over distinctive pairs of neighboring spins within the same (111) planes. The coupling between these spins is ferromagnetic, so J_2 is positive. The third and fourth terms in Eq. 7.1 describe the out-of-plane and in-plane ansotropies, respectively.

The last term is the laser's effective magnetic field interacting with the spins. However, experimentally $\mathbf{B}(t)$ is measured [Kampfrath et al. (2011)], so it is not easy to reproduce it theoretically here. All the other parameters are given in Table 7.2. With these parameters, the theory can reproduce the experimental results very well (see the solid line in Fig. 7.4).

7.3 Spin precession

Each electron in an atom has two kinds of angular momenta: orbital angular momentum $\mathbf{L}$ and spin $\mathbf{S}$. They contribute to the orbital magnetic dipole moment $\boldsymbol{\mu}_L$ ($\propto \mathbf{L}$) and spin magnetic dipole moment $\boldsymbol{\mu}_S$ ($\propto \mathbf{S}$), respectively. The magnetic polarity of a material depends on the directions of $\boldsymbol{\mu}_L$ and $\boldsymbol{\mu}_S$ of individual electrons (spin-orbit coupling), and how neighboring magnetic moments are oriented with respect to one another (exchange coupling). In ferromagnetic materials, neighboring magnetic moments align in the same direction ($\uparrow\uparrow\uparrow\uparrow\cdots$) to have a large magnetic polarization, whereas in antiferromagnetic materials they align in the opposite direction ($\uparrow\downarrow\uparrow\downarrow\cdots$) so that their net magnetization cancels out.

Suppose that an electron is subject to a static, uniform magnetic field in the z-direction: $\mathbf{B} = B_z\hat{z}$. The magnetic field interacts with the spin magnetic dipole moment of the electron, given by

$$\boldsymbol{\mu}_s = \frac{e}{m}\,\mathbf{S}, \tag{7.4}$$

where e and m are the charge and mass of an electron, and $\mathbf{S}$ is the spin angular momentum. Their interaction Hamiltonian is

$$\mathcal{H}_m^{(\mathrm{S})} = -\boldsymbol{\mu}_s \cdot \mathbf{B}. \tag{7.5}$$

In terms of Pauli matrices, the spin angular momentum can be defined as

$$\mathbf{S} = \frac{\hbar}{2}\left[\begin{pmatrix}0 & 1\\ 1 & 0\end{pmatrix}\hat{x} + \begin{pmatrix}0 & -i\\ i & 0\end{pmatrix}\hat{y} + \begin{pmatrix}1 & 0\\ 0 & -1\end{pmatrix}\hat{z}\right], \tag{7.6}$$

where $\{\hat{x}, \hat{y}, \hat{z}\}$ are the unit vectors in Cartesian coordinates. Therefore, the magnetic interaction Hamiltonian (Eq. 7.5) for a static, uniform magnetic field in the z-direction is given by

$$\mathcal{H}_m^{(\mathrm{S})} = -\boldsymbol{\mu}_s \cdot \mathbf{B} = -\frac{e}{m}\,\mathbf{S}\cdot(B_z\hat{z}) = -\frac{e\hbar B_z}{2m}\begin{pmatrix}1 & 0\\ 0 & -1\end{pmatrix}. \tag{7.7}$$

Since this Hamiltonian is diagonal, its eigenfunctions are simply

$$|+\rangle = \begin{pmatrix}1\\ 0\end{pmatrix}, \quad |-\rangle = \begin{pmatrix}0\\ 1\end{pmatrix} \tag{7.8}$$

with the corresponding eigenvalues

$$E_{\pm} = \mp\frac{e\hbar B_z}{2m} = \mp\mu_B B_z, \tag{7.9}$$

where $\mu_B \equiv e\hbar/(2m)$ is called the Bohr magneton. This means that the energy of spin-up and spin-down electrons is split by an amount $\Delta E = 2\mu_B B_z$ in the presence of a static, uniform magnetic field $\mathbf{B} = B_z\hat{z}$.[6]

Now, suppose further that the spin state of an electron at $t = 0$ is neither of the two eigenstates (Eq. 7.8) but their superposition, given by

$$|\chi(0)\rangle = a_+|+\rangle + a_-|-\rangle = \begin{pmatrix}a_+\\ a_-\end{pmatrix}, \tag{7.11}$$

with the normalization condition

$$|a_+|^2 + |a_-|^2 = 1. \tag{7.12}$$

[6]Note that the angular momentum $\mathbf{L}$ of an electron also interacts with an external magnetic field $\mathbf{B}$, such that

$$\mathcal{H}_m^{(\mathrm{L})} = -\boldsymbol{\mu}_{\mathrm{L}} \cdot \mathbf{B}, \tag{7.10}$$

where $\boldsymbol{\mu}_{\mathrm{L}} = (e/m)\mathbf{L}$ is the angular magnetic dipole moment. The combined effect of $\mathcal{H}_m^{(\mathrm{S})}$ and $\mathcal{H}_m^{(\mathrm{L})}$ causes the energy of $(2\ell + 1)$-degenerate states of an atom to shift in proportion to their magnetic quantum number (Zeeman effect). For simplicity, we will discuss only the interaction associated with spin degrees of freedom in this section.

Because $\{|+\rangle, |-\rangle\}$ are the eigenstates of the Hamiltonian (Eq. 7.7) with eigenvalues $E_\pm$, they evolve in time as

$$\begin{aligned} e^{-i\mathcal{H}_m^{(\mathrm{S})}t/\hbar}|+\rangle &= e^{-iE_+t/\hbar}|+\rangle = e^{i\Delta\omega t/2}|+\rangle, \\ e^{-i\mathcal{H}_m^{(\mathrm{S})}t/\hbar}|-\rangle &= e^{-iE_-t/\hbar}|-\rangle = e^{-i\Delta\omega t/2}|-\rangle, \end{aligned} \tag{7.13}$$

where $\Delta\omega \equiv (E_- - E_+)/\hbar = 2\mu_B B/\hbar$. Therefore, the spin state at some later time $t > 0$ is

$$|\chi(t)\rangle = \underbrace{a_+ e^{i\Delta\omega t/2}}_{\equiv a_+(t)}|+\rangle + \underbrace{a_- e^{-i\Delta\omega t/2}}_{\equiv a_-(t)}|-\rangle. \tag{7.14}$$

So, we can evaluate the expectation value of each spin component as a function of time, such that

$$\langle S_x(t)\rangle = \langle\chi(t)|S_x|\chi(t)\rangle = \frac{\hbar}{2}(a_+{}^*(t), a_-{}^*(t))\begin{pmatrix} 0 & 1 \\ 1 & 0 \end{pmatrix}\begin{pmatrix} a_+(t) \\ a_-(t) \end{pmatrix} \tag{7.15}$$

$$\langle S_y(t)\rangle = \langle\chi(t)|S_y|\chi(t)\rangle = \frac{\hbar}{2}(a_+{}^*(t), a_-{}^*(t))\begin{pmatrix} 0 & i \\ -i & 0 \end{pmatrix}\begin{pmatrix} a_+(t) \\ a_-(t) \end{pmatrix} \tag{7.16}$$

$$\langle S_z(t)\rangle = \langle\chi(t)|S_z|\chi(t)\rangle = \frac{\hbar}{2}(a_+{}^*(t), a_-{}^*(t))\begin{pmatrix} 1 & 0 \\ 0 & -1 \end{pmatrix}\begin{pmatrix} a_+(t) \\ a_-(t) \end{pmatrix}. \tag{7.17}$$

For example, let

$$a_+ = \cos\left(\frac{\alpha}{2}\right), \quad a_- = \sin\left(\frac{\alpha}{2}\right), \tag{7.18}$$

which satisfies the normalization condition (Eq. 7.12). Then, the expectation values of S_z at later time $t > 0$ are

$$\langle S_z(t)\rangle = \langle S_z(0)\rangle = \frac{\hbar}{2}\left(|a_+|^2 - |a_-|^2\right) = \frac{\hbar}{2}\cos\alpha, \tag{7.19}$$

which means that the spin is initially at an angle α from the z-axis (and remains at this angle for $t > 0$). The expectation values of the other two spin-components are

$$\langle S_x(t)\rangle = \frac{\hbar}{2}\sin\alpha\cos\left(\frac{\Delta\omega}{2}t\right), \tag{7.20}$$

$$\langle S_y(t)\rangle = \frac{\hbar}{2}\sin\alpha\sin\left(\frac{\Delta\omega}{2}t\right). \tag{7.21}$$

That is, in the presence of a static, uniform field $\mathbf{B}$, the spin angular momentum $\mathbf{S}$ precesses around the direction of $\mathbf{B}$ with a period of $T = 4\pi/\Delta\omega$, provided that they are not in parallel with one another initially ($\alpha \neq 0$). In particular, if $\mathbf{S} \perp \mathbf{B}$ initially ($\alpha = 90°$), then $\mathbf{S}$ rotates on a plane perpendicular to $\mathbf{B}$ and reverses its direction in every $T/2$. After the magnetic field turns off, the precession relaxes to the equilibrium where $\mathbf{S}$ aligns with $\mathbf{B}$ through magnetization damping.

In antiferromagnetic materials such as NiO, it is not necessary to change the magnetization direction by 180° but only by 90° for magnetic recording [Fiebig et al. (2008); Higuchi et al. (2011)]. Such change in magnetic anisotropy is achieved by a relatively weak magnetic field [Saito et al. (1980)]. The magnetization switching using the spin precession is therefore a very plausible approach for antiferromagnetic materials [Satoh et al. (2006); Duong et al. (2004)]. It is known that a circularly-polarized laser field **E** induces an effective magnetic field $\mathbf{H} \propto \mathbf{E} \times \mathbf{E}^*$ in a material, via the inverse Faraday effect (IFE) [Pershan et al. (1966)]. A linearly polarized laser field also induces a magnetic field, particularly if the laser field is not polarized in parallel to the crystal axes; this phenomenon is called the inverse Cotton-Mouton effect (ICME) [Pershan et al. (1966)]. Both IFE and ICME cause the spin precession in NiO but around the different axis and with a 90-degree phase difference [Satoh et al. (2010); Tzschaschel et al. (2017)].

7.4 Rabi oscillation

In the previous section, the external magnetic field was assumed to be time independent. In this section, we would like to address the effect of time-dependent magnetic field. Consider an electron subject to the following magnetic field:

$$\mathbf{B}(t) = B_z\hat{z} + B_0\left[\cos(\omega_o t)\hat{x} + \sin(\omega_o t)\hat{y}\right], \tag{7.22}$$

where B_z and B_0 are constant. That is, we have a time-dependent, circularly-polarized magnetic field on the xy-plane in addition to a static, uniform magnetic field along the z-axis.[d] From Eqs. 7.4, 7.5 and 7.6, the magnetic interaction Hamiltonian then becomes

$$\mathcal{H}_m^{(\mathrm{S})}(t) = -\boldsymbol{\mu}_s \cdot \mathbf{B}(t) = \begin{pmatrix} E_+ & W(t) \\ W^*(t) & E_- \end{pmatrix}, \tag{7.23}$$

where $E_\pm = \mp\mu_B B_z$ and

$$W(t) = \frac{e\hbar B_0}{2m}e^{i\omega_0 t} = W_0 e^{i\omega_0 t}, \tag{7.24}$$

with $W_0 \equiv \mu_B B_0$. The matrix (Eq. 7.23) can be diagonalized by the following unitary transformation:[e]

$$\tilde{\mathcal{H}}_m^{(\mathrm{S})} = S^{-1}\mathcal{H}_m^{(\mathrm{S})}S = \begin{pmatrix} \tilde{E}_+ & 0 \\ 0 & \tilde{E}_- \end{pmatrix}, \tag{7.25}$$

where

$$\tilde{E}_\pm = \frac{1}{2}\left[E_+ + E_- \pm \sqrt{(E_- - E_+)^2 + 4|W_0|^2}\right], \tag{7.26}$$

and

$$S = \begin{pmatrix} \cos\frac{\theta}{2} & \sin\frac{\theta}{2}e^{i\phi} \\ \sin\frac{\theta}{2}e^{-i\phi} & -\cos\frac{\theta}{2} \end{pmatrix}, \tag{7.27}$$

[d] A linearly polarized laser field along the z-axis, for example, would make an electron oscillate back and forth in the z-direction, whose motion would then induce a circularly-polarized magnetic field on the xy-plane (Ampere's Law).

[e] The matrix S is called unitary if its Hermitian conjugate is its inverse, i.e., $S^\dagger S = SS^\dagger = I$.

in which the mixing angles θ and ϕ are defined by

$$\tan\theta = \frac{2|W_0|}{E_- - E_+}, \text{ and } e^{i\phi} = \frac{W^*}{|W_0|}. \tag{7.28}$$

Accordingly, we can express the time-dependent spin state of a system with Hamiltonian (Eq. 7.23) as

$$|\chi(t)\rangle = a_1 e^{-iE_1 t/\hbar}|\phi_1\rangle + a_2 e^{-iE_2 t/\hbar}|\phi_2\rangle, \tag{7.29}$$

where $\{a_1, a_2\}$ are time-independent coefficients, and $|\phi_1\rangle$ and $|\phi_2\rangle$ are the eigenstates of the Hamiltonian (Eq. 7.23) with eigenvalues $E_1 = \tilde{E_+}$ and $E_2 = \tilde{E_-}$, respectively, given by

$$|\phi_1\rangle = \cos\frac{\theta}{2}|+\rangle + \sin\frac{\theta}{2}e^{i\phi}|-\rangle, \tag{7.30}$$

$$|\phi_2\rangle = \sin\frac{\theta}{2}e^{-i\phi}|+\rangle - \cos\frac{\theta}{2}|-\rangle. \tag{7.31}$$

The time-independent coefficients $\{a_1, a_2\}$ in Eq. 7.29 are determined from the initial condition. For example, suppose that the initial spin state is $|+\rangle$. Inverting Eqs. 7.30 and 7.31, we have

$$|+\rangle = \cos\frac{\theta}{2}|\phi_1\rangle + \sin\frac{\theta}{2}e^{i\phi}|\phi_2\rangle, \tag{7.32}$$

$$|-\rangle = \sin\frac{\theta}{2}e^{-i\phi}|\phi_1\rangle - \cos\frac{\theta}{2}|\phi_2\rangle. \tag{7.33}$$

Therefore, it follows from the initial condition $|\chi(0)\rangle = |+\rangle$ that

$$a_1 = \cos\frac{\theta}{2}, \quad a_2 = \sin\frac{\theta}{2}e^{i\phi}. \tag{7.34}$$

The transition amplitude to the $|-\rangle$ state is therefore

$$\begin{aligned}\langle -|\chi(t)\rangle &= \cos\frac{\theta}{2}e^{-iE_1 t/\hbar}\langle -|\phi_1\rangle + \sin\frac{\theta}{2}e^{i\phi}e^{-iE_2 t/\hbar}\langle -|\phi_2\rangle \\ &= \sin\frac{\theta}{2}\cos\frac{\theta}{2}\left(e^{-iE_1 t/\hbar} - e^{-iE_2 t/\hbar}\right)e^{i\phi} \\ &= i\sin\theta\sin\left(\frac{\Delta\omega}{2}t\right)e^{i\phi}, \end{aligned} \tag{7.35}$$

where $\Delta\omega \equiv (E_1 - E_2)/\hbar$ is the transition frequency between the quasistatic eigenstates $|\phi_1\rangle$ and $|\phi_2\rangle$. Thus, the probability for the spin to remain in the $|-\rangle$ state can be found as

$$P_-(t) = |\langle -|\chi(t)\rangle|^2 = \sin^2\theta\sin^2\left(\frac{\Delta\omega}{2}t\right). \tag{7.36}$$

Using Eqs. 7.26 and 7.28, it becomes

$$P_-(t) = \frac{\Omega^2}{\Omega^2 + \left(\frac{\Delta}{2}\right)^2}\sin^2\left(\sqrt{\Omega^2 + \left(\frac{\Delta}{2}\right)^2}t\right), \tag{7.37}$$

where $\Omega \equiv W_0/\hbar$ is called the Rabi frequency, and $\Delta \equiv \omega_0 - \Delta\omega$ (detuning).[f] In particular, when $\Delta = 0$ (i.e., the driving field frequency and the separation of spin-up and spin-down energy are in resonance), then $P_-(t) = \sin^2(\Omega t)$; i.e., the spin state oscillates between spin-up and spin-down states with a period of $T = 2\pi/\Omega$. Physically, this means that the spin direction reverses in every $T/2$ if a time-dependent, circularly-polarized magnetic field is applied on the plane perpendicular to the initial direction of the spin. This is called Rabi oscillation, after I. I. Rabi who is the father of magnetic resonance imaging (MRI).

[f] In Eq. 7.37 the frequency $\sqrt{\Omega + (\Delta/2)^2}$ is called the generalized Rabi frequency.

7.5 Spin-orbit coupling in an atom

Spin-orbit coupling refers to the interaction of an electron's spin with its orbiting motion inside a central potential $U(r)$. For example, when driven by a circularly-polarized laser field $\mathbf{E}_{\rm ext}(t)$, an electron in an atom would move in an orbit centered around the nucleus. This motion of the electron yields a circular current, which in turn generates an orbital magnetic moment $\boldsymbol{\mu}_L$ in the direction perpendicular to the polarization plane of the laser field. The average of these magnetic moments per unit volume would amount to an effective magnetic field $\mathbf{H} \propto (\mathbf{E}_{\rm ext} \times \mathbf{E}^*_{\rm ext})$[g] in a crystal, i.e., the inverse Faraday effect. The electron spin $\mathbf{S}$ would then interact with this orbit-induced magnetic field $\mathbf{H}$.

[g] This analogy ends if we have linearly polarized light.

Even in the absence of an external laser field, an electron in an atom is subject to a static electric field due to a radially symmetric nuclear potential $U(r)$, given by

$$\mathbf{E}(r) = -\frac{1}{e}\frac{dU}{dr}\frac{\mathbf{r}}{|\mathbf{r}|}. \tag{7.38}$$

If it has a momentum $\mathbf{p} = m\mathbf{v}$, then it also experiences a magnetic field, given by

$$\mathbf{B}(r) = -\frac{\mathbf{p}}{m} \times \mathbf{E}(r). \tag{7.39}$$

This field interacts with the spin magnetic dipole moment $\boldsymbol{\mu}_s = (e/m)\mathbf{S}$ of the electron. The interaction Hamiltonian is therefore [Sakurai (1994)]

$$\mathcal{H}_{SO} = -\boldsymbol{\mu}_s \cdot \mathbf{B}(r) = \frac{1}{m^2c^2}\frac{1}{r}\frac{dU}{dr}\mathbf{L}\cdot\mathbf{S}, \tag{7.40}$$

where where $\mathbf{L} = \mathbf{r}\times\mathbf{p}$ is the angular momentum of an electron. In reality, the amount of perturbation is only 1/2 of Eq. 7.40, due to a relativistic correction (Thomas precession) [Thomas (1926)].

If we assume separable eigenfunctions of the form

$$\psi_\pm(\mathbf{r}) = \frac{R_{j\pm}(r)}{r}\mathcal{Y}^{m_j}_{j\pm}(\theta,\phi), \tag{7.41}$$

where $j_\pm = \ell \pm 1/2$ and $|m_j| \leq j_\pm$, then the angular solutions are the *spin spherical harmonics*, given by

$$\mathcal{Y}_{j\pm}^{m_j}(\theta,\phi) = \pm\sqrt{\frac{\ell \pm m_j + \frac{1}{2}}{2\ell+1}}\, Y_\ell^{m_j-\frac{1}{2}}(\theta,\phi)\begin{pmatrix}1\\0\end{pmatrix} + \sqrt{\frac{\ell \mp m_j + \frac{1}{2}}{2\ell+1}}\, Y_\ell^{m_j+\frac{1}{2}}(\theta,\phi)\begin{pmatrix}0\\1\end{pmatrix}. \quad (7.42)$$

They are simultaneous eigenfunctions of J^2, L^2, S^2 and J_z.

We can use first-order perturbation theory to estimate the change of energy due to the spin-orbit coupling (Eq. 7.40), i.e., $\delta E = \langle \frac{1}{2}\mathcal{H}_{SO}\rangle$. Since $\mathbf{L}\cdot\mathbf{S} = \frac{1}{2}(\mathbf{J}^2 - \mathbf{L}^2 - \mathbf{S}^2)$, where $\mathbf{J} = \mathbf{L}+\mathbf{S}$ is the total angular momentum, we have

$$\langle \mathbf{L}\cdot\mathbf{S}\rangle = \frac{\hbar^2}{2}\left[j(j+1) - \ell(\ell+1) - \frac{3}{4}\right]. \quad (7.43)$$

Therefore, the change of energy is given by

$$\delta E_{n\ell}^{j} = \frac{1}{2m^2c^2}\left\langle \frac{1}{r}\frac{dU}{dr}\right\rangle_{n\ell} \frac{\hbar^2}{2}\begin{cases} \ell & \text{if } j = \ell + 1/2 \text{ or } \ell = 0 \\ [-(\ell+1)] & \text{if } j = \ell - 1/2 \end{cases} \quad (7.44)$$

where

$$\left\langle \frac{1}{r}\frac{dU}{dr}\right\rangle_{n\ell} = \int_0^\infty R_{n\ell}(r)\,\frac{1}{r}\frac{dU}{dr} R_{n\ell}(r)\,dr. \quad (7.45)$$

Equation 7.44 shows that the change in the energy due to the spin-orbit coupling $\mathcal{H}_{SO}$ of a single-electron atom (or a "hydrogen-like" atom) is (i) proportional to the orbital quantum number ℓ, and (ii) responsible for the energy difference between the spin-up and spin-down states (unless $\ell = 0$).

For multiple electrons in an atom, we must calculate their total angular momentum $\mathbf{J}$ as follows: For light atoms (Z<26), the spin-orbit interaction is small ($\sim 10^3$ cm^{-1}) as compared to the electron-electron interaction ($\sim 10^4$ cm^{-1}), so that the total orbital angular momentum $\mathbf{L} = \sum_i \mathbf{L}_i$ and total spin $\mathbf{S} = \sum_i \mathbf{S}_i$ are both good quantum numbers, and $\mathbf{J} = \mathbf{L} + \mathbf{S}$; this is called the Russel-Saunders or the LS coupling scheme. Then, we calculate

$$\langle \mathbf{L}\cdot\mathbf{S}\rangle = \frac{\hbar^2}{2}\left[J(J+1) - L(L+1) - S(S+1)\right]. \quad (7.46)$$

Following Eq. 7.40, we write the spin-orbit interaction Hamiltonian for the multielectron atom as

$$\mathcal{H}_{\text{SO}} = \lambda\,(\mathbf{L}\cdot\mathbf{S}), \quad (7.47)$$

where λ is called the spin-orbit coupling constant. In principle, λ can be calculated from a radial integral of type (Eq. 7.45) using a many-electron wave function, e.g., Slater determinant of

Table 7.3
Spin-orbit coupling constants λ for $3d$-transition metal ions in the ground state [Bersuker (1996)].

Ion	configuration	term	λ (cm^{-1})	λ (meV)
Ti^{3+}	d^1	^2D	154	19.1
V^{3+}	d^2	^3F	104	12.9
V^{2+}	d^3	^4F	55	6.8
Cr^{3+}	d^3	^4F	87	10.8
Cr^{2+}	d^4	^5D	57	7.1
Mn^{3+}	d^4	^5D	85	10.5
Mn^{2+}, Fe^{3+}	d^5	^6S	0	0.0
Fe^{2+}	d^6	^5D	-100	-12.4
Co^{2+}	d^7	^4F	-180	-22.3
Ni^{2+}	d^8	^3F	-335	-41.5
Cu^{2+}	d^9	^3D	-852	-105.6

single-electron wave functions (7.41). Or, λ can be determined from spectroscopic data as follows: Since the first-order correction is $\delta E = \lambda \langle \mathbf{L} \cdot \mathbf{S} \rangle$, the energy difference between the two states of the same L and S and a consequent J is estimated as

$$E_{J+1} - E_J = \lambda(J+1). \tag{7.48}$$

This is the rule of Landé intervals, which enables us to determine λ from spectroscopic data. Table 7.3 lists values of such an empirically determined λ for $3d$-transition metal ions. Notice that λ is negative when the d-shell is more than half-filled. (For d^5, $L = 0$ and therefore $\lambda = 0$.) For rare earth elements in particular, there is an approximate empirical formula [El'yashevich (1953)]:

$$\lambda\,[\mathrm{cm}^{-1}] = 200(Z - 55). \tag{7.49}$$

7.6 Magnetic resonance in NiO clusters

For transition metal ions with several unpaired d-shell electrons (e.g., Fe^{3+} and Ni^{2+}) in the condensed phase, their stationary-state energy levels depend on the crystal symmetry as well (crystal-field splitting). The combined effect of crystal field and spin-orbit coupling removes the degeneracy of most transition metal ions completely, even without application of any external field. For example, consider bulk and surface clusters of antiferromagnetic NiO shown in Figure 7.5. They belong to different symmetry groups: an octahedral complex (O_h) and the square pyramidal complex (C_{4v}), respectively.

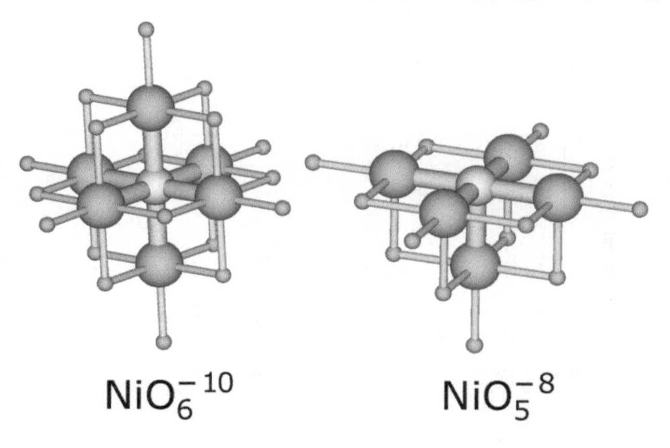

Figure 7.5
Clusters used for modeling the bulk (NiO_6^{-10}) and the surface (NiO_5^{-8}) of NiO. Large dark spheres represent oxygen atoms, the central grey sphere the nickel atom, and the small spheres the surrounding effective core potentials. The clusters are also embedded in a layer of effective core potentials and charge points in order to describe the next neighboring nickel atoms and the Madelung potential (not shown here) [Lefkidis and Hübner (2006); Li et al. (2014a)].

As discussed in Section 7.1, these clusters were the first ones to achieve spin-switching based on the 3-level scheme (Figure 7.2).

Without going into the technical details on the exact quantum chemical calculations, we present the final outcome in Figure 7.6, which is in a good agreement with the experimental results. The symmetry notations (A_{1g}, E_g, etc.) in the figure are for the irreducible representations of a point group, due to Mulliken [Mulliken (1933)]; one-dimensional representations are denoted by A and B, two- and three-dimensional representations by E and T, respectively. Numeric indices reflect an additional classification, such as symmetric or antisymmetric, etc. If the group includes the inversion operation, notations have an additional index, either g or u, indicating the parity: g for even (*gerade* in German) or u for odd (*ungerade*). In the literature, the Bethe notation ($\Gamma_1, \Gamma_2, \cdots$) [Bethe (1929)] is also widely used.[h] The following is the table of correspondence for the two types of notations.

[h] Adding $\pm$ in superscript denotes parity in the Bethe notation.

Mulliken	A	B	E	T_1	T_2
Bethe	Γ_1	Γ_2	Γ_3	Γ_4	Γ_5

The reason why NiO is particularly interesting is that, contrary to the ferromagnetic Ni, it possesses a spin density strongly localized on a Ni^{2+} ion and discrete energy levels which can be addressed individually by laser pulses. So this

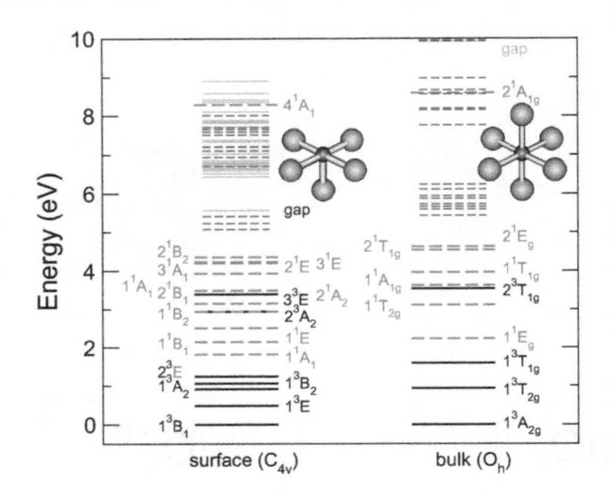

Figure 7.6
Energy levels of the NiO_5^{-8} (left) and NiO_6^{-10} (right) clusters, calculated using the complete-active-space self-consistent-field (CAS-SCF) method [Lefkidis and Hübner (2005)]. The wide solid lines are the triplet d-states, wide dashed lines the singlet d-states, narrow solid lines are the triplet charge-transfer states, and the narrow dashed lines are the singlet charge-transfer states. The gap is identified as the lowest triplet charge-transfer state.

system, although extended, allows for spin switching and not only partial demagnetization like the ferromagnetic Ni. One additional energetic detail is that flipping the spin of every atom of an antiferromagnet does not change the total net magnetization, which remains zero. In other words, the total energy cost is considerably lower than that of ferromagnets. Furthermore, NiO is an insulator with a characteristic sequence of discrete energy levels. The associated long coherence times are advantageous in magnetization control.

In the limit of the electric dipole approximation, the spin-up and spin-down states split by the spin-orbit coupling do not mix, because their orbital angular momentum is the same. This means that a direct transition between the spin-up and spin-down states is forbidden, unless the magnetic field of a driving laser pulse is considered, as we did in Section 7.4 (see also Refs. [Kampfrath et al. (2011)] and [Baierl et al. (2016)] for such experiments on NiO); this coupling mechanism (magnetic dipole transition), however, is six orders of magnitude weaker than the electric dipole transition.[i] A more practical way to couple the spin states of NiO by a laser field is through an intermediate excited state, as shown in Fig. 7.2 (see also Ref. [Fiebig et al. (2001)] and [Duong et al. (2004)] for experimental setup). If our initial and final states are in a d-shell, then the intermediate state needs to be in a p- or f-shell, due to the selection rule ($\Delta L = \pm 1$).

[i] The ratio between the radiated power of a magnetic dipole to that of an electric dipole is proportional to $(1/c)^2 = 1/137^2$ (atomic units) [Griffiths (1981)].

The Hamiltonian of the three-level system in Fig. 7.2 is

given by

$$\mathcal{H}(t) = \mathcal{H}_0 + V(t), \tag{7.50}$$

where

$$\mathcal{H}_0|\alpha\rangle = E_\alpha|\alpha\rangle, \quad (\alpha = 1, 2, 3) \tag{7.51}$$

and

$$V(t) = \sqrt{I_o}z\cos(\omega_o t), \tag{7.52}$$

where I_o and ω_o are the peak intensity and the frequency of a driving laser field, respectively. It is assumed that levels $|1\rangle$ and $|3\rangle$ have the same parity, whereas $|2\rangle$ has the opposite parity. In the interaction picture (or Dirac picture), the time-evolution of an arbitrary state $|\psi(t)\rangle = \sum_\alpha a_\alpha(t)|\alpha\rangle$ is found by solving

$$\frac{\partial}{\partial t}a_\alpha(t) = \frac{1}{i\hbar}\sum_\beta \langle\alpha V(t)|\beta\rangle a_\beta(t)e^{-i(E_\beta - E_\alpha)t/\hbar}. \tag{7.53}$$

In the limit of the rotating-wave approximation, the matrix element becomes [Ho and Chu (1985)]

$$\langle\alpha V(t)|\beta\rangle = \begin{pmatrix} 0 & W_{12} & 0 \\ W_{12}^* & \Delta_1 & W_{23} \\ 0 & W_{23}^* & \Delta_1 + \Delta_2 \end{pmatrix}, \tag{7.54}$$

where

$$W_{\alpha\beta} = \frac{\sqrt{I_o}}{2}\langle\alpha z|\beta\rangle, \tag{7.55}$$

$$\Delta_1 = (E_2 - E_1) - \hbar\omega_o, \tag{7.56}$$

$$\Delta_2 = (E_3 - E_2) + \hbar\omega_o. \tag{7.57}$$

In particular when $\Delta_1 = 0$ (resonant excitation), then $\Delta_2 = E_3 - E_1$, and quasistationary states which diagonalize the interaction Hamiltonian (7.54) are expressed as [Fleischhauer et al. (2005)]

$$\begin{aligned} |+\rangle &= \sin\Theta\sin\Phi|1\rangle + \cos\Theta\sin\Phi|2\rangle + \cos\Phi|3\rangle, \\ |-\rangle &= \sin\Theta\cos\Phi|1\rangle + \cos\Theta\cos\Phi|2\rangle - \sin\Phi|3\rangle, \\ |0\rangle &= \cos\Theta|1\rangle - \sin\Theta|3\rangle. \end{aligned} \tag{7.58}$$

The mixing angles Θ and Φ are given by

$$\begin{aligned} \Theta &= \tan^{-1}\left(\frac{W_{12}}{W_{23}}\right), \\ \Phi &= \frac{1}{2}\tan^{-1}\left(\frac{\sqrt{W_{12}^2 + W_{23}^2}}{E_1 - E_3}\right). \end{aligned} \tag{7.59}$$

The corresponding quasieigenvalues are

$$\varepsilon_\pm = \frac{\hbar}{2}\left(\pm\sqrt{\Omega_{12}^2 + \Omega_{23}^2 + \Delta^2} \mp \Delta\right), \quad \varepsilon_0 = 0. \tag{7.60}$$

Table 7.4
Energies, symmetries and total angular momenta of the 12 lowest states for the NiO_5^{-8} and NiO_6^{-10} clusters [Lefkidis and Hübner (2007)]. States marked in bold can be used for switching. For the bulk states 1 and 3 in the table below are considered spin-up and spin-down, and for the surface states 2 and 3 (in both cases the energy differences are minimal). Used with permission.

	bulk			surface		
State	Symmetry type	Energy (eV)	$\langle J_z \rangle$	Symmetry type	Energy (eV)	$\langle J_z \rangle$
12	$1A_{1g}$	**0.9674**	**0.00**	$3E$	0.9256	0.69
11	$1T_{2g}$	0.9551	0.68		0.9256	-0.69
10		**0.9551**	**0.00**	$2B_1$	0.9076	0.00[a]
9		0.9551	-0.68	$2B_2$	**0.5795**	**0.00**
8	$2T_{1g}$	0.9248	0.60	$1B_1$	**0.5570**	**0.00**
7		**0.9248**	**0.00**	$2E$	0.5162	1.28
6		0.9248	-0.60		0.5162	-1.28
5	$1E_g$	**0.9166**	**0.00**	$1A_2$	**0.4480**	**0.01**
4		**0.9166**	**0.00**	$1A_1$	**0.4372**	**-0.01**
3	$1T_{1g}$	0.0000	1.20	$1E$	0.0032	1.31
2		0.0000	0.00		0.0032	-1.31
1		0.0000	-1.20	$1B_2$	0.0000	0.00

[a] The state has $\langle J_z \rangle = 0$ but originates (before SOC splitting) from a $^3A_2$ state for which the transition from the ground state is forbidden by symmetry.

where $\Omega_{\alpha\beta} \equiv W_{\alpha\beta}/\hbar$ and $\Delta \equiv \Delta_2/\hbar = (E_3 - E_1)/\hbar$. This splitting of the unperturbed energy by $\varepsilon_\pm$ is known as the Autler-Townes effect. Physically, it means that we have Rabi oscillation between the states $|1\rangle$ and $|3\rangle$ at a frequency $\tilde{\Omega} = \sqrt{{\Omega_{12}}^2 + {\Omega_{23}}^2 + \Delta^2} \pm \Delta$, which is clearly greater than the transition frequency $(E_3 - E_1)/\hbar = \Delta$ of the spin states. This is an advantage of the 3-level scheme, since the greater frequency means a shorter period of Rabi oscillation. If we could set the duration of a driving laser pulse to coincide with 1/2 of the Rabi-oscillation period $2\pi/\tilde{\Omega}$, then the oscillation will terminate, and we would achieve complete population transfer from one spin state to the other.

At this point it is perhaps time to elaborate a little bit on the numerical calculation of the transition matrix elements $\langle \alpha V(t)|\beta\rangle$ in Eq. 7.53. The quality of our results depends on two factors. The one is the energy of the levels (which is then compared against the experimentally measured peaks), and also the accuracy of the wavefunctions $|\alpha\rangle$. The latter is important whenever expectation values of any operator are needed. For our calculation of NiO clusters, which depend mainly on the intragap d-character states and not the charge-transfer states, we choose the complete-active-space self-consistent-field (CAS-SCF) method (results in Table 7.4). This is es-

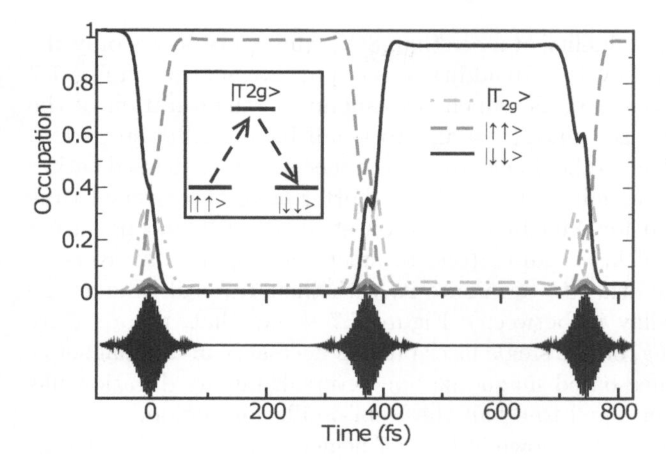

Figure 7.7
Population transfer between spin-up and spin-down ground states in bulk NiO. There is mainly one spin-mixed excited state involved in the process (T_{2g}), which always exhibits two peaks around the center of the laser pulse. The transfer needs about 100 fs. The pulse has FWHM 140 fs, energy 0.98 eV, and B_{max}=125000 A/m. The polarization is linear, and perpendicular to the externally applied static magnetic field $B_{\text{ext}} = 12.5$ A/m [Lefkidis and Hübner (2007)]. Used with permission.

pecially designed for static correlations, and therefore yields excellent energies. The main characteristic of CAS-SCF is that for the different configurations, different sets of molecular orbitals are used which are not necessarily orthogonal to each other.

For the 3-level magnetic resonance in the bulk of NiO, we use the triplets 1T_{g} as initial spin-up and final spin-down states. Note that since we have added spin-orbit coupling, mathematically speaking the multiplicity is not well defined anymore (since our wavefunctions are no longer eigenfunctions of the $\mathbf{S}^2$ operator), and therefore not given in the table. Still, one can calculate the expectation value of $\mathbf{J} = \mathbf{L} + \mathbf{S}$, which gives us information about the spin direction. As intermediate states we choose 1E_{g}. As it turns out, there are several other states which can be used as well (marked bold in Table 7.4).[7]

Figure 7.7 gives the time-dependent populations of the initial (solid line) and final (dotted line) states, as well as the participating intermediate states. The first remarkable observation is of course the very high fidelity of the process, and the second is its repeatability. Even after ten cycles the process

[7]If the electronic (de)excitations occur due to the coupling to the electric field of the laser pulse, there is a selection rule stating that the parity of the wavefunction must change, in other words it must be $g \leftrightarrow u$, and therefore the transitions $\text{T}_{\text{g}} \leftrightarrow \text{E}_{\text{g}}$ are forbidden. However, here the coupling is to the *magnetic field* of the laser, and hence they are allowed.

has still a fidelity of more than 85% (the figure shows only the first three cycles). In addition, the spin reversal time in Fig. 7.7 is less than 100 fs, which corresponds to the duration of the driving laser pulse (and also to a half Rabi-oscillation period, according to the 3-level resonance mechanism discussed earlier in this section). This is much shorter than the reversal time reported for NiO in Ref. [Satoh et al. (2006)] (9.5 ps) using a pair of laser pulses (one to start the population inversion and the other to terminate it, with their reversal time as the time delay in between). Figure 7.7 shows that, theoretically speaking, only a single laser pulse is necessary in the magnetic-resonance-based magnetization reversal, but its duration has to be set equal to a half the Rabi oscillation period.

The results shown in Fig. 7.7 neglect the effects of exchange interactions between the neighboring spins and magnetization damping. One way to incorporate such effects is to solve the semiclassical equation of the magnetization vector $\mathbf{M}(t)$, called the Landau-Lifshitz-Gilbert (LLG) Equation [Koopmans et al. (2005)]. One can start with the Hamiltonian [Wienholdt et al. (2012)]

$$\mathcal{H} = \sum_{i \neq j} \mathcal{J}_{ij} \mathbf{S}_i \cdot \mathbf{S}_j - \gamma \mathbf{B}(t) \cdot \sum_{i}^{N} \mathbf{S}_i. \tag{7.61}$$

The first term is Heisenberg's exchange interaction Hamiltonian with exchange constants $\mathcal{J}_{ij}$ taking into account the nearest neighbors (or more, depending on the models).[j] The second term describes the interaction of a collection of N spins with an external magnetic field $\mathbf{B}(t)$, and γ is the gyromagnetic ratio defined by

$$\mathbf{M}_i = \gamma \mathbf{S}_i. \tag{7.62}$$

Then, we define the effective magnetic fields $\mathbf{H}^{\text{eff}}$ as [Tzschaschel et al. (2017)]

$$\begin{aligned} \mathbf{H}_i^{\text{eff}} &= -\frac{\partial \mathcal{H}(\mathbf{M}_1, \mathbf{M}_2, ..., \mathbf{M}_N)}{\partial \mathbf{M}_i}, \\ &= -\left[\hat{x}\frac{\partial \mathcal{H}}{\partial M_x} + \hat{y}\frac{\partial \mathcal{H}}{\partial M_y} + \hat{z}\frac{\partial \mathcal{H}}{\partial M_z}\right] \end{aligned} \tag{7.63}$$

and solve the Landau-Lifshitz-Gilbert equations for $\mathbf{M}_i$. That is,

$$\frac{\partial \mathbf{M}_i}{\partial t} = -\gamma \left(\mathbf{M}_i \times \mathbf{H}^{\text{eff}} + \eta \, \mathbf{M}_i \times \frac{\partial \mathbf{M}_i}{\partial t} \right), \tag{7.64}$$

where η is a magnetic damping constant.

[j] There is another type of exchange interaction called the Dzyaloshinskii-Moriya interaction or the antisymmetric exchange interaction, which is proportional to $\mathbf{D}_{ij} \cdot (\mathbf{S}_i \times \mathbf{S}_j)$, where $\mathbf{D}_{ij}$ is a vector constrained by crystal symmetry.

7.7 Exercises

1. Prove Eqs. 7.20 and 7.21.
2. Show that the matrix S given by Eq. 7.27 is unitary, i.e., $S^\dagger S = SS^\dagger = I$.
3. Find the amount of difference in energy between the ^1P (singlet) and ^3P (triplet) excited states of the hydrogen atom using Eq. 7.44.
4. Consider the term ^{5}D for a multielectron atom.
   (a) What are the possible values of the total angular momentum J?
   (b) Find the shift of energy δE of the ^{5}D level due to the spin-orbit coupling (Eq. 7.47), in terms of a spin-orbit coupling constant λ.
5. Consider a system with two spins for the Hamiltonian in Eq. 7.1. Assume a Gaussian shape of the magnetic field, and compute the time evolution of spins. Hints: one can use the Matlab code in Section 6.4.2.
6. How do the degenerate energy levels of a free ion with angular momentum quantum numbers $J = 1,\ 2,\ 3,\ 4$ in a crystal field with cubic symmetry $\mathbf{O}$ split?
7. The transition metal monoxide NiO forms an ionic antiferromagnetic crystal with the NaCl crystal lattice structure. The transition metal atom is in the crystal field of $\mathbf{O}_h$ symmetry and has the electronic configuration Ni^{2+} with 8 electrons in the d-shell (Ar[$3d^8$]). Classify all 45 electronic states of the $3d$ electrons for the case of free Ni^{2+} ions. How do they split in the crystal field symmetry?
8. Calculate the effect on the linear-optics response of a quantum harmonic oscillator, which is perturbed (i) by a quadratic and (ii) by a cubic potential.
9. The multipole of transition matrix elements can be calculated by expanding the vector potential of the laser into a Taylor series and summing over the terms of the same order (see Appendix A.5). In the expansion, identify the local and nonlocal contributions both in real and in reciprocal (k) space. (Note: the pertinent operator is $\hat{\mathbf{r}}$ in real space and $\hat{\mathbf{p}}$ in the reciprocal space.) How can we transform the elements of the one to the other representation?

10. **O** and $\mathbf{D}_4$ represent two symmetry groups whose operations are illustrated in the following figure:

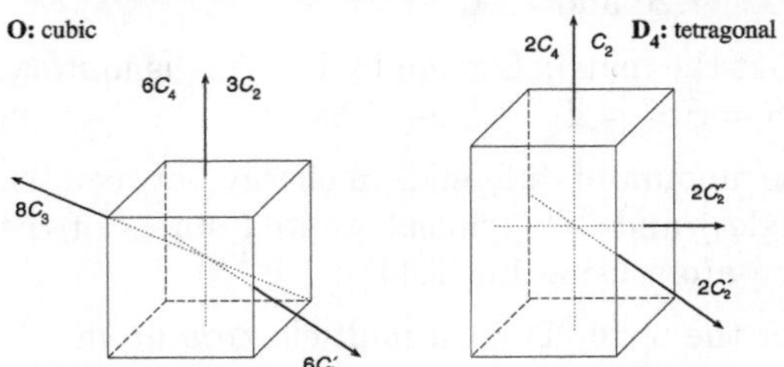

Their character tables of groups $\mathbf{D}_4$ and **O** are,

$\mathbf{D}_4$	E	C_2	$2C_4$	$2C_2'$	$2C_2''$
A_1	1	1	1	1	1
A_2	1	1	1	-1	-1
B_1	1	1	-1	1	-1
B_2	1	1	-1	-1	1
E	2	-2	0	0	0

O	E	$3C_2$	$6C_2'$	$8C_3$	$6C_4$
A_1	1	1	1	1	1
A_2	1	1	-1	1	-1
E	2	2	0	-1	0
T_1	3	-1	-1	0	1
T_2	3	-1	1	0	-1

Determine the further splitting by an additional perturbation of tetragonal symmetry ($\mathbf{D}_4$). Which classes of $\mathbf{D}_4$ and **O** are associated with each other and how do they decompose from **O** to $\mathbf{D}_4$? Sketch the energy level diagram.

8 Magnetic molecules and magnetic logic

8.1 Λ processes in molecular systems
8.2 A closer look into electronic correlations
8.3 Molecular vibrations
8.4 Magnetic logic on molecules
8.5 First steps towards magnetic logic gates
8.6 Concluding remarks
8.7 Exercises

The advent of femtosecond spin dynamics was in 1996, when it was first shown that laser pulses can induce demagnetization in extended ferromagnetic systems [Beaurepaire et al. (1996); Hübner and Zhang (1998)]. Very quickly the term ultrafast was coined to describe subpicosecond dynamics, a time scale extremely interesting for the computer industry, since it opens a door to faster magnetic storage media [Hohlfeld et al. (1997); Aeschlimann et al. (1997); Scholl et al. (1997)].

Within the next couple of years several mechanisms were proposed to explain this new phenomenon. The Elliot-Yafet mechanism highlights the coupling of the spins to the lattice [Koopmans et al. (2005)], a mechanism that has been disputed for the demagnetization of metallic ferromagnets in several cases [Radu et al. (2009); Meier et al. (2009)]; the *sp-d* model emphasizes the coupling of the localized spins to the carrier spins [Cywiński and Sham (2007)], as well as classical Gilbert-damping-based mechanisms [Simanek and Heinrich (2003)]. Finally, some authors suggested that at such short time scales the only plausible mechanism is the interaction between the spin and laser as it is mediated through spin-orbit coupling [Zhang and Hübner (2000); Vonesch and Bigot (2012); Bigot et al. (2009)].

Today research has come a long way from considering magnetic materials solely for information storage, but even in this particular research area smaller clusters are more appealing than larger ones. And there are several good reasons for nanotechnology: One can pack more information in smaller spaces making any relevant apparatus smaller and handier, and one needs less power to manipulate small clusters (especially when using lasers to do so). Furthermore, if one thinks of extending their applications towards magnetic logic operations, small clusters exhibit less energy dissipation problems (i.e., heating up the apparatus).

This chapter is roughly divided into two parts. At first we apply our techniques analytically to some simple cases in order to gain a deeper insight into the physics at hand. In the remaining part, we present actual scientific results which are obtained through real quantum mechanical calculations and go far beyond analytically solvable models. Insofar, we benefit from all the aforementioned methodologies and tools to yield rules describing which laser-induced processes are possible. The educational content (and hence the reason for including them in this book) consists exactly in identifying those rules through the extensive use of computers, in close analogy to the field of experimental mathematics: "the computer provides us with a 'laboratory' in which [one] can perform experiments: analyzing examples, testing out new ideas, or searching for pattern" [Borwein and Bailey (2004)].

8.1 Λ processes in molecular systems

Quantum mechanically, to a first approximation, the light couples only to the orbital angular momentum (see Eq. 3.82), and can induce electronic excitations with $\Delta l = \pm 1$ (see selection rules, Eq. 3.52). This means that a direct transition from a spin-up d state to a spin-down d state is forbidden.[a] The way to go is through an intermediate spin-mixed excited state, which couples both to the spin-up and to the spin-down state. Naturally not all molecules have these spin mixed states, so we must find them in the first place. If our initial and our final states are of d character, we need an intermediate state of either p or f character (the second one being a candidate only in atoms with f electrons in the first place). Spin mixing can come from spin-orbit coupling. The whole process is called a Λ process (see Fig. 8.1) and consists of two stages, an excitation stage during which a photon is absorbed and a deexcitation stage during which a photon is emitted, which, however, are triggered by the *same* laser pulse.

[a] Unless the magnetic field of the laser pulse is considered; this coupling mechanism, however, is six orders of magnitude weaker!

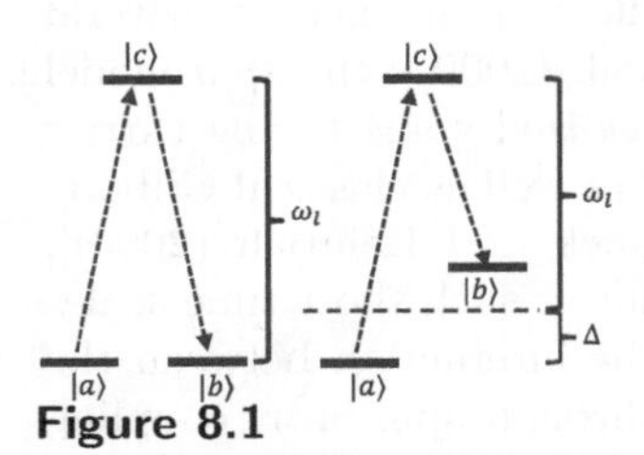

Figure 8.1
A Λ process necessitates three energy levels $|a\rangle$, $|b\rangle$, and $|c\rangle$, out of which the intermediate state $|c\rangle$ is higher in energy than the other two. There are two allowed transitions forming the Greek capital letter Λ. A single laser pulse excites the electronic population from the initial state first to the intermediate state and subsequently deexcites the system to the final state. The initial and the final states can be degenerate (left) or non-degenerate (right) [Lefkidis and Hübner (2013)].

The astute reader might have noticed that we almost always deal with triplet states.[1] It is clear that singlet states (being non-magnetic) are per definition not interesting for magnetic logic, but why not doublets or quartets? Strictly speaking doublets could also do the job, but one does not benefit much from them. The reason is that the electrons have spin $\frac{1}{2}$, in other words *each* electron is a doublet (in either an up or a down state). So, manipulating doublet spins practically means manipulating individual electrons. So there is no real advantage compared to manipulating the charge of individual electrons, since the carrier in both cases is the same, namely the electron itself. The mass of the electron remains the same, no matter which of its properties is addressed. Triplets, however, consist of *pairs* of electrons, and therefore their dynamics can differ drastically from the dynamics of the individual constituent electrons. In other words, triplets represent the minimal case of spin- and charge-dynamics separation. Higher multiplicities, of course, also offer this possibility, albeit up to

[1] If a system has two electrons, there are four possible spin configurations (if we do not take into account the spatial degrees of freedom). $|\uparrow\uparrow\rangle$, $|\downarrow\downarrow\rangle$, $|\uparrow\downarrow\rangle$ and $|\downarrow\uparrow\rangle$. Of these only the first two are eigenstates of the $\hat{S}^2$ operator, while for the last two we must take linear combinations. So we get again four spin states: a singlet $(|\uparrow\downarrow\rangle - |\downarrow\uparrow\rangle)/\sqrt{2}$, and a triplet with the three substates $|\uparrow\uparrow\rangle$, $(|\uparrow\downarrow\rangle + |\downarrow\uparrow\rangle)/\sqrt{2}$, and $|\downarrow\downarrow\rangle$.

a point. Not only does the dynamics become too complicated and therefore the results inconclusive, but accumulation of too many single electrons actually leads to macrospins. This means that the magnetic entity ultimately behaves like a classical nanomagnet, thus losing all interesting (not to say necessary) quantum effects!

8.1.1 Degenerate case

The model Λ system is extremely educational since it can be solved analytically. We build our Λ system as follows. First we consider the degenerate case, that is, the case in which the initial and the final states have the same energy $E_a = E_b$ (left side of Fig. 8.1). The initial ground state and the final ground state are the spin-down and spin-up triplets of an S state, respectively. So we have $|a\rangle = |0,0\rangle \otimes |\downarrow\downarrow\rangle$ and $|b\rangle = |0,0\rangle \otimes |\uparrow\uparrow\rangle$. The intermediate state is the $|j=1, m_j=0\rangle$ term ($^3\mathrm{P}_1$) resulting from the consideration of spin-orbit coupling (SOC) in an excited triplet P state.[b] Using Clebsch-Gordan coefficients (or diagonalizing the respective SOC Hamiltonian), we find that

$$\begin{aligned} |c\rangle &= |j=1, m_j=0\rangle \\ &= \frac{1}{\sqrt{2}}\left(|1,1\rangle \otimes |\downarrow\downarrow\rangle - |1,-1\rangle \otimes |\uparrow\uparrow\rangle\right). \end{aligned} \tag{8.1}$$

[b] SOC, as defined in Eq. 7.40, splits the triplet $^3\mathrm{P}$ into a quintet $^5\mathrm{P}_2$, a triplet $^3\mathrm{P}_1$ and a singlet $^1\mathrm{P}_0$ in spectroscopic notation $^{2s+1}\mathrm{L}_j$; j refers to the total angular momentum when spin and orbital angular momentum are coupled.

Then we calculate the transition matrix elements between the states for linearly polarized light along the x-axis:

$$\begin{aligned} \langle a|\hat{x}|c\rangle &= \frac{1}{\sqrt{2}}\langle 0,0|\hat{x}|1,1\rangle = \frac{1}{2}\Big(\langle s|\hat{x}|p_x\rangle + i\langle s|\hat{x}|p_y\rangle\Big) = 2\mu \\ \langle b|\hat{x}|c\rangle &= -\frac{1}{\sqrt{2}}\langle 0,0|\hat{x}|1,-1\rangle = -\frac{1}{2}\Big(\langle s|\hat{x}|p_x\rangle - i\langle s|\hat{x}|p_y\rangle\Big) = -2\mu \\ \langle a|\hat{y}|c\rangle &= \frac{1}{\sqrt{2}}\langle 0,0|\hat{y}|1,1\rangle = \frac{1}{2}\left(\langle s|\hat{y}|p_x\rangle + i\langle s|\hat{y}|p_y\rangle\right) = i2\mu \\ \langle b|\hat{y}|c\rangle &= -\frac{1}{\sqrt{2}}\langle 0,0|\hat{y}|1,-1\rangle = -\frac{1}{2}\left(\langle s|\hat{y}|p_x\rangle - i\langle s|\hat{y}|p_y\rangle\right) = i2\mu, \end{aligned}$$

where we set $\langle s|\hat{x}|p_x\rangle = 4\mu$ in order to simplify the equations later, while $\langle s|\hat{x}|p_y\rangle = 0$ because of the selection rules, see Eq. 3.52. In the interaction picture for the degenerate case, the total wavefunction of the system is expressed as

$$|\Psi\rangle = a(t)e^{-\frac{E_a}{\hbar}t}|a\rangle + b(t)e^{-\frac{E_b}{\hbar}t}|b\rangle + c(t)e^{-\frac{E_c}{\hbar}t}|c\rangle, \tag{8.2}$$

where E_a, E_b, and E_c are the energies of the three states. $a(t)$, $b(t)$, and $c(t)$ are time-dependent coefficients. The interaction with a resonant laser pulse with frequency $\omega_l = (E_c - E_a)/\hbar = (E_c - E_b)/\hbar$ is given by the perturbation Hamiltonian $\mathcal{H}' = \cos(\omega_l t)\hat{x}$. Applying time-dependent perturbation theory leads to the system of differential equations (working in atomic units it is also $\hbar = 1$)

$$\begin{cases} \dot{a}(t) = -i\langle a|\mathcal{H}'|c\rangle\, c(t)e^{-i(E_c - E_a)} \\ \dot{b}(t) = -i\langle b|\mathcal{H}'|c\rangle\, c(t)e^{-i(E_b - E_a)} \\ \dot{c}(t) = -i\Big(\langle c|\mathcal{H}'|a\rangle\, a(t)e^{-i(E_a - E_c)} + \langle c|\mathcal{H}'|b\rangle\, b(t)e^{-i(E_b - E_c)}\Big) \end{cases}$$

Putting everything together, writing $\cos(\omega_l t) = \frac{1}{2}(e^{i\omega_l t} + e^{-i\omega_l t})$, neglecting all terms containing double frequencies $e^{\pm 2i\omega_l t}$ (which is called the rotating-wave approximation, RWA), and using the initial conditions $a(0) = 1$, $b(0) = c(0) = 0$ (since we start with all the population in the initial state), we finally get

$$\begin{cases} \dot{a}(t) = -i\mu\, c(t) \\ \dot{b}(t) = i\mu\, c(t) \\ \dot{c}(t) = -i\mu\,[a(t) - b(t)] \end{cases} \Rightarrow \begin{cases} a(t) = \frac{1}{2}\left[1 + \cos(\sqrt{2}\mu\, t)\right] \\ b(t) = \frac{1}{2}\left[1 - \cos(\sqrt{2}\mu\, t)\right] \\ c(t) = -\frac{i}{\sqrt{2}}\sin(\sqrt{2}\mu\, t) \end{cases} \quad (8.3)$$

and ultimately the induced polarization in the material,

$$\begin{aligned} P_x(t) &= \mathrm{Tr}\left[\rho(t)\mathbf{D}_x\right] = -\sqrt{2}D\sin(2\sqrt{2}D\, t)\sin(\omega_l t) \quad (8.4) \\ P_y(t) &= \mathrm{Tr}\left[\rho(t)\mathbf{D}_y\right] = -2\sqrt{2}D\sin(\sqrt{2}D\, t)\cos(\omega_l t). \end{aligned}$$

Here $\mathbf{D}_x$ and $\mathbf{D}_y$ are the 3×3 matrices containing the $\hat{x}$ and $\hat{y}$ transition elements, respectively.

For a sinusoidal laser pulse, the maximum fidelity[c] is given by $\frac{d|b(t)|^2}{dt} = 0$ and happens at times $t_{\max} = \frac{k\pi}{\sqrt{2}\mu}$, where k is an odd integer. The third component P_z is zero.

[c] The **fidelity** is defined as the maximum population transfer from the initial state to the final state.

In the case of a spin-switching Λ process (say from spin-up to spin down), during the excitation a photon of one helicity is absorbed, while a photon of the opposite helicity is emitted. This way the angular momentum is conserved [Lefkidis and Hübner (2007); Lefkidis et al. (2009); Lefkidis and Hübner (2013)].

From a physical point of view the fate of the lost angular momentum is an extremely important point (and also a subject of intense scientific research over decades [Lefkidis and Hübner (2005); Zhang et al. (2002b); Bigot et al. (2009)]). Therefore we will spend some time rigorously proving that in the Λ process the laser light acts as an angular momentum bath. Furthermore, by doing so, we will see that this reversible quantum model (since we do not include any kind of relaxation in it) already contains the grain of macro-irreversibility because it allows us to macroscopically distinguish between the absorption and the emission phase.

Experimentally the expectation value of the spin changes [for ferromagnetic Ni, see Beaurepaire et al. (1996)], so the lost angular momentum must clearly go to a different degree of freedom, since it is a conserved quantity. In extended systems some obvious (but not unique) candidates to act as an angular-momentum bath are the phonons. Obviously in molecules we

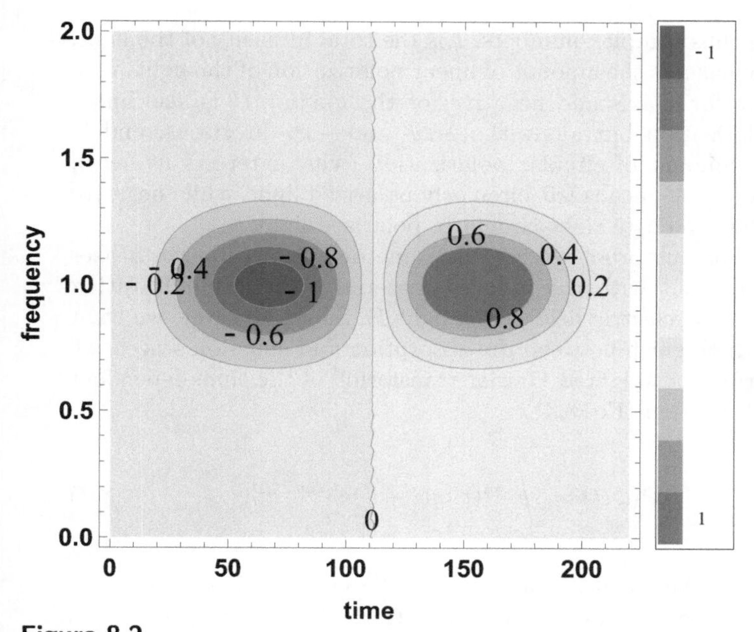

Figure 8.2
Fourth component $V(\omega, t)$ of the Stokes vector indicating the helicity of the induced polarization in the system (frequency $\omega_l = 1$, standard deviation of the pulse envelope $\sigma = 4$, transition-matrix elements $D = 0.01$) during a spin-flipping Λ process for the degenerate case (Fig. 8.1 left panel). The different helicities between absorption (negative value) and emission (positive value) phases are evident. Both axes are in arbitrary units [Lefkidis and Hübner (2013)].

have fewer options for such "baths," and therefore in general the conservation laws act much more strictly, and other candidates such as the laser pulse must be considered. Although a Λ process microscopically explains the transfer of angular momentum from the spin to the light and vice versa, the question remains, how to experimentally detect this. A possible way is to exploit the helicity difference of the two participating photons, in combination with the fact that photon emission and absorption are intrinsically connected to the material's induced polarization $\mathbf{P}(t)$. In classical optics the polarization of the light is described with the Stokes vector

$$\mathbf{S} = (I, Q, U, V), \tag{8.5}$$

where the four components are defined as

$$\begin{aligned} I &= E_x^2 + E_y^2 \\ Q &= E_x^2 - E_y^2 \\ U &= 2\mathrm{Re}\left(E_x E_y^*\right) \\ V &= -2\mathrm{Im}\left(E_x E_y^*\right). \end{aligned} \tag{8.6}$$

Here E_x and E_y are expressed in the frequency domain and are thus complex numbers. I is the total intensity of the light, Q measures the amount of linear polarization of the light (positive for x-axis and negative for the y-axis), U is also linear polarization but along the $+xy$- and $-xy$-directions, and V the amount of circular polarization (what interests us here). Positive V means left-circularly polarized light, while negative values indicate right-circularly polarized light.

For our proof we use the same definitions for the Stokes vector but with the induced material polarization $\mathbf{P}$ rather than the electric field of the light $\mathbf{E}$. Since, however, we want to distinguish between the absorption and emission stages, we perform a weighted Fourier transform[d] of the time-dependent polarization (Eq. 8.4),

[d]Through the **Fourier transform** of the light-induced polarization in the material with a **weighting time function**, one can observe in time the change in the helicity of the absorbed or emitted light.

$$\tilde{\mathbf{P}}(\omega, t) = \int_{-\infty}^{\infty} \mathbf{P}(t')g(t-t')e^{-i\omega t'}\,dt', \qquad (8.7)$$

where the weight function $g(t-t')$ ensures that temporally distant points do not contribute to the frequency spectrum as much as close ones. A typical example is a Gaussian function $g(t-t') = \frac{1}{\sigma\sqrt{2\pi}}e^{-\frac{1}{2}\left(\frac{t-t'}{\sigma}\right)^2}$, where the standard deviation σ decides as a parameter the width of the weighing window. This way we can follow in time the frequency change of the polarization. Figure 8.2 depicts the time- and frequency-resolved V component of the Stokes vector for a spin-switching process on a model system consisting of triplet s and p orbitals. We nicely see the difference in the helicities of the polarization at different times (corresponds to left system of Fig. 8.1). In the case of the non-degenerate system depicted in the right panel of Fig. 8.1, the picture remains similar, with the only difference that now the absorbed and the emitted photons have slightly different frequencies.

8.1.2 Chirped lasers

The frequency of a laser pulse can also vary dynamically (chirp), an effect which can actually be exploited in order to control the fidelity of a given Λ process. For a linear chirp α the laser frequency is $\omega_{\text{laser}} = \omega_0 + \alpha t$ and the equations of motion acquire a $e^{-i\alpha t^2}$ time dependency, so Eq. 8.3 becomes

$$\begin{cases} \dot{a}(t) = -iD\,c(t)\,e^{-i\alpha t^2} \\ \dot{b}(t) = iD\,c(t)\,e^{-i\alpha t^2} \\ \dot{c}(t) = -iD\,[a(t) - b(t)]\,e^{+i\alpha t^2}. \end{cases} \qquad (8.8)$$

The $e^{-i\alpha t^2}$ phase factor is what remains from $e^{-i\omega_{\text{laser}}t}$ after compensation from the energy-levels resonance $e^{-i\,\Delta\omega\,t}$ within RWA.

After a lengthy and quite complicated integration of the

above system of coupled differential equations one arrives at the analytical solutions [Lefkidis and Hübner (2015)]:

$$
\begin{aligned}
a(t) =& \frac{1}{2\mu\Gamma(A)}\Big\{2\mu K_3\Gamma(A) - (-1)^{\frac{3}{4}}\sqrt{\alpha}\Big[M\Big(-\frac{A}{2},\frac{1}{2},-i\alpha t^2\Big)-1\Big] \\
& K_1\Gamma\left(\frac{A+1}{2}\right) - 2\alpha t M\left(\frac{1-A}{2},\frac{3}{2},-i\alpha t^2\right) \\
& \Big[K_1\Gamma\left(1+\frac{A}{2}\right) + K_2\Gamma(1+A)\Big]\Big\}, \\
b(t) =& \frac{1}{2\mu\Gamma(A)}\Big\{2\mu K_4\Gamma(A) + (-1)^{\frac{3}{4}}\sqrt{\alpha}\Big[M\Big(-\frac{A}{2},\frac{1}{2},-i\alpha t^2\Big)-1\Big] \\
& K_1\Gamma\left(\frac{A+1}{2}\right) - 2\alpha t M\left(\frac{1-A}{2},\frac{3}{2},-i\alpha t^2\right) \\
& \Big[K_1\Gamma\left(1+\frac{A}{2}\right) + K_2\Gamma(1+A)\Big]\Big\}, \\
c(t) =& K_1 H_{-A}\left(\sqrt[4]{-1}\sqrt{\alpha}t\right) + K_2 M\left(\frac{A}{2},\frac{1}{2},i\alpha t^2\right).
\end{aligned}
\tag{8.9}
$$

Here $A = i\mu^2/2$, μ is half the strength of the transition matrix elements $\langle a|\hat{x}|b\rangle = -\langle b|\hat{x}|c\rangle = -i\langle a|\hat{y}|c\rangle = i\langle b|\hat{y}|c\rangle = 2\mu$, $H_n(z)$ are the Hermite polynomials, $M(a,b,z)$ are the Kummer's confluent hypergeometric functions, $\Gamma(x)$ is the Gamma function, and the constants K_i depend on the initial conditions. Another factor is the time of resonance t_{res} with respect to the period of the process T.[e] If $t_{\text{res}} = \frac{T}{4}$, the laser pulse has overall the least amount of detuning during the process, and hence the fidelity gets optimized. Therefore the sign of the chirp determines the best detuning policy: For $\alpha > 0$ ($\alpha < 0$) it is better to start at slightly lower (higher) frequencies. As we will see, this explains why a combination of detuning and chirp breaks the time-inversion symmetry of the Λ-process in a controllable manner, thus selectively driving the magnetization towards a desired orientation.

[e] This is when exactly the frequency of the pulse becomes resonant with the energy difference of the involved states.

8.1.3 Spectral broadening of the laser pulse

One final point to address before investigating real molecules is the tolerance of our suggested processes with respect to the laser-pulse broadening (since this is an experimentally relevant component). Here we repeat the mathematical derivation given by Jin et al. (2014). The electric field of a laser pulse with a Gaussian envelope normalized to its amplitude is

$$
E(t) = \frac{1}{2\sqrt{2\pi}\sigma_\omega} e^{-\frac{t^2}{2\sigma_{\text{env}}^2}} \cos(\omega_e t) e^{-\frac{(\omega_e-\omega_0)^2}{2\sigma_\omega{}^2}}, \tag{8.10}
$$

where σ_ω is the standard deviation of the laser frequency in the frequency domain, σ_{env} is the standard deviation of the envelope of the laser pulse in the time domain, ω_e is the frequency of the Gaussian-shaped laser, and ω_0 is the laser frequency, optimized with a specially developed genetic algorithm [Hartenstein et al. (2008)]. The third factor, which stands for the Gaussian distribution of the laser frequency in the continuous-wave limit (i.e., infinite pulse width), contributes a normalization factor of $\frac{1}{\sqrt{2\pi}\sigma_\omega}$. The coefficient of $\frac{1}{2}$ stems from the term $\cos(\omega_e t)$ [i.e., $\frac{1}{2}(e^{i\omega_e t}+e^{-i\omega_e t})$]. Performing a Fourier transform to the frequency domain gives

$$\tilde{E}(\omega) = \frac{\sigma_{\rm env}}{2\sqrt{2\pi}\sigma_\omega}\left[\exp(-\frac{\sigma_\omega^2\sigma_{\rm env}^2(\omega-\omega_e)^2+(\omega_0-\omega_e)^2}{2\sigma_\omega^2})+ \right. \\ \left. +\exp(-\frac{\sigma_\omega^2\sigma_{\rm env}^2(\omega+\omega_e)^2+(\omega_0-\omega_e)^2}{2\sigma_\omega^2})\right]. \tag{8.11}$$

Integrating with respect to the laser energy ω_e over all frequencies gives

$$\langle\tilde{E}(\omega)\rangle = \frac{1}{2\sigma}\left[e^{-\frac{(\omega-\omega_0)^2}{2\sigma^2}]} + e^{-\frac{(\omega+\omega_0)^2}{2\sigma^2}}\right], \tag{8.12}$$

finally yielding the total broadening

$$\sigma = \sqrt{\frac{1}{\sigma_{\rm env}^2}+\sigma_\omega^2}. \tag{8.13}$$

Equation 8.13 thus incorporates both the spectral broadening due to the finite laser pulse and the actual broadening due to the experimental uncertainty of the central frequency of the laser even at the continuous-wave limit.

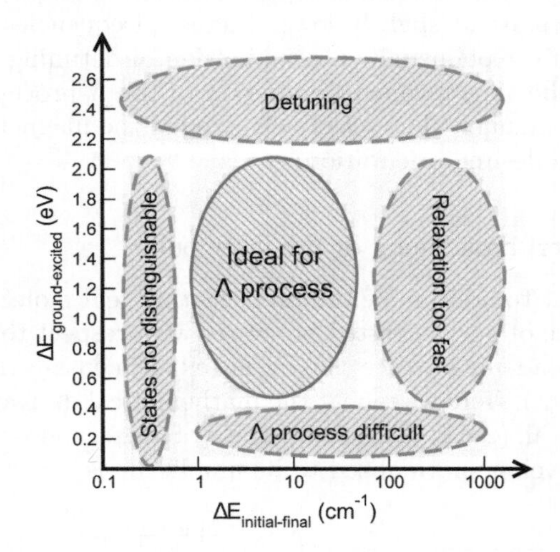

Figure 8.3
Sketch of the required energy differences between initial and final states (abscissa axis in logarithmic scale) and between the initial state and the intermediate excited state (ordinate axis) in the Λ process [Li et al. (2011)].

Obviously, in order for the Λ process to have any experimental value, we must identify the energy regions, for which it can prove successful. For this, two energy differences are important: the difference between initial and final states $\Delta E_{\text{initial-final}}$ (Zeeman splitting), and the energy between initial and intermediate states $\Delta E_{\text{ground-excited}}$ (more or less resonant to the laser frequency). Neither of them should be too small or too large (Fig. 8.3). If $\Delta E_{\text{initial-final}}$ is too small, then thermal excitations very quickly populate equally both of them; if it is too large, then a direct relaxation is possible, making the highest of the two short-lived. If $\Delta E_{\text{ground-excited}}$ is too small, then it is not possible for the laser pulse to be resonant both with the excitation and the deexcitation; if it is too large, then the same laser pulse will address a multitude of other states in the energetic vicinity of the intermediate state, making the whole process extremely nonlinear, and thus possibly unstable.

8.2 A closer look into electronic correlations

Now that we have elaborated on the laser-induced Λ processes and before delving deeper into the mechanisms and processes that are important for the design of clusters suitable for magnetic logic, it is a good idea to spend a few lines on explicitly considering the importance of correlations. In order to do so, we take two two-center clusters as our real examples.

8.2.1 Correlations and interatomic distances

We first calculate the lowest many-body electronic state in Ni_2 as a function of the interatomic distance using high-level n-electron valence-state perturbation theory [NEVPT2, see Angeli et al. (2001)] as well as the equation-of-motion coupled cluster (EOM-CCSD) method [Krylov (2008)]. Our results are shown in Fig. 8.4. One main observation is that the equilibrium distance of the electronic ground state very much depends on the basis set and correlation method used: The NEVPT2 method yields 2.2 Å [Chaudhuri et al. (2015), Fig. 8.4], which is very close to the experimental value of 2.15 Å, while SAC-CI and EOM-CCSD computations, depending on the exact basis set used, result in slightly longer interatomic distances, as long as 2.4 Å [Dong et al. (2013); Chaudhuri et al. (2014, 2015), see also Fig. 8.4]. Subsequently we derive successful spin-flip scenarios at several distances. Because Ni_2 is highly symmetric, the spin density for all low-lying electronic states is equidistributed among both Ni atoms; hence we derive *global* and not *local* spin-flips. For each interatomic distance, we find a different optimized laser pulse.

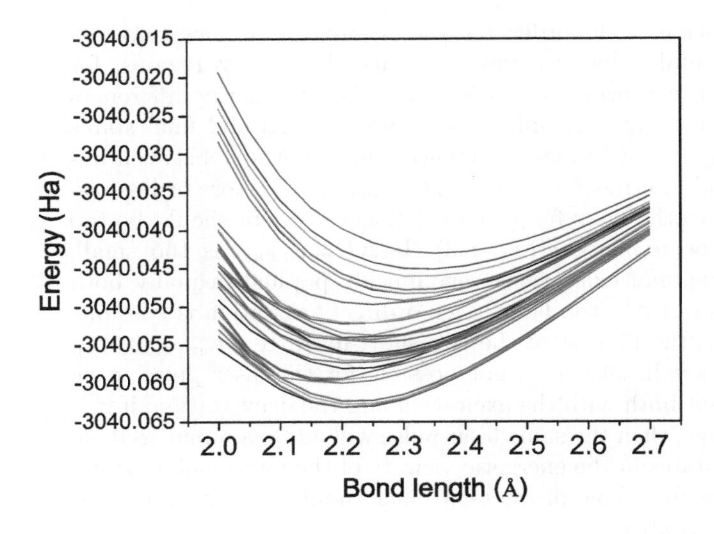

Figure 8.4
Level scheme of the electronic states of the Ni_2 dimer as a function of the bond length, computed with the NEVPT2 method and the Def2-QZVPP [24s18p10d3f1g/11s6p5d3f1g] basis set [Chaudhuri et al. (2015)]. Used with permission from Springer Nature.

If we propagate our system, for a specific interatomic distance, using the laser pulses which are optimized for *other* distances, then the fidelity is lower. One might think that the only reason for this is that due to the energetic shifting of the electronic states, the laser pulse is off-resonant. To test this hypothesis, we manually fit the laser frequency to match the energy separation between initial, intermediate, and final states for the different distances, and repeat the propagation without changing the other laser parameters (namely direction, duration, and amplitude). Clearly, there is more to a successful spin-flip scenario than just resonance, which is due to the correlations.

The main effect of the **electronic correlations** is that they shift the energies of the electronic states. Additionally, they can change the strength of the transition-matrix elements $\langle i|\hat{O}|j\rangle$ between states $|i\rangle$ and $|j\rangle$ and therefore the strength of several phenomena. However, they *do not* alter the selection rules. What is forbidden remains forbidden.

8.2.2 Correlations and ultrafast spin dynamics

In order to clearly attribute the change in the Λ process to correlations, we must check how the correlations alter the whole process, which can be done by selectively switching off correlation channels and monitoring the result. Allowing a small digression to a cluster which does not contain Ni, we check this on the CoMgCo-CO cluster. After finding a successful spin-switching scenario, we gradually remove configurations (i.e., Slater determinants from the many-body wavefunction, see Section 3.3) and repeat the propagation.

We find out that the fidelity of the whole Λ process deteriorates as correlations are removed. It is interesting that this does not happen in a linear but rather in a step-like way. The fidelity drops abruptly every time a certain Slater determinant, which allows for a needed transition, is removed (Fig. 8.5).

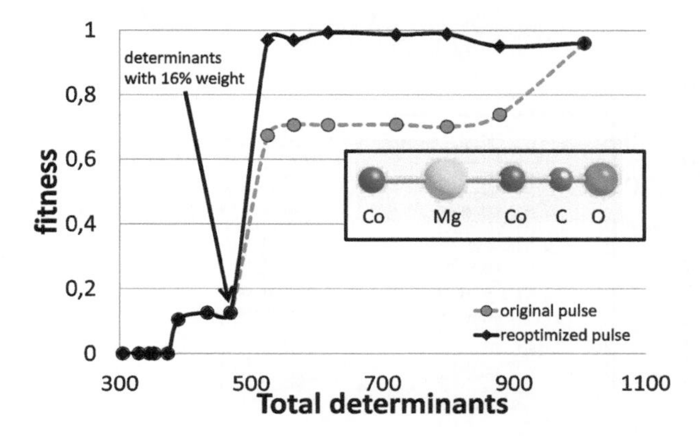

Figure 8.5
Dependence of the fitness of the Λ process on the total number of Slater determinants included in the calculation with the original laser pulse (dashed line) and when reoptimizing the pulse every time (solid line). The arrow points at the 16%-weight threshold. Inset: CoMgCo-CO structure [Zhang et al. (2011)]. Used with permission from the American Institute of Physics.

8.3 Molecular vibrations

Quantum chemical calculations can do more than describe the experimental electronic absorption spectra. In fact, quantum chemical theories can also very successfully describe infrared spectra, and thus help characterize structures and isomers, which is a prerequisite to identify suitable ultrafast phenomena. As we will see later in this chapter, for nanospintronics one must go beyond the trivial case of dimers: at least three active magnetic centers are needed. Therefore, here as an example we mention for the case of a Co cationic trimer synthesized with nozzle expansion, to which an ethanol molecule (EtOH) is attached, and then the whole structure isolated through mass spectrometry [Jin et al. (2014)]. Figure 8.6(a) shows the theoretically optimized geometry of Co_3^+(EtOH), as well as electronic many-body states both before and after the inclusion of spin-orbit coupling. The quantum chemical computations were performed at the SAC-CI level [Nakatsuji (1979)]. Figure 8.6(b) shows the energy spectrum with and without SOC.

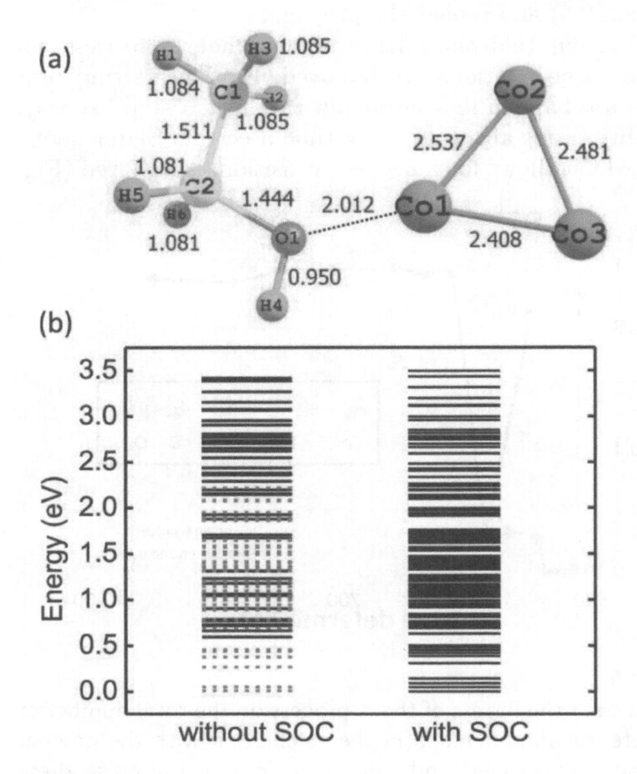

Figure 8.6
(a) Optimized structure of Co_3^+(EtOH) together with the bond lengths (in Å). (b) Triplet (dotted lines) and singlet (solid lines) energy levels (without SOC) and levels with SOC [Jin et al. (2014)].

The electronic absorption spectrum is given in Fig. 8.7. For each peak of the spectrum two parameters are important. One is the position of each peak (which is given by the energy difference $E_b - E_a$ of the two electronic states a and b among which the transition takes place), and the other is the intensity[f] of the peak given through the oscillator strength of each transition,

[f] The intensity of the absorption peaks are given through the **oscillator strength**.

$$f_{a\to b} = \frac{2}{3}\,(E_b - E_a)\,\langle a|\mathbf{r}|b\rangle^2, \tag{8.14}$$

where $\mathbf{r}$ is the electric dipole transition operator (see the discussion on their derivation, Eq. A.46).

The oscillator strength obeys the Thomas-Reiche-Kuhn sum rule[g]

[g] In some books, and depending on the normalization of the many-body wavefunctions, the rule is given as $\sum_b f_{a\to b} = N$, where N is the number of electrons in the system.

$$\sum_b f_{a\to b} = 1, \tag{8.15}$$

and can be proven by combining two quantum mechanical

commutators, namely $i[\hat{H}, \hat{x}] = \hat{p}_x$ and $i[\hat{x}, \hat{p}_x] = i\hbar$, with Eq. 8.14.

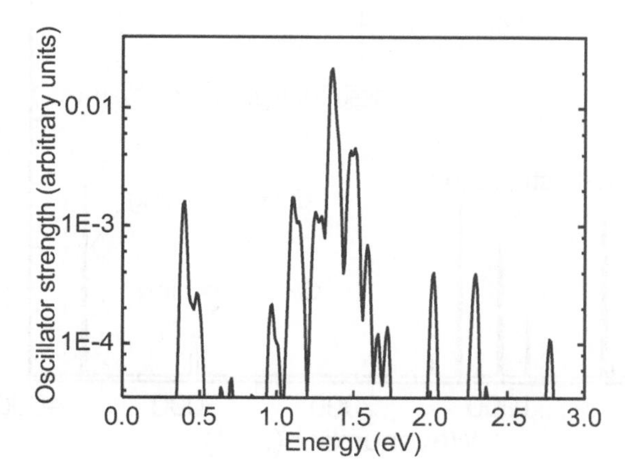

Figure 8.7
Theoretical oscillator strength (in logarithmic scale with arbitrary units) with a Gaussian broadening $\sigma = 0.015$ eV for the ground state of Co_3^+(EtOH) [Jin et al. (2014)].

The infrared spectrum of a molecule gives information about its geometrical structure, since it is mostly due to the molecular vibrations. For complicated structures, the quantum chemical calculations can become extremely challenging, because very small energy scales are involved and are based on molecular normal-mode analysis. The analysis is a special case of the Hellmann-Feynman theorem,[2] which relates the derivative of the total energy of a system to the expectation value of the derivative of the Hamiltonian with respect to a parameter λ, in this case the atomic coordinates [Feynman (1939)]:

$$\begin{aligned}
\frac{dE_\lambda}{d\lambda} &= \frac{d\langle\Psi|\hat{H}(\lambda)|\Psi\rangle}{d\lambda} \\
&= \left\langle \frac{\partial\Psi}{\partial\lambda} \middle| \hat{H}(\lambda) \middle| \Psi \right\rangle + \left\langle \Psi \middle| \hat{H}(\lambda) \middle| \frac{\partial\Psi}{\partial\lambda} \right\rangle + \left\langle \Psi \middle| \frac{\partial\hat{H}(\lambda)}{\partial\lambda} \middle| \Psi \right\rangle \\
&= E\left\langle \frac{\partial\Psi}{\partial\lambda} \middle| \Psi \right\rangle + E\left\langle \Psi \middle| \frac{\partial\Psi}{\partial\lambda} \right\rangle + \left\langle \Psi \middle| \frac{\partial\hat{H}(\lambda)}{\partial\lambda} \middle| \Psi \right\rangle \\
&= E\underbrace{\frac{\partial\langle\Psi|\Psi\rangle}{\partial\lambda}}_{=\frac{\partial 1}{\partial\lambda}=0} + \left\langle \Psi \middle| \frac{\partial\hat{H}(\lambda)}{\partial\lambda} \middle| \Psi \right\rangle = \left\langle \Psi \middle| \frac{\partial\hat{H}(\lambda)}{\partial\lambda} \middle| \Psi \right\rangle. \qquad (8.16)
\end{aligned}$$

[2]The **Hellmann-Feynman theorem** lets us calculate the dependence of the electronic Hamiltonian on some parameter. Often, such parameters are the interatomic distances, and so we can derive the frequencies of the molecular vibrations.

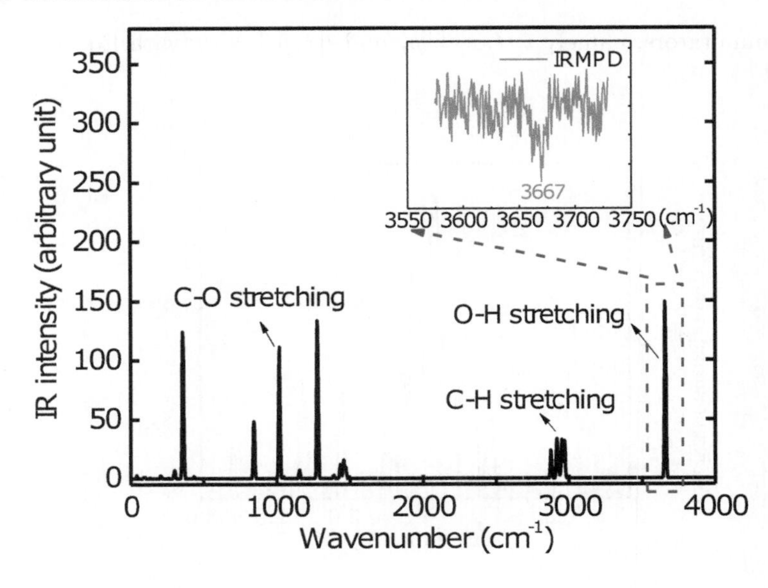

Figure 8.8
Theoretical and experimental (inset) infrared-multi-photon-dissociation (IRMPD) spectrum of the cluster Co_3^+(EtOH) [Jin et al. (2014)].

This means that one must compute the force matrix of the system (which is the Hessian matrix[h] of the total energy with respect to the atomic dislocations). Although there are some techniques with which one can derive the matrix analytically for a Hartree-Fock calculation, this is unfortunately not the case for most post-Hartree-Fock methods. Since in a molecule with N atoms there are $3N$ atomic coordinates, this is also the number of single-point energy calculations that must be performed. In reality, quite often even more calculations are needed due to numerical instabilities. And of course all the calculations refer typically only to the electronic ground state. If electron-phonon coupling constants are desired, then one must also recalculate the Hessian matrix for several electronic levels.

[h] The **Hessian** matrix of a function which depends on several variables, e.g., $f(x, y, z)$, is defined as the matrix of all combinations of second derivatives:

$$H_f = \begin{pmatrix} \frac{\partial^2 f}{\partial x^2} & \frac{\partial^2 f}{\partial x\,\partial y} & \frac{\partial^2 f}{\partial x\,\partial z} \\ \frac{\partial^2 f}{\partial y\,\partial x} & \frac{\partial^2 f}{\partial y^2} & \frac{\partial^2 f}{\partial y\,\partial z} \\ \frac{\partial^2 f}{\partial z\,\partial x} & \frac{\partial^2 f}{\partial z\,\partial y} & \frac{\partial^2 f}{\partial z^2} \end{pmatrix}$$

Once calculated, the normal modes must also be visualized with some specialized software, in order to identify the different modes. Often, this is easier said than done, since the molecule can have many complicated modes, and a specific bond can participate in several of them.

As an example we take the case of Co_3^+(EtOH) where we are interested in the O-H stretching mode, which is a fingerprint for alcohol in IR spectroscopy (Fig. 8.8). The theoretically calculated frequency lies only 11 cm^{-1} higher than the experimentally measured one (3667 cm^{-1}, at the HF/6-31G(d) level and with a scaling factor of 0.895 [Merrick et al. (2007)]). This deviation is consistent with the one calculated for the Co_3^+(MeOH) cluster (methanol instead of ethanol), for which the theoretically predicted value is 10 cm^{-1} higher than the experimentally measured one (3649 cm^{-1}). In other words, when going from MeOH to EtOH as an attached ligand, theory predicts a blue shift of 18 cm^{-1} as compared to the measured value of 17 cm^{-1}. The chemical reason behind this is the stronger positive inductive effect (+I) of the ethyl group with respect to the methyl group.

8.3.1 Electron-vibron coupling

In the previous section we discussed the molecular vibrations in a classical manner, that is, we treated the vibrations as classical oscillations of the atoms, and solved the electronic Hamiltonian within the Born-Oppenheimer approximation.[i] Of course this is just an approximation, which disregards the (sometimes) strong interaction between the motion of the electrons and this of the nuclei. In order to remedy this, one also needs to quantize the vibrations, a procedure yielding the so-called vibrons, in close analogy to the phonons in extended systems.

[i]The **Born-Oppenheimer** approximation states that the positions of the nuclei are fixed - only the electrons are treated quantum mechanically. This approximation is based on the fact that the nuclei are thousands of times heavier than the electron, and therefore they move more slowly. Thus in a first approximation one can decouple the two motions.

One way to do this is the following: After calculating the vibronic modes of the molecule, one canonically quantizes them, i.e., considers them as quantum harmonic oscillators. Now the total wavefunction of the system (also called the *dressed* system) consists of two parts: the electronic part and the vibronic part. We can write

$$|\Psi_{\text{tot}}\rangle = |\alpha, \mathbf{N}_i\rangle, \tag{8.17}$$

where α refers to the electronic state, and $\mathbf{N}_i$ is the number of vibronic quanta for the i-th vibrational mode. Now we want to calculate the dynamics of the dressed system. There is one more ingredient, however, namely we need to introduce electron-vibron coupling, as the funnel. This can be computed indirectly with the help of the Hellman-Feynman theorem just introduced. The coupling constants are

$$\lambda^i_{\alpha,\beta} = \langle \alpha, \mathbf{0}_i | \frac{\partial \hat{H}}{\partial q} | \beta, \mathbf{1}_i \rangle, \tag{8.18}$$

where $\mathbf{0}_i$ and $\mathbf{1}_i$ mean no and one vibron in the i-th mode, while α and β index the electronic part of the wavefunction Ψ. q is the vibronic coordinate. Without any coupling the total Hamiltonian takes a diagonal form, with each entry as the sum of the electronic energy E_{e} and vibronic energy E_{ph},

$$\hat{H}^0_{\text{e}} = \begin{pmatrix} E_{\text{e}} & 0 & \\ 0 & E_{\text{e}} + E_{\text{ph}} & \\ & & \ddots \end{pmatrix}. \tag{8.19}$$

Including the interactions $\lambda^i_{\alpha,\beta}$, this becomes

$$\hat{H}_{\rm e} = \begin{pmatrix} E_{\rm e} & \lambda^i_{\alpha,\beta} & \\ \lambda^i_{\alpha,\beta} & E_{\rm e} + E_{\rm ph} & \\ & & \ddots \end{pmatrix}. \tag{8.20}$$

If we knew all $\lambda^i_{\alpha,\beta}$'s we would simply diagonalize our total Hamiltonian $\hat{H}$. However, this is not possible. What we can do is to calculate the total electronic energy in the presence of vibrons, by repeating the frozen-vibron calculations and averaging over many points along the vibron coordinate q. In other words, we take snapshots of the electronic energy for the different geometries as the atomic coordinates in the molecule oscillate around their equilibrium positions. This yields

$$\hat{H}_{\text{ab-initio}} = \begin{pmatrix} E_{\rm e} & 0 & \\ 0 & E_{\rm e} + E_{\rm ph} + \Delta E^i_{\alpha,\beta} & \\ & & \ddots \end{pmatrix}, \tag{8.21}$$

which is diagonal. In fact, the entries of $\hat{H}_{\text{ab-initio}}$ are the eigenvalues of $\hat{H}_e$ averaged over a whole vibronic period plus the energy of the quantized vibration! For some very simple cases it is mathematically possible to demand that $\hat{H}_{\rm e}$ and $\hat{H}_{\text{ab-initio}}$ yield exactly the same eigenvectors, which finally leads to

$$\lambda^i_{\alpha,\beta} = \pm\sqrt{\frac{(\Delta E^i_{\alpha,\beta})^2 + 2E_{\rm ph}\Delta E^i_{\alpha,\beta}}{4}}. \tag{8.22}$$

Ultimately, if we restrict ourselves to vibronic excitations within the same mode, we are able to calculate the coupling constants.

8.3.2 Molecular vibrations and spin dynamics

When it comes to the realization of such spin-flip and spin-transfer scenarios, one of the most important questions is how to experimentally detect them. Of course, one can always hope to directly detect the magnetic state of an atom (using for instance some kind of magnetic scanning tunneling microscopy), but in practice this is too cumbersome and time consuming for real-life magnetic-nanologic applications. An easier, also non-destructive detection method is through optical, infrared, or ultraviolet spectroscopy.[j] To this end, the material must possess some characteristic, easily detectable absorption peaks, which can give information on its magnetic state. Since metallic clusters usually have too complicated electronic and vibrational spectra, it is a good idea to rely on some attached ligands with well-known, if possible fairly isolated peaks, such as carbonyls. We call such ligands chromophores. CO has a strong, sharp stretching frequency peak in the range between

[j] In some molecules the vibrational (IR) frequency of a particular group of atoms is very characteristic and also sensitively depends on the molecular electronic state. Then the exact IR frequency can be used as an experimental tool to indirectly detect the electronic (or magnetic) state of the molecule. If this is the case, we call this group of atoms an **IR marker**.

1650 and 1850 cm^{-1} (between 2.05 and 2.30 eV), which is very sensitive to its immediate physical and chemical environment. However, the interaction of the carbonyl with the magnetic clusters of interest is always bidirectional: not only the C-O stretching frequency is influenced by the metallic cluster, but also the metallic cluster is influenced by the ligand, notably since the latter reduces the overall symmetry.

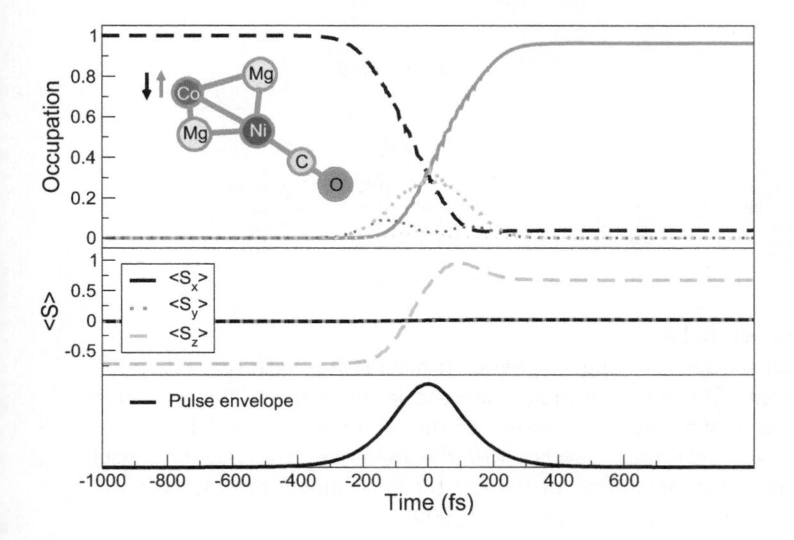

Figure 8.9
Local magnetic switch at the Co end of the charged cluster $[CoMg_2Ni\text{-}CO]^+$ (with SOC). (a) The occupations of the initial (dashed), target (solid), and intermediate (dotted) states vs. time. The inset shows a sketch of the optimized cluster. (b) Time-resolved expectation values of the spin components. (c) The laser pulse envelope that induces the switching [Li et al. (2009)].

One of the first examples investigated in this way was a spin-flip scenario on the $CoMg_2Ni\text{-}CO^+$ cluster (Fig. 8.9), in which the spin localization is induced by two factors, the difference of the two magnetic centers (Co vs. Ni) and the presence of the CO-ligand. Due to the strong electron-vibron coupling of the cluster, we find that the exact electronic (and hence also magnetic) state of the system is also reflected in the infrared vibrational frequency of the CO chromophore. In fact, there are two effects. One is that the equilibrium distance of the CO bond is different for different electronic states (i.e., it is 1.15 Å for the ground state, while it is enlarged to 1.17 and 1.16 Å for the first and second excited states, respectively (see Fig. 8.10). This means that any vertical electronic excitation also induces vibronic excitations. The second effect is that the curvature of the energy hypersurfaces for the different electronic states is different, which leads to different vibrational frequencies. And since every state has a different spin localization, in the end the vibrational frequency indirectly reveals information about

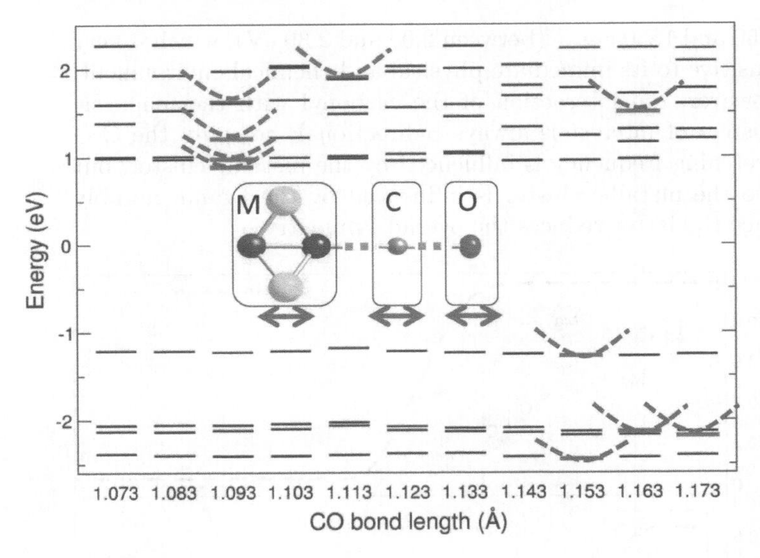

Figure 8.10
Calculated low-lying magnetic (triplet) states with respect to different CO bond lengths. The solid levels indicate the minimum energy of a state with respect to the CO bond lengths (the dashed curves near these levels show the minimum trends). The inset shows the separation of the cluster into three blocks for the frequency calculations [Li et al. (2009)].

where the spin sits! The vibrational energy differences are on the order of 20 to 100 cm^{-1} and thus experimentally easily detectable.

Perhaps at this point we should again remind the reader how these calculations are performed. First we optimize the geometry of the cluster at the Hartree-Fock level. Then we reoptimize it at the post-Hartree-Fock level (typically SAC-CI or EOM-CCSD) for the electronic ground state, as well as for the relevant excited states. Finally, we calculate the normal modes of the system for *all* obtained electronic states, and perform at least three single-point calculations for each of them (one at equilibrium geometry and two for distortions along the normal mode; remember that for the harmonic approximation one needs at least three points to determine the curvature of the parabola). In all, for 10 electronic states, we need 10 geometry optimizations, 10 normal mode analyses, and 30 post-Hartree-Fock single-point calculations, making the whole process a rather demanding endeavor! Note also, that for this purpose it is not necessary to quantize the normal modes, since we do not explicitly need the electron-phonon or spin-phonon coupling constants.

8.3.3 Geometry change as a tool: Mechanical strain

One of the greatest experimental challenges regarding such delicate processes like the laser-induced Λ processes is probably the protection of small magnetic structures (molecular magnets or clusters) from the possibly destructive influence of the immediate environment. At the same time, one wishes to leave room for interaction with the outside world, so that a laser pulse can still address its electronic and magnetic properties. One very promising way to do so is to enclose the magnetic entities in small organic cages, namely fullerenes. Fullerenes (especially spherical fullerenes) are very stable carbon molecules which can carry in their inner cavity a multitude of small magnetic clusters, and thus protect the latter from the chemical surroundings. Additionally, since the fullerenes themselves can be ordered nicely on surfaces (e.g., with the tip of a scanning tunneling microscope [Cuberes et al. (1996)]), they can also be used to create large two-dimensional arrays of non-interacting magnetic clusters.

One such endohedral fullerene (also called endofullerene) is $Co_2@C_{60}$, where the so-called buckminsterfullerene C_{60} contains the homodinuclear magnetic cluster Co_2. Our usual high-level quantum chemical calculations reveal that as anticipated the largest part of the spin-density is localized on the Co atoms. This of course immediately raises the question of whether in such a structure it is still possible to induce laser-driven spin dynamics, since the electric field of the laser will mainly interact with the cage only. However, a closer inspection of the electronic cloud reveals that there is a certain degree of hybridization between the Co and C atoms, meaning that the wavefunctions of the enclosed entity and the cage are not fully decoupled from each other, therefore rendering our familiar Λ processes possible.

A perhaps counterintuitive observation is that despite the rather weak coupling, the cage significantly influences the spin-density distribution of the magnetic moiety. Since the latter is symmetrical, one would rather expect the spin density to be equidistributed among both Co atoms, at least for the energetically low-lying electronic states. This, however, is not the case. The presence of the cage induces a strong spin localization, which in turn means a strong electron-vibron coupling (later we will come back to this very important phenomenon once more, when we will take a closer look at another cluster, which serves as a paradigm for magnetic logic, namely Ni_3Na_2). Therefore, a remarkable effect occurs if one exerts a tensile strain on the cage (that is, if one uses mechanical pressure along the dimer's axis). Despite the weak overlap of the electronic clouds of the cage and the dimer, the spin-density of the latter redistributes and can vary from a slight change in the absolute Mulliken population up to a complete shift of the spin density from one magnetic atom to the other. For the

P-type endofullerene the tensile strain does not significantly affect the spin distribution of the ground state (it always remains on Co_2). In state 3, however, a dramatic change takes place: the spin density starts from Co_1 and with increasing strain slowly moves over to Co_1. This movement is even more pronounced for states 4 and 5 (in the sense that the spin is already completely localized on the other Co atom for a tensile strain of 0.5% only). In the HP-type endofullerene, in which the spin density is equidistributed among the two Co atoms in the first place, no perceivable shift takes place.

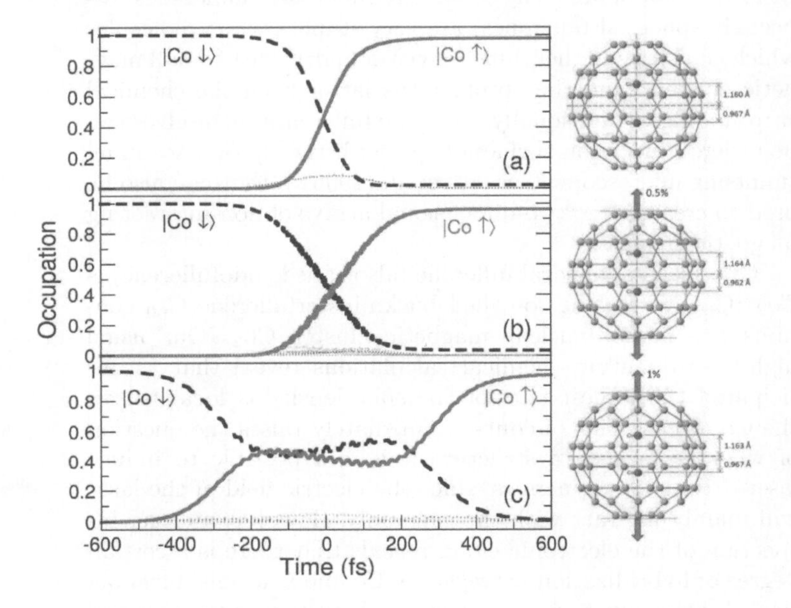

Figure 8.11
Local spin switching at one Co site of the endohedral fullerenes $Co_2@C_{60}$ via Λ processes with the tensile strain of (a) 0, (b) 0.5%, and (c) 1.0%, respectively. The initial state (dashed), final state (solid), and intermediate states (dotted) for each spin-switching process are indicated. On the right the strain states of the fullerenes are shown [Li et al. (2015)].

As one can immediately suspect, the different spin-density localizations also give rise to different spin-manipulation scenarios. In Fig. 8.11 we see three such spin-flip scenarios for three different tensile strengths. Note that each laser pulse is separately optimized, so we can be sure that the effect is not due to a simple resonance change. As it turns out, the smoothest process takes place with a slight strain (0.5%, middle panel), while a stronger one induces an almost chaotic (yet successful) behavior (lower panel).

The speed of the processes is also affected, the longest one being for the strongest strain. Comparing the three time-dependent occupations of the respective initial and final states, we detect some aftermath of the typical spin-charge dynamics separation in our systems: although the magnitude of the transition matrix elements becomes stronger (evident in the fast oscillations for 1% strain), the spin dynamics (i.e., the speed of the overall flipping of the spin) becomes slightly slower. Additionally, too many oscillations can also lead to a chaotic behavior and thus be detrimental to the applicability of the process. Of course, one cannot generalize for all cases, but in our experience this is always the tendency whenever designing laser-induced scenarios using triplet channels. We remind the reader once again that doublet states can be built from a single unpaired electron and thus usually exhibit similar spin and charge dynamics, contrary to triplet states which necessitate two unpaired electrons.

8.4 Magnetic logic on molecules

Our ultimate goal is to use the magnetic molecules as alternatives for the semiconductor based electronic circuits in conventional computers. The reasons for this are obvious. First of all, the spin of the electron consumes less energy than its charge when it is altered. Second, one can pack more densely molecular spins than semiconductor electron-doping charges. In a molecular magnet one can have up to three or four distinct spins in each molecule (at least theoretically), while in today's semiconductor technology the number is on the order of one charge every 10^4 atoms. And third, if one uses coherent lasers as the driving force, then the speed of the prospective magnetic-logic device can be in the subpicosecond regime.

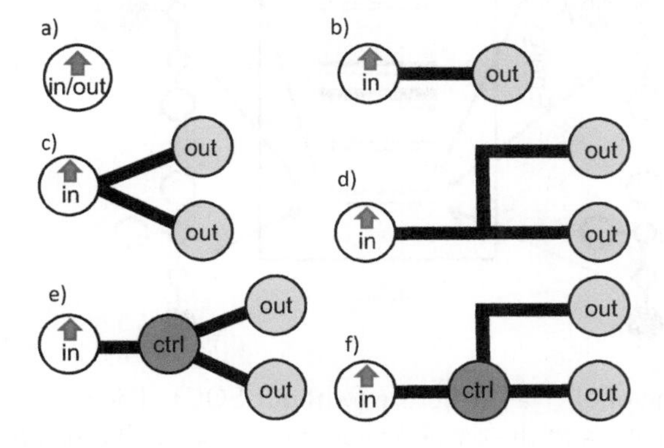

Figure 8.12
Several possible structures for magnetic logic, with one bit (a), two bits (b), three bits symmetric (c) and asymmetric (d), and four bits symmetric (e) and asymmetric (f). The unfilled circles with the word "in" indicate the input bit, and the grey ones with the word "out" the spatially separated output bits. With four or more bits one can imagine control bits as well (dark circles with word "ctrl") [Hübner et al. (2009)].

So we must derive concepts for laser-induced spin manipulation scenarios which can ultimately lead to the realization of such magnetic-logic building blocks. More specifically, we will shortly discuss examples for the three main functionalities needed in a computer: the logic operations (Boolean logic), the initialization of the device ("write" operation), and the transfer of information ("read" operation).[k]

[k] For full-fledged **magnetic logic**, one needs molecules with at least three active centers.

8.4.1 How many magnetic centers do we need?

In order to exploit the spin as information carrier, the structure must contain at least three magnetic atoms. Only one magnetic atom ("bit") just allows for information storage (switching the bit from 0 to 1 and vice versa), two atoms additionally allow for information transfer, and three atoms already allow for conditional branching, thus enabling logic capabilities. Of course, it is possible to conceive more complicated situations, e.g., an asymmetric three-magnetic-center cluster, where the two transfer paths are inequivalent, or even adding a fourth atom, which can act as a control or carrier bit (Fig. 8.12).

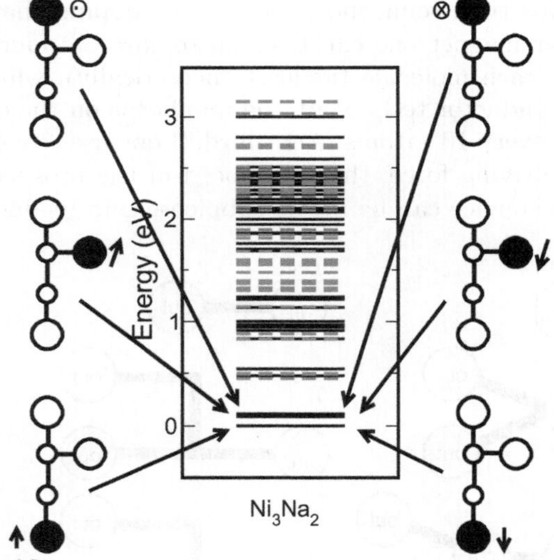

Figure 8.13
Level scheme of Ni_3Na_2 structure (without SOC). The solid lines are spin triplets, and the dashed ones spin singlets. The six structures next to the level scheme show the spin localization. Large circles represent Ni atoms, small circles magnetically inert Na atoms. The solid circles indicate the spin localization of each state (the arrows next to the sphere show its easy-axis direction). The upper two states have the spin perpendicular to the molecular plane (xy-plane) [Hübner et al. (2009)].

For our clusters we typically find that the lowest-lying many-body states originate from triplets or quintets, with their spin densities usually localized at a single magnetic center only. For energies below 1 eV we always find at least one "spin-up" and one "spin-down" state localized at any given magnetic center, plus several non-magnetic ones (see for example Fig. 8.13). Note that after inclusion of SOC, "spin-up" or "spin-down" merely means that, although S is not a good quantum number, the expectation value of its projection along the respective easy axis $\hat{q}$ is relatively high. Moreover, the names do not refer to the spin of individual electrons but to the expectation value of the spin-density operator acting on the whole many-body wavefunction.

8.4.2 The Ni_3Na_2 paradigm

Aiming at magnetic logic, we design our first candidate structure with three Ni atoms (our three terminals), which are interconnected through Na atoms (Fig. 8.14). The structure could be synthesized, e.g., by atomic deposition on a Cu (001) surface, which is also the reason for choosing an interatomic distance of 3.615 Å, which matches the lattice constant of the substrate. Generally, for the proof-of-principle, the Cu substrate can be considered chemically inert. It is not magnetic, and its electronic energy levels lie well below those of the cluster. Of course, it may slightly shift them, but it neither induces any redistribution of the spin density nor affects the optical selection rules [Pal et al. (2009, 2010)]. The selection of Na as bridging atoms is because they are metallic but not magnetic (and thus mediate the interactions between the Ni atoms without altering their magnetic state), and they are also cheap to compute (since they do not add any d electrons to the system). The geometry of the structure is such that the three Ni atoms differ slightly due to the local chemical environment. This is enough to trigger spin localization. Using different atomic species for the different sites is quite often not helpful, because then the spin dynamics becomes too local: spin-transfer is extremely difficult between different species.

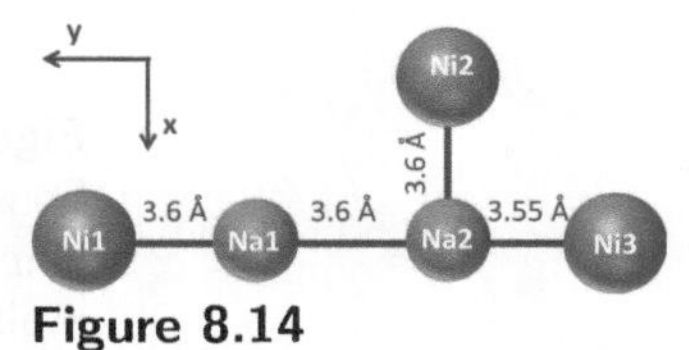

Figure 8.14
Geometrical structure of the three-magnetic-center Ni_3Na_2 cluster. The numbers show the interatomic distances in Å [Chaudhuri et al. (2014)].

Figure 8.13 shows the electronic level scheme of Ni_3Na_2, together with sketches indicating the localization of the spin densities for some of the lowest electronic states. The localization, in fact, comes perhaps as a mild surprise. Figure 8.15 shows some of the d-character molecular orbitals, where mostly the electronic density is localized on only one Ni center, and the different d-type orbitals of every atom (e.g., d_{xz}, d_{yz}, d_{z^2} etc.) come in groups. Since the static correlations lead to virtual electronic excitations between molecular orbitals with the same energy, one surely expects that in the many-body wavefunctions the spins remain well separated as well, which indeed is the case. However, some molecular orbitals, such as no. 44 and no. 46, have contributions from two or three Ni atoms, which, contrary to what one might expect, not only do not

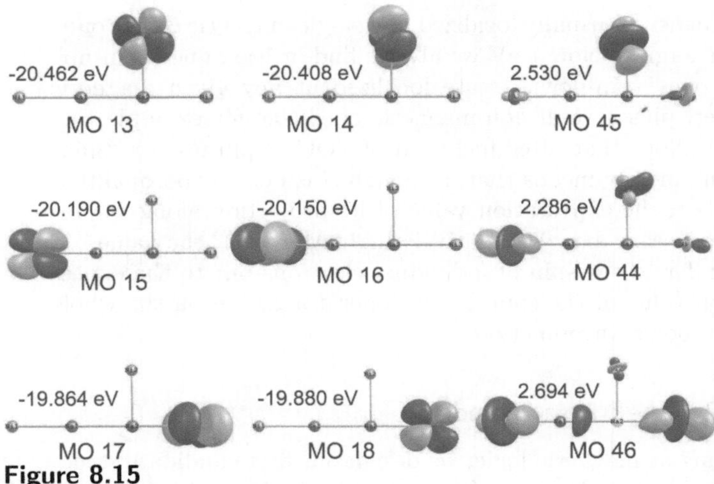

Figure 8.15
Some of the strongly localized, single-electron *d*-character molecular orbitals in the three-magnetic-center Ni_3Na_2 cluster. The multiplicity of the cluster for the Hartree-Fock step is triplet. The isosurfaces of the molecular orbitals are dark grey and light grey for positive and negative values, respectively [Chaudhuri et al. (2014)].

share spin density, but do not even have spins with parallel orientation!

The direction of the spin is indeed very interesting and has a dramatic effect on the spin transferability. While for two of the atoms the easy axis is more or less along the long molecular chain, for the third one (the top nickel in Fig. 8.13), the easy axis is perpendicular to the molecular plane. In fact this is not really a surprise. Since every atom has a different local geometry we do not expect them to get "oriented" in the same way. Indeed, we can inspect the *d*-type atomic orbitals involved in the many-body states (Fig. 8.15), and realize that every atom has a different local *z*-axis, which tends to be parallel to its bonds. So for MO 46 the (slightly distorted) d_{z^2} orbitals of nickels 1 and 3 are along the molecular axis, while for MO 45 for nickel 2 it is perpendicular to it. This finding is an indication of the importance of correlations, which may not only fine-tune but often even significantly alter the magnetic properties of such nanoclusters.

8.4.3 Elementary laser-induced processes

What are the most elementary laser-induced processes one can seek? Based on the already discussed Λ processes, one concentrates on the spin properties of the initial and the final states involved. These properties are obviously the localization and the orientation. If both the initial and the final states of a Λ process have the spin density localized on the same mag-

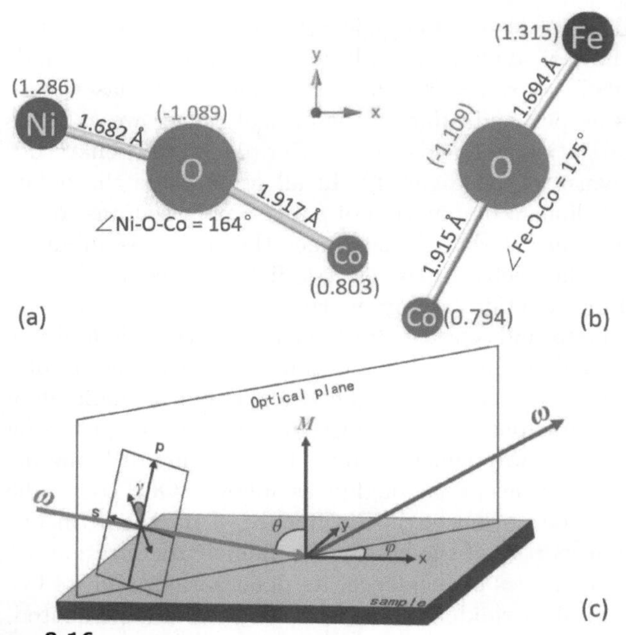

Figure 8.16
Optimized structures of the two quasilinear molecular ions (a) $[Ni\text{-}O\text{-}Co]^+$ and (b) $[Fe\text{-}O\text{-}Co]^+$. The bond lengths and angles are also shown. The numbers in the parentheses give their atomic charge densities [Li et al. (2014b)].

netic center and the spin directions point up and down, respectively, we are dealing with a local spin-flipping process. If the spin orientation is different but for both cases the spin density is spread throughout the whole molecule, we are studying a global spin switching process. If the spin orientation is the same but the spin density is localized on different atoms, then it is a laser-induced spin-transfer process. In every case the underlying mechanism is the laser-induced electronic transition from a suitably chosen initial to a suitably chosen final state. Of course, not every laser pulse can do the job.

At this point, let us digress a little and explain the meaning of the laser parameters, which are implemented in our genetic algorithm (see below) in the example of two oxygen-bridged systems (namely $[Ni\text{-}O\text{-}Co]^+$ and $[Fe\text{-}O\text{-}Co]^+$, see Fig. 8.16). In total there are six parameters: amplitude, duration (half-width-at-half-maximum, for sech^2-shaped time-dependence of the laser pulses), frequency ω, and the geometry of incidence given by the angles θ, ϕ, and γ. The meaning of those angles is visualized in Fig. 8.16(c). The optical plane is defined by two vectors: the propagation of the laser before hitting the target surface and after being reflected by it (two arrows). Here we use the typical physical convention in which θ is the polar angle and ϕ the azimuthal angle. The angle γ is needed for the definition of the polarization plane when the laser is linearly polarized. For $\gamma = 0$ the electric field of the laser pulse oscillates in the optical plane (we call this a p polarization; p stands for "parallel"). For $\gamma = 90°$ the electric field of the laser pulse oscillates as perpendicular to the optical plane (we call this an s polarization; s stands for the German word "senkrecht" which means "perpendicular"). In all other cases the polarization is a linear combination of p and s polarizations. If the laser pulse is circularly polarized, then the angle γ is unimportant, since the vector of the electric field rotates around the propagation direction (two grey arrows).

In order to find a successful laser pulse (which is a highly nonlinear, higher-dimensional and thus analytically not a solvable problem), we rely on a genetic algorithm. Genetic algorithms[l] are computational algorithms capable of optimizing procedures in a way which mimics the evolution of living organisms. The concept is roughly as follows: One codes the variables to be optimized in a sequence of numbers (in our case, the properties of the laser pulse, namely its duration, its intensity, its angles of incidence, its polarization, and its frequency). First, a random set of such sequences is generated. Every sequence is called an "individual" and each set is called a "generation". Each individual of the first randomly created generation is checked to see what it does on the system (in other words, the time-dependent Schrödinger equation for this particular pulse is solved). The state of the system after the laser pulse is projected on the *desired* final state, producing a number between 0 and 1. Here 0 means the desired state is not reached at all, while 1 means that the laser pulse leads the system (our magnetic molecule) completely to the desired final state. This number is called fidelity f. The most successful laser pulses (the ones with the highest fidelities) are combined with each other in the same way the DNA of living species is combined during reproduction: some bits are randomly mutated and some individuals exchange parts of their genetic code. In this way we create the next generation (children) through *natural selection*. Each individual of this new generation is once again checked for its fidelity, and the best ones chosen for the next generation. The procedure is repeated until in one generation there are individuals (in our cases, laser pulses) with a high fidelity. For our purposes a fidelity is considered high enough if it surpasses 85%, in other words, if the desired final state is at least 85% populated [Hartenstein et al. (2008)].

[l] **Genetic algorithms** are powerful computational algorithms capable of optimizing procedures in a way which mimics the evolution of living organisms. (For an exemplary implementation in Mathematica of such an algorithm on a few-level system which optimizes the frequency and the duration of a laser pulse, see Appendix A.10).

We try to find local spin-switching, spin-transfer, and combined spin-switch-and-transfer scenarios in our paradigm system Ni_3Na_2. It turns out that we can flip the spin on *all* three atoms, and we can transfer the spin (without flipping it) between the two nickels with parallel easy axes. Simultaneous transfer and flip are possible but with lower fidelities.

Interestingly, we cannot find laser-induced spin-transfer processes when the two local easy axes are perpendicular to

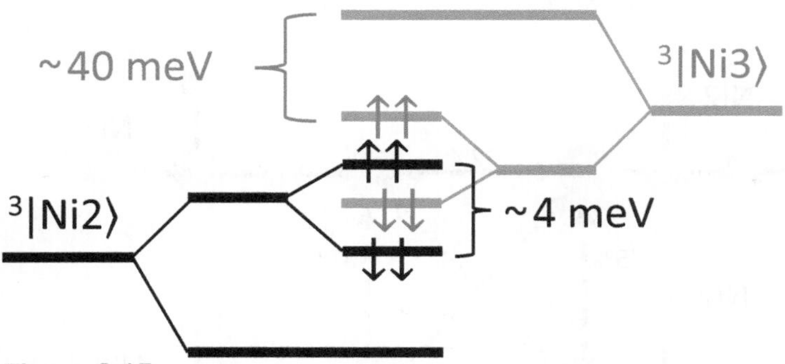

Figure 8.17
Sketch of interlocked triplet states in Ni_3Na_2. Two triplet states, localized on Ni_2 (black) and Ni_3 (grey), are split due to SOC and the Zeeman effect in such a way that the order of the substates is mixed.

each other, and we cannot flip and transfer the spin simultaneously (unless some very specific conditions are met, which will be discussed later). This observation, together with the fact that no state has the spin localized on two out of the three atoms, tells us that in Ni_3Na_2 it is not possible to design magnetic logic by using the spin as input bit alone.

Although our logic gate profits from the fact that (at first) we did not find a simultaneous spin-flip-and-transfer scenario, the fundamental question remains as to whether this is possible at all. As it turns out, it is if the two triplet states (with their spin density localized on the two different atoms) are energetically so close that the special level ordering, which we call interlocking, occurs: The lowest substate of the upper triplet state has lower energy than the highest substate of the lower triplet state (same as interlocking the fingers of both hands, see Fig. 8.17). In this particular case there is enough electronic overlap between *all four* substates which have $m_s \neq 0$, and, at the same time, little energetic difference, so that a Λ process is possible. So ultimately we get additional manipulation scenarios, however with only modest fidelities (Fig. 8.18). We can say that interlocking leads to a substantially increased coupling between the spin and the spatial degrees of freedom, thus allowing the spin to "go around the corner" as a special case of the Goodenough-Kanamori rules (see also below).

8.4.4 Spin transferability

Evidently, spin transferability also heavily depends on the spatial degrees of freedom (geometry). We already mentioned that generally we do not find spin-transfer scenarios, unless the easy axes on both magnetic atoms are parallel. These two have to influence each other somehow, since the exchange interaction is mediated through the molecular orbitals. For the readers with a more solid chemical background, we again refer to the

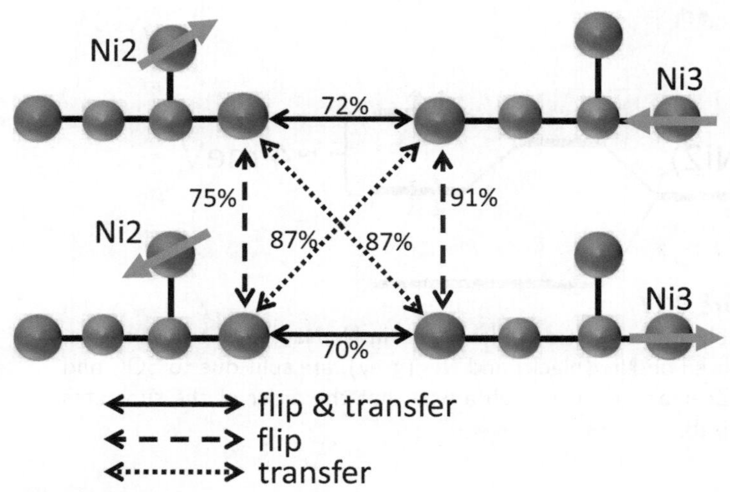

Figure 8.18
Schematic for various laser-induced processes obtained for the structure with singlets and triplets in the excited states. The directions of the arrows on the magnetic centers represent the direction of the spins. From the pictorial representation it is clear that simultaneous spin-flip and transfer can be achieved between Ni_2 and Ni_3 with modest fidelity, and local spin flips and transfers can be achieved with high fidelities [Chaudhuri et al. (2014)].

aforementioned Goodenough-Kanamori rules, which state that in molecular magnets the exact geometry ultimately decides whether two magnetic ions interact ferro- or antiferromagnetically [Goodenough (1958); Kanamori (1959); Kahn (1993)].

Of course, the first spatial degree of freedom that comes to mind are the molecular vibrations. Without going too much into technical details, we investigate the vibron- (phonon) spin interaction by computing the normal modes and then recompute all electronic and magnetic states for every mode. Our results indicate that the shrinking and stretching of the d1 bond [see Fig. 8.19 (a)] has a huge impact on the orientation of the easy axis on the spin, if it sits on Ni_3. This might be surprising at first; however, it becomes plausible if we once again take a look at the molecular orbitals 44 and 46 in Fig. 8.15. In fact, depending on the normal mode in question, we can distinguish between several **magnetic phases** [Fig. 8.19 (b)]. The characterization "phases" comes from the fact that the orientation of the spin on Ni_1 changes abruptly and not gradually with the interatomic distance d1, which is also an indicator of strong vibron-spin coupling [Xiang et al. (2012, 2013)]. In fact the energy change that accompanies this phase transition allows us to derive the coupling strength [for more details see Xiang et al. (2013)]. Looking for the transfer possibilities reveals that they are possible only if the spin on Ni_3 is

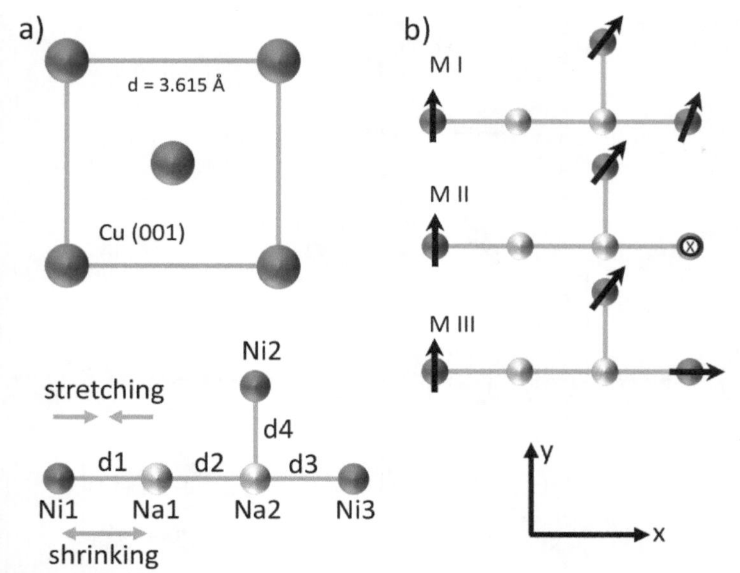

Figure 8.19
(a) The geometric structure of the Cu(001) surface and Ni_3Na_2 with the atoms and the interatomic distances. Stretching and shrinking are understood with respect to the original value of 3.6 Å (lattice constant of Cu). (b) Three of the five magnetic phases due to different interatomic distances, with respect to spin orientation (see text). The arrows indicate the spin direction when the spin density is localized on the respective atoms. C.f. Fig. 8.14. Adapted from [Xiang et al. (2013)].

parallel to the x- or the y-axis, in other words, within magnetic phases II and III (see Figs. 8.19(b) and 8.20). It is important to remind the reader that the spin density always sits on *one single* Ni atom. The sketch indicates the direction of the spin *whenever* it sits on that particular atom within the same magnetic phase.

Spin transfer from and to Ni_1 was not found. We attribute this to the larger distance to the other magnetic atoms, as well as to the fact that the interaction is mediated through more than one Na atom, thus justifying our design for this prototypical model. Already the different local geometrical details of the three Ni atoms suffice to produce rich cooperative phenomena.

8.4.5 Mapping quantum dynamics onto classical trajectories

All real molecules undergo molecular vibrations. For some magnetic molecules this fact is much more important than simply producing infrared spectra - it can actually induce spin dynamics if the spin-orbit and the electron-vibron couplings are strong enough. Our paradigm molecule Ni_3Na_2 is such a

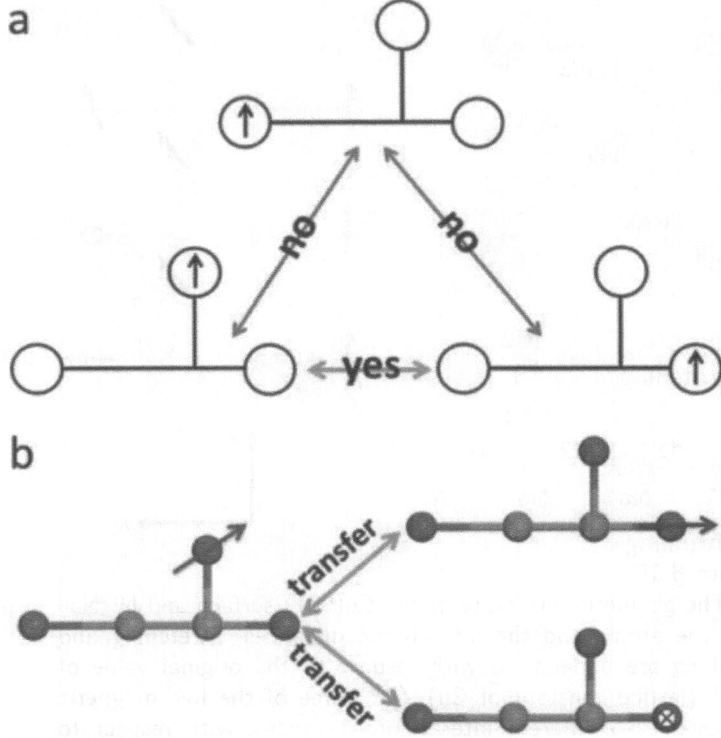

Figure 8.20
Possible spin transfer scenarios in Ni_3Na_2. (a) Sketch map, showing that only the spin transfer between Ni_2 and Ni_3 connected by one Na atom can be achieved. Spheres indicate the magnetic centers and arrows the localization and direction of the spin. (b) Orientation dependence of the spin transfer between Ni_2 and Ni_3 atoms. It shows that the spin can be transferred from Ni_3 to Ni_2 only if its direction, when localized on Ni_3, is parallel either to the x- or the y-axis (long and short molecular axis, respectively) [Xiang et al. (2013)]. Used with permission from Springer Nature.

system. Let us repeat the methodology introduced in Section 8.3.1, to calculate the vibron-electron coupling.

We perform normal mode calculations (that is we find the vibrational modes of the molecule) by calculating the force constant matrix. The force constants along the x- and y- axes (as defined in Fig. 8.19) are numerically computed from the second derivatives of the total electronic energy with respect to the atomic dislocations,

$$k_{ab} = \frac{\partial^2 E}{\partial a\, \partial b}, \text{ with } a, b = x_i \text{ and } y_i.\ i = 1, 2, \ldots 5 \qquad (8.23)$$

where x_i and y_i represent the dislocation of atom i. We manually displace the atoms and recalculate for each frozen geometry the many-body energy of all electronic states up to 1 eV. Then we formally quantize the normal modes, and we use the

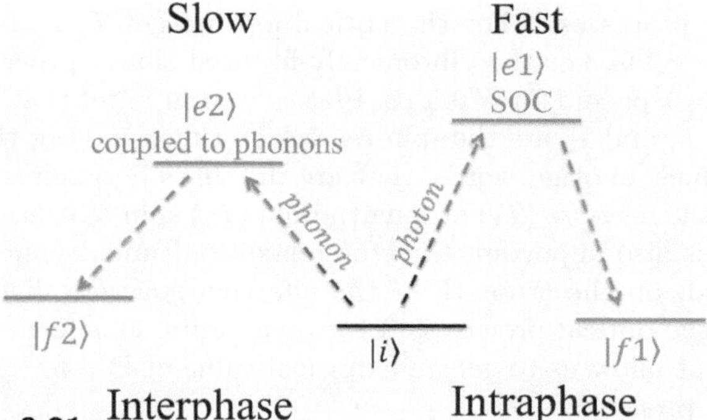

Figure 8.21
Unified illustration of the two microscopic mechanisms for the slow (ps), interphase, and the ultrafast (fs) intraphase spin-flip process. $|e_1\rangle$ stands for one or more spin-mixed electronic states (due to spin-orbit coupling, SOC) while $|e_2\rangle$ stands for one or more phonon-mixed electronic states. Both processes start from an initial state $|i\rangle$ and end in a final state with different spin orientation ($|f_1\rangle$ or $|f_2\rangle$) via photon (ultrafast) or phonon (slow) scattering [Xiang et al. (2012)].

formalism of Eqs. 8.18-8.22. After getting the coupling constants $\lambda^i_{\alpha\beta}$ for each magnetic (rather than electronic) state α and β and vibronic mode i, we can construct the Hamiltonian and diagonalize it. It turns out that not all vibronic modes mix with all magnetic states. In fact, each mode (characterized by its own infrared frequency) couples only to specific spin directions, and thus five rise to what we call magnetic phases [Fig. 8.19(b)]. There are altogether five magnetic phases. Surprisingly, whenever the energy range of the vibronic motion changes, the spin undergoes an abrupt reorientation. Conceptually this is a Λ process in which the spin couples to the phononic rather then the photonic degree of freedom, thus yielding a unified picture for the two paths (Fig. 8.21).

In order to better compare those two spin-flipping paths, we can project the quantum-mechanical trajectory of the spin onto a classical one, as derived from the Landau-Lifschitz-Gilbert equation of motion of the magnetization $\mathbf{M}$ [Kittel (1996)]

$$\frac{d\mathbf{M}}{dt} = \gamma(\mathbf{M} \times \mathbf{B}_{\text{eff}}) + \frac{1}{T}(\mathbf{M}_0 - \mathbf{M}), \tag{8.24}$$

where

$$T = \begin{pmatrix} T_2 & 0 & 0 \\ 0 & T_2 & 0 \\ 0 & 0 & T_1 \end{pmatrix}. \tag{8.25}$$

$\mathbf{M}_0$ is the equilibrium magnetization (here the final magnetic state), $\mathbf{B}_{\text{eff}}$ an effective magnetic field, and γ the gyromagnetic ratio. Although this is just an empirical equation, we can fit our results and thus extract the experimentally relevant longitudinal and transversal relaxation times T_1 and T_2 for both kinds of Λ processes.[m] For the optical one we get $T_1 = 100$ fs and $T_2 = 80$ fs. For the vibronically-induced slower process we get $T_1 = 1$ ps and $T_2 = 0.5$ ps. Please keep in mind that in both cases T_1 and T_2 are the spin *relaxation* times and not the spin-flip times [in other words, they are the times between two subsequent transverse (T_1) or longitudinal (T_2) spin scattering events]. It is also important that the relaxation time depends, as expected, on the strength of the effective magnetic field, which for the optical process is 0.1 atomic units. Our formalism does not allow us to get an empirical value of $\mathbf{B}_{\text{eff}}$ for the vibronic Λ process.

[m] By quantum mechanically calculating the time dependence of the spin density with the time dependent Schrödinger equation and then fitting it to a classical spin model, one can extract **phenomenological spin-relaxation times**.

8.4.6 Higher multiplicities

Up to now we have restricted ourselves to triplets, although Ni_3Na_2 can also have higher multiplicities. As already mentioned earlier, going to much higher multiplicities changes the observed behavior of most metallic clusters: At some point they start forming a macrospin which, in many cases, can even be described classically. However, it is still interesting to find out if by gradually doing so, one also opens additional manipulation channels.

First let us discuss the quintet states of the cluster. At a minimum level this opens up the possibility of demagnetization. If you remember the angular momentum bookkeeping of the Λ process, we said that since one photon is absorbed and one emitted, the initial and the final state must differ by exactly $\Delta S_z = \pm 2$.[n] For triplets this means going from $S_z = +1$ to $S_z = -1$ (or the other way around). This scenario can obviously be found for quintets as well. Ni_3Na_2 has an interesting characteristic: both triplets and quintets have an even number of unpaired electrons (two and four, respectively), a number incommensurable with three atoms. This means that practically always the spin density is divided unequally among the atoms, and hence most of the spin-flipping scenarios are local.

[n] But of course this selection rule can be overcome by tilting the direction of the laser pulse or by using elliptically polarized light.

Quintets exhibit a behavior which is somewhat different from that of the triplets, with regard to the easiness to derive different scenarios. Generally it is easier to find transitions when the ${}^5|\psi\rangle_0$ and ${}^5|\psi\rangle_{\pm 1}$ substates are involved than when the ${}^5|\psi\rangle_{\pm 2}$ must contribute. This admittedly strange phenomenon has an origin similar to the fact that strong correlations also exhibit a richer dynamics, and can be understood when looking into the Slater-determinant decomposition of the many-body wavefunctions.

Starting from a closed-shell configuration (singlet), one creates the quintet substates with a double virtual excitation with a detailed explanation of the notations below:

$$
\begin{aligned}
{}^5|\psi\rangle_2 &= |\psi^{ab}_{\bar{i}\bar{j}}\rangle, \\
{}^5|\psi\rangle_1 &= \frac{1}{2}\left(|\psi^{\bar{a}b}_{\bar{i}\bar{j}}\rangle + |\psi^{a\bar{b}}_{\bar{i}\bar{j}}\rangle - |\psi^{ab}_{i\bar{j}}\rangle - |\psi^{ab}_{\bar{i}j}\rangle\right), \\
{}^5|\psi\rangle_0 &= \frac{1}{\sqrt{6}}\left(|\psi^{\bar{a}\bar{b}}_{\bar{i}\bar{j}}\rangle + |\psi^{ab}_{ij}\rangle - |\psi^{a\bar{b}}_{\bar{i}j}\rangle - |\psi^{a\bar{b}}_{i\bar{j}}\rangle - |\psi^{\bar{a}b}_{\bar{i}j}\rangle - |\psi^{\bar{a}b}_{i\bar{j}}\rangle\right), \\
{}^5|\psi\rangle_{-1} &= \frac{1}{2}\left(|\psi^{a\bar{b}}_{ij}\rangle + |\psi^{\bar{a}b}_{ij}\rangle - |\psi^{\bar{a}\bar{b}}_{\bar{i}j}\rangle - |\psi^{\bar{a}\bar{b}}_{i\bar{j}}\rangle\right), \\
{}^5|\psi\rangle_{-2} &= |\psi^{\bar{a}\bar{b}}_{ij}\rangle.
\end{aligned}
\tag{8.26}
$$

The above equations describe virtual double excitations from the Hartree-Fock configuration where i and j are the occupied molecular orbitals, a, and b are unoccupied molecular orbitals. An orbital without an overbar is an α (spin-up) MO, while an overbar indicates a β (spin-down) MO. As an example, the Slater determinant $|\psi^{ab}_{\bar{i}\bar{j}}\rangle$ results from the Hartree-Fock Slater configuration with two virtual, one-electron excitations, namely two spin-flipping excitations (from β to α) from MOs i and j to MOs a and b. It is also noteworthy that since at least two electrons must be virtually excited from the closed-shell Hartree-Fock configuration in order to get four unpaired electrons, quintets are generally more correlated than triplets (which can be built with single excitations). What we see is that the lower the magnitude of S_z the more Slater determinants are needed to construct the substate, and thus more interaction with other states are possible. In a simplified physical picture we can say that whenever a superposition of more states is needed, the dynamics is richer (compare with Fig. 8.5 which refers to correlations).

8.4.7 More complicated M and nonlinear M processes

Like most optical (i.e., electrodynamic) processes, whenever one order of an effect is not present, the next one might be. When discussing the Λ process in the first place, we explicitly insisted that the intermediate excited state be optically addressable from both the initial and the final states. One question arises: What if such a state is not found? Are there still any chances of a successful scenario? The answer is that this may be the case if other channels which require more than two electronic transitions can be found.

One particular case is the M process, where we need altogether four transitions (Fig. 8.22). M processes are generally slower than Λ processes, and require a more strict condition than their Λ counterparts with respect to the magnitude of the transition matrix elements: The matrix elements of the first and the last transitions must be of comparable magnitude, and the same holds for the second and the third ones. In the opposite case the electronic population is not "deflected"

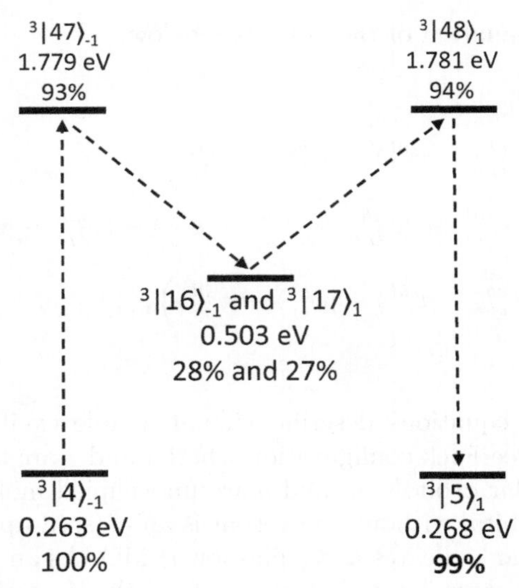

Figure 8.22
Schematic of the spin-flipping M process in Ni_2. Depicted are the involved states, their energies (in eV), as well as their maximal transient occupation during the whole process. The occupation of the final state $^3|5\rangle_1$ (99%) is printed in bold, because it is also the fidelity of the whole process. States $^3|16\rangle_{-1}$ and $^3|17\rangle_1$ are quasi-degenerate [Chaudhuri et al. (2014)].

to the next state but returns back, lowering the fidelity of the process.

In fact, things can become even more complicated. If our genetic algorithm cannot find any suitable M processes, sometimes it suggests a nonlinear M process in which six electronic transitions are needed (Fig. 8.23, where quintet rather than triplet states are involved). It is nonlinear, because the laser pulse frequency ω_l is not resonant (or at least almost resonant) with all electronic transitions. In fact, the second and the fifth transitions are at $2\omega_l$ resonance. This process also requires a similar symmetry as the M process with respect to the magnitude of the transition elements: they all come in pairs (the first must be equal to the last one, the second to the fifth, and the third to the fourth one). The processes are usually even slower than the M processes, and this is not only because of the additional transitions, but also due to the nonlinearity.

Although we call these M and Λ processes and discuss the number of transitions (or photons involved), it is always *one single laser pulse* that drives them, and that the transitions occur more or less simultaneously. The same universality between the three processes applies to the necessity of spin-orbit coupling, as a mechanism which ultimately funnels the electrons from the spin-down to the spin-up reservoir (or vice versa). This becomes immediately evident if one examines the constitution of the middle intermediate state in Figs. 8.22 and

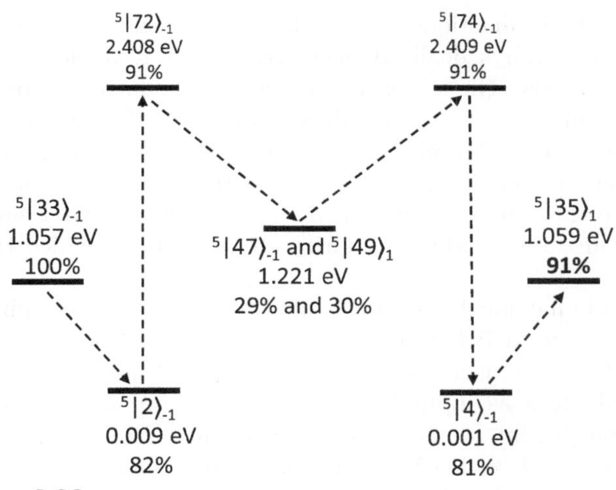

Figure 8.23
Schematic representation of a nonlinear M process and simplified pictorial representation of the population transfer process in Ni_3Na_2. Indicated are the involved quintet states, their energies and their maximum transient occupations during the whole process. The energy of the laser pulse is 1.163 eV. The occupation of the final state $^5|35\rangle_1$ (91%) is printed in bold because it is also the fidelity of the whole process [Chaudhuri et al. (2014)].

8.23. Both are superpositions of a spin-up and a spin-down state, while the other transitions do not usually alter the value of S_z (of course, there can be many exceptions to this rule). The reader should generally keep in mind that *all* calculated states enter our time-propagation computations; here we mention merely the ones which predominantly participate in the Λ and M processes.

8.5 First steps towards magnetic logic gates

Armed with the different elementary laser-induced processes we are now in a position to start looking for magnetic logic, which means for processes which reproduce the truth tables of various logic gates. The idea is to use the spin density of our magnetic atoms to code bits of information and suitable laser pulses as driving forces for the logic operations (in analogy to the current flowing through the transistors and chips of conventional computers).[o] Note that although our molecules are quantum objects, the resulting logic can still be classical (normal Boolean gates).

[o] **Spintronics** uses the electronic spin as information carrier (bits) to perform logic operations. If the system is on the nanoscale (molecules or nanoclusters), we speak of **nanospintronics**.

8.5.1 Boolean logic on Ni_3Na_2: NAND gate

Let us first continue discussing our paradigm molecule Ni_3Na_2, which historically also represents the first successful magnetic logic gate. As we mentioned before, the spin density of all low-lying states is always localized on one Ni atom (it never spreads across two atoms). Higher excited states can also have the spin density equidistributed among all Ni atoms, but these are obviously much less relevant for magnetic logic, since there is no distinction between possible input and output bits. Therefore, we turn to the external magnetic field as one of the two input bits, and investigate which scenarios are possible for which **B** field.

We find that local spin flip is always possible, while spin transfer requires a **B** field parallel to the main molecular axis (Fig. 8.14). So by using *both* the magnetic field *and* the spin on one Ni atom as input bits, and the spin on another Ni atom as output bit, we are able to create three Boolean logic gates: OR, AND, and NAND, depending on which magnetic state is interpreted as bit zero and which as bit one [Hübner et al. (2009, 2010)]. Table 8.1 gives the truth table for the magnetic NAND gate, which is the most interesting of the three, because with this one it is possible to build any other Boolean-logic gate; therefore it is called a universal gate.

This is definitely a first clear-cut success: a molecule can act as a logic circuit in the subpicosecond regime (the time is given by the duration of the suitably chosen laser pulses). It should be noted, however, that the speed bottleneck of a real device based on this type of logic functionality does not come from the laser-induced process, but from the speed with which one can turn the external magnetic field (which is a macroscopic object). This is a not an unsolvable problem. As it turns out, this deficiency can be cured by adding a fourth magnetic center. This can lead to *pure-spin* logic gates (i.e., gates in which *all* bits are solely coded by the magnetic state of the respective active magnetic centers and not the external magnetic field, which simply serves as a spectator). So the operational speed depends only on the ultrafast optical Λ or M processes [Chaudhuri et al. (2017)].

Table 8.1
Universal NAND gate. The spin at Ni_3 and the orientation of the **B** field are input bits, and the Ni_2 "down" state is the output bit. We exploit the fact that transfer is only possible if the magnetic field is parallel to the long molecular axis (x-direction as defined in Fig. 8.19) [Hübner et al. (2010)].

input 1 (spin)	input 2 (**B**-field)	output (spin+position)
1 ($Ni_3\uparrow$)	1 (parallel)	0 ($Ni_2\uparrow$)
0 ($Ni_3\downarrow$)	1 (parallel)	1 ($Ni_2\downarrow$)
1 ($Ni_3\uparrow$)	0 (tilted)	1 ($Ni_3\uparrow$)
0 ($Ni_3\downarrow$)	0 (tilted)	1 ($Ni_3\downarrow$)

8.5.2 ERASE functionality

The implementation of spin-based logic in a real device necessitates one more ingredient, not discussed until now: Preparing the system for the computation ("write" operation). This means that the system must be put to a specific electronic and/or magnetic state, regardless of the state it is in. With respect to the information content of a possible magnetic computer, overwriting a state translates into loss of information and poses a challenge, since a laser pulse induces a unitary transformation of the electronic wavefunction and therefore preserves information. In this section, we are specifically looking into (quasi-)irreversible processes which constitute the so-called ERASE functionality.

We know that a monochromatic laser pulse which drives a system from state $|A\rangle$ to state $|B\rangle$ also drives it backwards from $|B\rangle$ to $|A\rangle$. Although this statement is true for many pulses and can also be found in many textbooks, it is not strictly mathematically true. In fact, it holds only for symmetric pulses (i.e., with a laser field $\mathbf{E}(t) = \mathbf{E}(-t)$, for pulses centered around $t_0 = 0$), and when neglecting higher order terms (magnetic dipole transitions, electric quadrupole transitions, etc.). This is a consequence of the invariance of the time-dependent Schrödinger equation

$$i\hbar\frac{\partial\psi(t)}{\partial t} = \hat{H}(t)\psi(t) \tag{8.27}$$

under the substitutions $t \mapsto -t$ and $\psi(t) \mapsto \psi^*(-t)$. Given the relatively high sensitivity of most Λ processes to the exact parameters of the laser pulse, one obvious way to practically induce irreversibility in the process is to use non-symmetric laser pulses, $\mathbf{E}(t) \neq \mathbf{E}(-t)$.

One can argue that the success of an optical process is not necessarily connected to its simplicity. Generally, at least from a mathematical point of view, one can often find laser pulses which drive the system from an initial electronic or magnetic state to a desired one, and with a suitable combination of such elementary processes, one can design magnetic logic elements. Intuitively one might think that if these processes have a high tolerance degree (with respect, e.g., to laser detuning, angle of incidence, or duration), the process is always technologically advantageous. This, however, is not necessarily the case for optical processes. The reason is that all coherent processes are typically reversible;[3] in other words, the same laser pulse (if it is time-symmetric) drives the system also in the opposite direction, leaving open the problem of how to prepare the system in a desired state. In the following we will introduce two

[3] Optical processes are **coherent processes** and therefore reversible. In order to introduce irreversibility to a process, one must find a way to **break the time-reversal symmetry** of the whole setting (molecule plus light), for example using asymmetric laser pulses or taking advantage of quantum interference.

candidates that can tackle this problem, namely laser chirp and quantum interference.

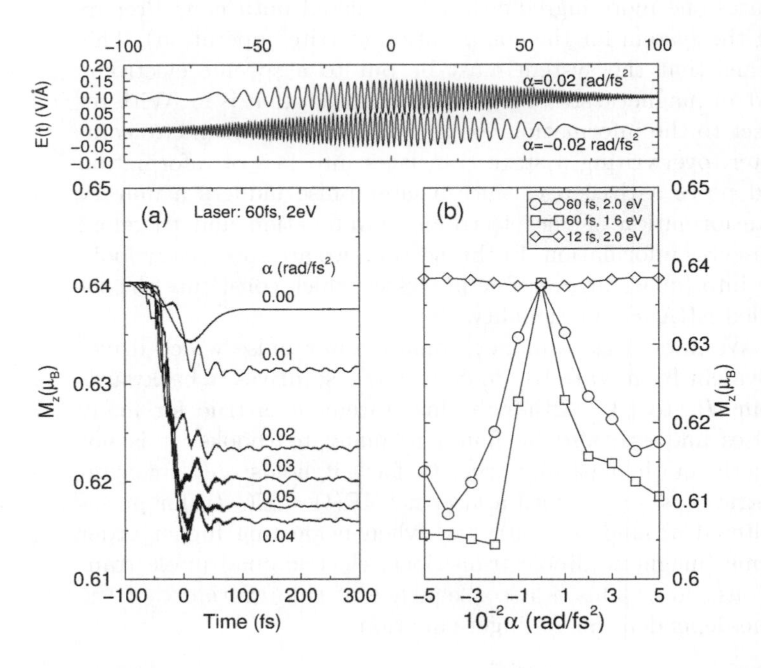

Figure 8.24
(a) Effect of laser chirp on spin moment change in ferromagnetic Ni. Laser pulse duration is 60 fs, and its photon energy $\hbar\omega = 2.0$ eV. (b) Final spin moment as a function of chirp for three different sets of pulse duration and photon energy: 60 fs/2.0 eV (empty circles), 60 fs/1.6 eV (empty squares), and 12 fs/2.0 eV (empty diamonds). Top: Laser pulse field with two different chirps $\alpha = 0.02$ rad/fs^2 (top) and $\alpha = -0.02$ rad/fs^2 (bottom). The field for the positive chirp is shifted by 0.1 for clarity [Zhang et al. (2012)]. Used with permission from the American Institute of Physics.

8.5.2.1 Laser chirp

The simplest method is to use a chirped pulse with a time-dependent frequency $\omega(t) = \omega_0 + \alpha t$. This was discussed for the model Λ system (see Eqs. 8.8 and 8.9). The effect is observable both in extended and in molecular systems. Figure 8.24 illustrates what happens in bulk Ni. In panel (a) we see that with no chirp ($\alpha = 0$) only a minor transient demagnetization of the sample takes place. As α increases, the sample demagnetizes more strongly. Panel (b) summarizes the final spin moment for different pulse durations and chirps.

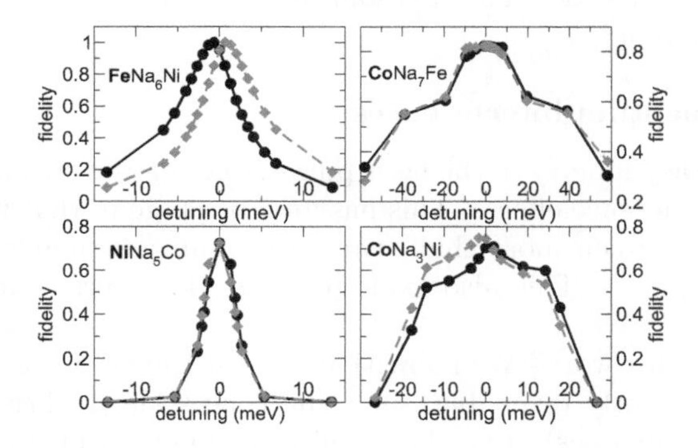

Figure 8.25
Fidelity vs. detuning (in meV) for spin-flipping Λ processes in four different magnetic molecules: (a) $FeNa_6Ni$; (b) $CoNa_6Fe$; (c) $NiNa_5Co$; and (d) $CoNa_3Ni$. The boldfaced elements are the atoms on which the spin density is localized and changes direction after the laser pulse. The solid lines correspond to flipping from $|\uparrow\uparrow\rangle$ to $|\downarrow\downarrow\rangle$, and the dashed lines from $|\downarrow\downarrow\rangle$ to $|\uparrow\uparrow\rangle$ [Zhang et al. (2012)].

Following the same train of thought, we investigate the effect of the laser chirp on spin-flipping Λ processes in real molecular systems. As first candidates we choose magnetic atoms attached to the ends of Na chains. Figure 8.25 shows the fidelity of the processes as a function of the relative detuning of the laser pulse.[p] The solid and dashed lines correspond to flips from spin-up to spin-down, and from spin-down to spin-up, respectively. We see that the three metals (Fe, Co and Ni) behave quite differently. Fe is the best case. Both spin-flip directions have nice distributions which have two distinct peaks. In other words, there are regions in the relative detuning in which the one flip direction is twice as effective as the opposite one, hence reaching a good selectivity.

[p]As relative detuning we define the ratio $\Delta_{\text{rel}} = \frac{\omega_0 - \omega_{\text{HWHM}}}{\omega_0}$, with ω_0 being the frequency at the peak of the laser pulse and ω_{HWHM} its frequency when it attains half of its maximum peak.

In the case of Ni, the peaks for the two directions are much narrower and also so close to each other that the selectivity is extremely high for both flip directions. The reason is that the level scheme of Ni is less dense than that of Fe (having two rather than four holes in the d shell) and thus stricter resonance conditions for the Λ processes. Co lies in-between. In fact, Co also exhibits an additional characteristic: the fidelity of the Λ process does not immediately drop with increasing chirp as with the other two magnetic atoms, but first attains a stable plateau somewhere between 60 and 70%. We believe that this is also due to its level scheme: it has enough many-body electronic levels so that the Λ process has more than one channel, leading to an increased stability (that is resonance tolerance) of the process. We suggest at this point that the reader also reviews the analytical solution Eq. 8.9 for a model three-level system.

8.5.2.2 Quantum interferences

Besides time asymmetry of the laser pulses, one can use strong quantum interference effects. This has an advantage in that it can operate between more than two states and uses simpler symmetrical pulses. This also leads to a less-than-perfect fidelity.

How does this work? We know that optical transitions are generally coherently reversible, since the time scale (at least for all of our purposes) of the laser pulses is orders of magnitude shorter than the typical molecular vibration time scales. However, this time symmetry is present only if one drives the system back from *exactly* the same state to which the laser pulse brought it. If our process is given by $|A\rangle \longrightarrow |B\rangle$, then the same pulse also induces $|B\rangle \longrightarrow |A\rangle$. More often than not, the final state of the forward process is not $|B\rangle$ but rather $\frac{1}{\sqrt{1+\epsilon^2}}(|B\rangle + \epsilon|C\rangle)$, where ϵ hopefully is a negligibly small number. Here $\frac{1}{\sqrt{1+\epsilon^2}}$ is just the normalization factor. The fidelity of the process (which is defined as the projection of the achieved state on the desired state) is

$$\begin{aligned} f =& \langle B|\frac{1}{\sqrt{1+\epsilon^2}}(|B\rangle + \epsilon|C\rangle) \\ =& \frac{1}{\sqrt{1+\epsilon^2}}\langle B|B\rangle + \frac{\epsilon}{\sqrt{1+\epsilon^2}}\langle B|C\rangle = \frac{1}{\sqrt{1+\epsilon^2}}. \end{aligned} \tag{8.28}$$

Now the opposite process is $\frac{1}{\sqrt{1+\epsilon^2}}(|B\rangle + \epsilon|C\rangle) \longrightarrow |A\rangle$, which, of course has the same fidelity. If, however, one starts from the pure state $|B\rangle$ instead, it can be that the small difference in the initial condition leads to a completely different result. Sometimes the effect of the absence of the small contamination $\epsilon|C\rangle$ can even go so far as to suppress the whole process. In this case we speak of quantum-interference irreversibility. Strictly mathematically speaking, this is not irreversibility, but practically has the same effect. In fact, there is an additional detail which can make this effect even more pronounced, namely the phase difference between states $|B\rangle$ and $|C\rangle$. If the outcome of an optical process is not a pure state but a superposition of states (a mixed state), then it is not an eigenstate of the Hamiltonian, and thus not a steady state. In fact, if we express the total wavefunction not in the interaction picture but in the Schrödinger picture, the wavefunction transforms into

$$|\Psi\rangle = \frac{1}{\sqrt{1+\epsilon^2}}\left(e^{-iE_Bt/\hbar}|B\rangle + \epsilon e^{-iE_Ct/\hbar}|C\rangle\right), \tag{8.29}$$

where E_B and E_C are the eigenenergies of the eigenstates $|B\rangle$ and $|C\rangle$, respectively. This means that the phase factors in the linear combination rotate with different speeds, and thus even the mere waiting can lead to different off-diagonal elements in the density matrix, which, in turn, results in different initial conditions for the laser pulse. If our process is sensitive enough to the presence (or absence) of the quantum "dirt" $\epsilon|C\rangle$, then it is also sensitive to the exact time-dependent phase difference to the desired state $|B\rangle$, and consequently the process exhibits an even higher degree of irreversibility. In practice we test this by explicitly propagating the system under the influence of the same laser pulse as the forward process, but starting from the final rather than the initial state.

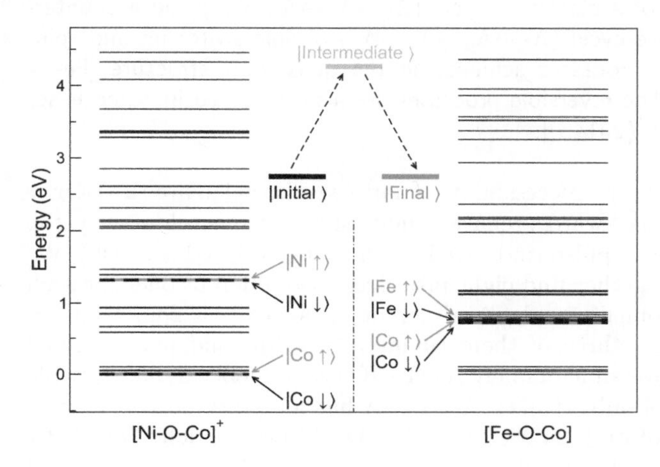

Figure 8.26
Lowest energy levels of the nonlinear structures $[\text{Ni-O-Co}]^+$ and $[\text{Fe-O-Co}]^+$ including SOC. In the left panel the initial (dashed line) and final (solid line) states for the spin-switching processes on the Ni and the Co ends of $[\text{Ni-O-Co}]^+$ are indicated, and in the right panel the initial and final states for the spin-switching or spin-transfer processes on $[\text{Fe-O-Co}]^+$ are shown. Inset shows a sketch of the Λ process [Li et al. (2014b)].

Now let us see how to implement this idea on a real material. In order to profit from both local spin flip and spin transfer, we want a cluster with two magnetic centers, which differ enough in order to induce spin-density localization. Two possible candidates are $[\text{NiOCo}]^+$ and $[\text{FeOCo}]^+$ (we have an even number of electrons, so we only have singlets and triplets rather than doublets and quartets). Figure 8.26 shows the calculated energy levels, as well as four states with a distinct spin-density localization and direction. In $[\text{FeOCo}]^+$ these four states are also depicted schematically in Fig. 8.27, together with all the possible processes we were able to find.

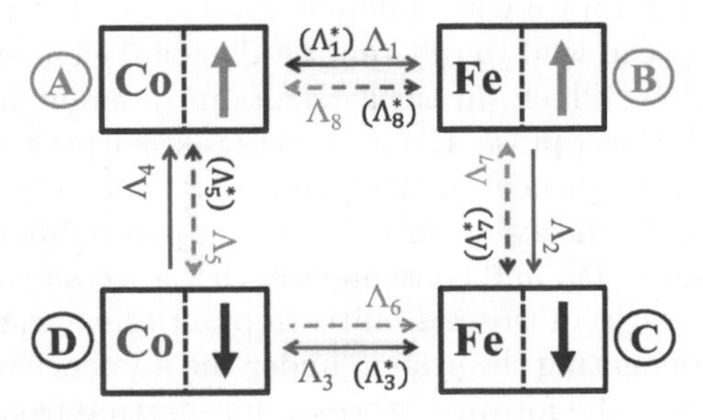

Figure 8.27
Sketch of a clockwise cycle ($\Lambda_1 \to \Lambda_2 \to \Lambda_3 \to \Lambda_4$) and a counter-clockwise cycle ($\Lambda_5 \to \Lambda_6 \to \Lambda_7 \to \Lambda_8$) of spin-switching and spin-transfer processes achieved in the quasilinear structure $[Fe\text{-}O\text{-}Co]^+$. The reversible processes are also indicated in parentheses [Li et al. (2014b)].

For every successful Λ scenario we also calculate the fidelity of the backward scenario under the influence of exactly the same laser pulse (indicated by the superscripted asterisk). We can altogether find eight processes: two different ones for each local spin flip and two different ones for each spin transfer. However, three of them (namely two flips and one transfer) are irreversible, namely Λ_2 (B$\to$C has a fidelity of 96.5% while C$\to$B of only 16.4%), Λ_4 (D$\to$A has a fidelity of 97.5% while A$\to$D of only 24.9%), and Λ_6 (D$\to$C has a fidelity of 87.2% while C$\to$D of only 7.8%), and Λ_2 (B$\to$C has a fidelity of 96.5% while C$\to$B of only 16.4%).

So we have a system consisting of four states, and we can use certain laser pulses to unidirectionally drive some states into other ones. In fact, a suitable sequence of four laser pulses can drive the system to state C *regardless* of which state we start from. This sequence is $\Lambda_2 \to \Lambda_6 \to \Lambda_5 \to \Lambda_6$, and the results after every pulse for all starting states are summarized in Table 8.2. This sequence is clearly an ERASE process, since it erases the information encoded in the system and initializes it to a known state (analogous to preparing the input of a logical operation in a conventional computer).

A nice mathematical way to depict the information loss of the process is the Shannon entropy S. It is defined as

$$S = -\sum_i p_i \log_2 p_i \tag{8.30}$$

and is a measure of the uncertainty in the system. Here p_i is the probability of the system being in state $|i\rangle$. If the system is *definitely* in a specific state, then this state has a probability of 1 and $S = -1 \log_2 1 = 0$ (in other words there is no uncertainty in the system). The maximum uncertainty for a total of n states arises when they all have the same probability $p_i = \frac{1}{n}$, and so $S = -n\left(\frac{1}{n}\log_2\frac{1}{n}\right) = \log_2 n$. If we start the ERASE function without any prior knowledge, our

Table 8.2
ERASE functionality on the quasilinear structure $[Fe\text{-}O\text{-}Co]^+$. The first and last columns represent the initial states and the final states, and the rest of the columns show the subsequent evolutions of the states under the influences of the corresponding Λ process shown in the first row. The items in parentheses show the final states if *all* processes were reversible. The last row shows how the Shannon entropy decreases after every irreversible pulse if we start from an unknown state. [Li et al. (2014b)]

	Λ_2		Λ_6		Λ_5		Λ_6	
A	$\rightarrow$	A	$\rightarrow$	A	$\rightarrow$	D	$\rightarrow$	C (C)
B	$\rightarrow$	C	$\rightarrow$	C	$\rightarrow$	C	$\rightarrow$	C (A)
C	$\rightarrow$	C	$\rightarrow$	C	$\rightarrow$	C	$\rightarrow$	C (B)
D	$\rightarrow$	D	$\rightarrow$	C	$\rightarrow$	C	$\rightarrow$	C (D)
$S = 2$		$S \approx 1.56$		$S \approx 1.07$		$S \approx 0.81$		$S \approx 0.13$

system can be in any of the states A,B, C, or D, each with a probability of 25%. As we gradually eliminate uncertainty with each pulse, the Shannon entropy decreases. An exception is the third pulse (Λ_5) which is only minimally irreversible, and therefore reduces the entropy considerably less than the other three pulses. In the end, we do not reach exactly zero, because the last pulse does not drive our system with 100% fidelity to state C. However, this could not be the case, because it would mean that the final state would not contain any quantum of "dirt," and thus the process could not be in this sense irreversible in the first place!

8.5.3 Charge-spin gearbox

One last thing to address before closing this chapter is the "read" functionality, in other words some means of transferring the information (encoded in the spin-bits) to a physically different place without altering it. One very simple and yet ingenious way to do that was proposed by Mahan not too long ago [Mahan (2009)]. His idea, which he called a spin-shift register, consists of a half-filled linear chain of spins (a spin wire with exactly one electron per site). Each site of this chain is occupied either by a spin-up ($|\uparrow\rangle$) or a spin-down ($|\downarrow\rangle$) electron. As it turns out, all bits can be shifted by one position if an additional electron runs through the wire.

Let us look at a concrete example in the five-member chain $|\uparrow\uparrow\downarrow\downarrow\uparrow\rangle$. Now an extra electron runs through the wire from left to right. When this electron first starts its journey from the left-most site, which is occupied by a spin-up electron, it can only assume the spin-down position $|\Updownarrow\uparrow\downarrow\downarrow\uparrow\rangle$ (here the symbol $\Updownarrow$ indicates double occupation, by a spin-up *and* a spin-down electron). When the extra electron jumps to the second position, the spin state of the wire becomes $|\uparrow\Updownarrow\downarrow\downarrow\uparrow\rangle$. When it jumps

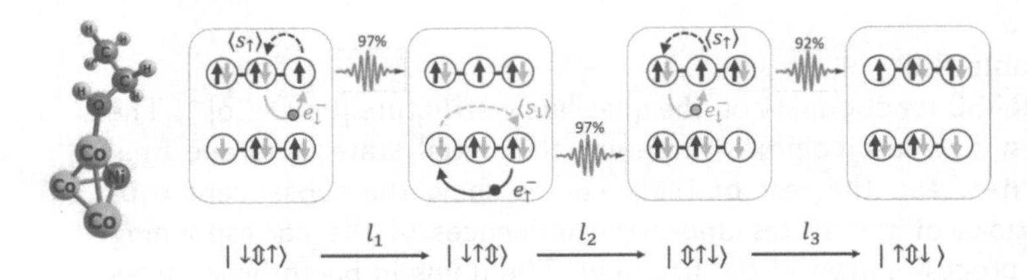

Figure 8.28
On the far left is the $Co_3Ni(EtOH)^+$ cluster. The following four rectangular boxes represent the spin-charge gearboxes. The operation starts with the first gearbox, where circles denote the Co atoms with spin-up (vertically up black arrow) and spin-down (vertically down grey arrow) densities. Under laser excitation, a spin-down ($e_\downarrow^-$) electron in one spin channel hops from the second Co atom to the third one (solid grey arrow), accompanied by the up-spin density transfer $\langle s_\uparrow \rangle$ (dashed arrow). The fidelity of charge and spin transfer is 97% (see the number above the laser pulse). The second gearbox is the result of the first operation. In addition, a spin-up ($e_\uparrow^-$) electron on the third Co atom in the other spin channel hops to the first Co atom (solid black arrow), accompanied by the spin-down density transfer $\langle s_\downarrow \rangle$ in the reverse direction (dashed arrow). The remaining two gearboxes are similar. Below each gearbox is its respective spin configuration. $\Updownarrow$ indicates a doubly occupied site ($\langle s \rangle = 0$). l_1, l_2, and l_3 represent three laser pulses used [Lefkidis et al. (2020)]. Used with permission from the American Chemical Society.

to the third position, an interesting thing happens: because the third position is occupied by a spin-down electron, due to the Pauli exclusion principle, it is the spin-up electron on the second site that must jump from the second to the third position. Thus the chain's spin state now becomes $|\uparrow\downarrow\Updownarrow\downarrow\uparrow\rangle$. The next jump leads to $|\uparrow\downarrow\downarrow\Updownarrow\uparrow\rangle$, and the one after that to $|\uparrow\downarrow\downarrow\uparrow\Updownarrow\rangle$. Since a spin-down electron entered the chain (remember originally the first site was occupied by a spin-up electron), spin conservation dictates that also a spin-down electron must leave the chain. So finally we get the spin-state $|\uparrow\downarrow\downarrow\uparrow\uparrow\rangle$. Comparing this final state with the initial one, we observe that all spins were shifted by one position to the left in a cyclic manner. In other words, five charge jumps to the right induce one spin jump (for every site) to the left, thus realizing a molecular "charge-spin gearbox"[q] with a translation ratio $5 : (-1)$. In other words, the charge must complete exactly five cycles in order for the spin to complete one cycle in the opposite direction.

[q] A **charge-spin gearbox** is a linear or cyclic chain, each site of which is occupied by a single electron: either spin-up or spin-down. When an additional electron runs through the complete chain in one direction, all spins get shifted by one position in the opposite direction in a cyclic manner.

When trying to adopt this abstract idea in a real magnetic molecule, one is confronted with two challenges. The first is that in molecules with a few active centers the spin density is either localized on one magnetic atom only, or equidistributed among many of them, so spin states like the ones we are interested in are practically never eigenstates of the molecule's

Hamiltonian. We can solve this problem by using suitable superpositions of states (both classical ensembles and quantum superpositions are possible). The second is that experimentally it is not easy to simply let an electron run through the molecule. So we devise spin states which we manipulate with our trusted laser pulses instead.

More specifically, we apply the idea of the spin-shift register to the synthesized and characterized molecule $[Co_3Ni(EtOH)]^+$, and use the three cobalt atoms as the active magnetic centers: Note that three is the minimal number of sites on which this idea is possible [Dutta et al. (2018); Lefkidis et al. (2020)]. As we can see in Fig. 8.28 we need two molecules, one containing one atom in the spin-up state, and the other containing a different atom in the spin-down state. These superpositions directly correspond to the different spin states necessary for the whole cycle. In this realization the extra electron is never removed from the system (we always have half-plus-one-filling), where the cyclic shifting of the spins is practically the same as flipping both spins (since the spins are shifted between two sites, while the third site is double occupied throughout). Repeating the cycle twice brings the spins back to their initial position (the spins of the two unpaired electrons are flipped once more), yielding a gearbox translation ratio of $2 : (-1)$.

8.6 Concluding remarks

In this chapter we reviewed some concepts which are pertinent to the ultrafast spin dynamics of molecular magnets, in particular laser-induced spin dynamics. We kept an eye on prospective nanospintronic devices and tried to discover elementary spin-manipulation processes which can be harvested to perform logic operations.

Everything presented here aims at elucidating the underlying physics and can help us (and others!) design molecules which can be used for future nanologic magnetic applications. Obviously there is no claim of completeness with respect to the innumerable ways to perform *ab initio* calculations on realistic systems, or to the vast pool of spin processes. We merely restricted ourselves to a small survey of physical concepts, mathematical methods, and models, decorated with some real results, in order to give to the readers a taste of this exciting field!

8.7 Exercises

1. **Spin operators in three dimensions**
The extension of the Pauli matrices in three dimensions are the spin operators $\hat{S}_x$, $\hat{S}_y$, and $\hat{S}_z$:

$$\hat{S}_x = \frac{1}{\sqrt{2}} \begin{pmatrix} 0 & 1 & 0 \\ 1 & 0 & 1 \\ 0 & 1 & 0 \end{pmatrix}, \quad \hat{S}_y = \frac{1}{\sqrt{2}} \begin{pmatrix} 0 & -i & 0 \\ i & 0 & -i \\ 0 & i & 0 \end{pmatrix},$$
$$\hat{S}_z = \begin{pmatrix} 1 & 0 & 0 \\ 0 & 0 & 0 \\ 0 & 0 & -1 \end{pmatrix}. \tag{8.31}$$

(a) Show that these matrices yield the correct commutation relations for an angular momentum (note that $\hbar$ is not included in their definitions here).
(b) Find their eigenvalues and eigenvectors.

2. **Spin operators in higher dimensions**
One way to build higher dimensional spin operators is to use the ladder operators $\hat{S}_+$ and $\hat{S}_-$ for which we have

$$\hat{S}_\pm |s, m_s\rangle = \sqrt{s(s+1) - m_s(m_s \pm 1)}\ |s, (m_s \pm 1)\rangle, \tag{8.32}$$

and then find $\hat{S}_x$ and $\hat{S}_y$ through the relations

$$\hat{S}_x = \frac{\hat{S}_+ + \hat{S}_-}{2} \quad \text{and} \quad \hat{S}_y = \frac{\hat{S}_+ - \hat{S}_-}{2}. \tag{8.33}$$

$\hat{S}_z$ is simply a diagonal matrix with entries $m_s, m_s - 1, \ldots, -m_s + 1, -m_s$. Give the explicit form of the spin matrices for $s = 2, \frac{3}{2}, 3, \frac{5}{2}$.

3. **Spin-orbit coupling**
Consider the Pauli Hamiltonian

$$\hat{H}_{\text{Pauli}} = \frac{1}{2m} \left[(\mathbf{p} - q\mathbf{A})^2 \right] - q\hbar\boldsymbol{\sigma} \cdot \mathbf{B}. \tag{8.34}$$

$\mathbf{p}$ is the momentum of the electron, q is its charge, $\mathbf{A}$ is the vector potential of the surrounding electromagnetic field in which the electron moves, $\mathbf{b}$ is the magnetic field, and $\boldsymbol{\sigma} = \{\sigma_x, \sigma_y, \sigma_z\}$ is a vector containing the Pauli matrices σ_x, σ_y, and σ_z as its elements.

(a) By expanding this in powers of $\frac{1}{c^2}$, explicitly derive the second-order correction terms

$$\hat{H}^{(2)} = -\frac{\mathbf{p}^4}{8m^3c^2} - \frac{q\hbar}{4m^2c^2}\boldsymbol{\sigma} \cdot [\mathbf{E} \times \mathbf{p}] - \frac{q\hbar^2}{8m^2c^2} \boldsymbol{\nabla} \cdot \mathbf{E}. \tag{8.35}$$

(b) Show that for a spherically symmetric electric field the second term becomes

$$\frac{\hbar^2}{2m^2c^2r}\frac{dV}{dr}\mathbf{l}\cdot\mathbf{s}. \tag{8.36}$$

Here $\mathbf{l}$ is the angular momentum operator, $\mathbf{s} = \frac{1}{2}\boldsymbol{\sigma}$ is the spin operator, and $V = q\Phi$ is the potential energy of the electron in the electric field.

4. **Many-electron wavefunctions**
The total wavefunction of a system with N electrons distributed along equally many one-electron wavefunctions needs to be antisymmetric with respect to electron exchange (since electrons are fermions), and can be described with the so-called Slater determinants (also called configurations)

$$\Psi = \frac{1}{\sqrt{N!}}\begin{vmatrix} \phi_1(\mathbf{r}_1)\chi_1(s_1) & \cdots & \phi_N(\mathbf{r}_1)\chi_N(s_1) \\ \vdots & \ddots & \vdots \\ \phi_1(\mathbf{r}_N)\chi_1(s_N) & \cdots & \phi_N(\mathbf{r}_N)\chi_N(s_N) \end{vmatrix}. \tag{8.37}$$

Here ϕ_i and χ_i are the spatial and spin parts of the i-th one-electron wavefunction (orbital), respectively, while $\mathbf{r}_i$ and s_i are the spatial and spin coordinates of the i-th electron, respectively. So if the second electron with spin down (β electron) of the He atom is excited into the fourth orbital, which is the $1p_z$ orbital, the term becomes $\phi_4(\mathbf{r}_2)\chi_4(\mathbf{r}_2) = p_z(\mathbf{r}_2)\beta(\mathbf{r}_2)$.

Note: Often in quantum chemistry one uses a shorter notation, namely $\phi_i(j)$ for the orbital which is up-spin (α orbital) and $\overline{\phi}_i(j)$ for an orbital which is down-spin (β orbital). Here j collectively gives the coordinates of the j-th electron. So, for instance, $p_z(\mathbf{r}_2)\beta(\mathbf{r}_2) \equiv \overline{p}_z(\mathbf{r}_2)$. Furthermore, we do not write the whole Slater determinant but only the diagonal elements, and thus we also do not need to write the electron index or the prefactor $\frac{1}{\sqrt{N!}}$. So, as an example, for two electrons distributed in two states $\phi_2(i)\alpha(i)$ and $\phi_1(i)\beta(i)$, we have

$$\begin{aligned}\Psi =& \frac{1}{\sqrt{2}}\begin{vmatrix} \phi_1(1)\alpha(1) & \phi_2(1)\beta(1) \\ \phi_1(2)\alpha(2) & \phi_2(2)\beta(2) \end{vmatrix} = \\ =& \frac{1}{\sqrt{2}}\begin{vmatrix} \phi_1(1) & \overline{\phi}_2(1) \\ \phi_1(2) & \overline{\phi}_2(2) \end{vmatrix} \equiv \left|\phi_1\overline{\phi}_2\right|. \end{aligned} \tag{8.38}$$

(a) Consider two spin orbitals α and β and two electrons (here we neglect the spatial part). Show that the only permissible wavefunctions are

$$\begin{aligned}\Psi_1 =& \alpha(1)\alpha(2), \Psi_2 = \beta(1)\beta(2)\\ \Psi_3 =& \frac{1}{\sqrt{2}}\Big(\alpha(1)\beta(2) + \beta(1)\alpha(2)\Big)\\ \Psi_4 =& \frac{1}{\sqrt{2}}\Big(\alpha(1)\beta(2) - \beta(1)\alpha(2)\Big). \end{aligned} \tag{8.39}$$

Which of these wavefunctions are symmetric and which is antisymmetric with respect to particle exchange? What does this mean for the spatial part of the wavefunction (not considered here)?

Note: The spin functions considered here are simple products, *not* Slater determinants.

(b) Consider the specific example of a two-electron system (e.g., the helium atom He) with two spatial orbitals a and b. Show that due to the required antisymmetry condition and the Pauli exclusion principle, the only permissible combinations are

$$\begin{aligned}\Psi_1 =& |a\bar{a}|, \Psi_2 = |b\bar{b}|, \Psi_3 = |ab|, \Psi_4 = |\bar{a}\bar{b}|\\ \Psi_5 =& \frac{1}{\sqrt{2}}\Big(|a\bar{b}| + |\bar{a}b|\Big), \Psi_6 = \frac{1}{\sqrt{2}}\Big(|a\bar{b}| - |\bar{a}b|\Big). \end{aligned} \tag{8.40}$$

Note: Here Slater determinants are considered.

5. **Splitting of p orbitals in Na**
Use the angular momentum matrices for $S = \frac{1}{2}$ and $L = 1$ to explain the splitting of the $3p$ orbital in the Na atom, which is manifested by the yellow emission lines of the sodium vapor lamp with frequencies 589.00 and 589.59 nm:

$$\begin{aligned}S_x =& \frac{\hbar}{2}\begin{pmatrix}0 & 1\\ 1 & 0\end{pmatrix}, \; S_y = \frac{\hbar}{2}\begin{pmatrix}0 & -i\\ i & 0\end{pmatrix},\\ S_z =& \frac{\hbar}{2}\begin{pmatrix}1 & 0\\ 0 & -1\end{pmatrix},\\ L_x =& \frac{\hbar}{\sqrt{2}}\begin{pmatrix}0 & 1 & 0\\ 1 & 0 & 1\\ 0 & 1 & 0\end{pmatrix}, \quad L_y = \frac{\hbar}{\sqrt{2}}\begin{pmatrix}0 & -i & 0\\ i & 0 & -i\\ 0 & 1 & 0\end{pmatrix},\\ L_z =& \hbar\begin{pmatrix}1 & 0 & 0\\ 0 & 0 & 0\\ 0 & 0 & -1\end{pmatrix}. \end{aligned} \tag{8.41}$$

Note: Calculate the spin-orbit coupling $\xi\hat{\mathbf{L}}\cdot\hat{\mathbf{S}}$ for the six levels by constructing a six-dimensional Hilbert space $\mathcal{H_L} \otimes \mathcal{H_S}$ ($3 \times 2 = 6$, three dimensions due to the orbital angular momentum and two due to the spin) and calculate the interactions between all basis vectors. The matrices in six dimensions are the Kronecker products of I_3 with S_x, S_y, and S_z, and of L_x, L_y, and L_z with I_2 (I_2 and I_3 are the two- and three-dimensional identity matrices). Since the atom is isotropic, the factor $\xi\hbar \approx 1.4$ meV is direction-independent.

6. **Normal and anomalous Zeeman effect in Na**
   Recalculate Exercise 5, both with and without spin-orbit coupling but this time also considering a weak external magnetic field **B** (normal and anomalous Zeeman effect). Take into account the relevant Landé factors for each case (this includes the spin, the orbital and the total angular momenta).

7. **Paschen-Back effect in Na**
   In Exercise 6 it was assumed that the splitting does not lead to new electronic state mixing. However, if the external field is strong enough, it can lead to saturation of one spin direction and, accordingly, to spin-orbit decoupling (Paschen-Back effect). How can we see this through the basis of the calculated matrices?

8. **Carbon monoxide molecule**
   Calculate the electronic structure of CO in the Hartree-Fock approximation. Then calculate the corrections of the ground state energy with the SAC-CI method. Calculate the energies of the lowest excited states. *Note:* The SAC-CI calculations can, among others, be done with the free-of-charge quantum chemistry package GAMESS.

9. **Vibrational modes of the carbon monoxide molecule**
   For the system of Exercise 8 calculate at the SAC-CI level the normal modes of vibration and then quantize them. To numerically calculate the constant of the normal modes, you need to numerically differentiate the potential energy curve (i.e., the total energy of the molecule as a function of the interatomic distance). In order to do so, you need to repeat the SAC-CI calculations for different interatomic distances around the equilibrium distance (in the harmonic regime, that is approximated by a parabola). Formal quantization of this harmonic potential will give you the phononic energies. Then also look at the energy of the electronic excited states. The difference between these energies and the sum of the electronic state energy at the equilibrium distance plus the phononic energy will give you the electron-phonon coupling.

10. **Boolean logic operations**
Consider the truth tables of the OR, AND, and NAND logic gates.

AND			OR			NAND		
In	In	Out	In	In	Out	In	In	Out
0	0	0	0	0	0	0	0	1
0	1	0	0	1	1	0	1	1
1	0	0	1	0	1	1	0	1
1	1	1	1	1	1	1	1	0

(a) Show that both the OR and the AND gates can be constructed by combinations of more NAND gates. *Note:* For this reason the NAND gate is called a universal gate.
(b) Show that by proper combinations of the OR and the NAND gates one can also construct *all* 16 possible logic gates (the number 16 stems from the all 2^4 possibilities for the output bits).

A Appendices

A.1 KLI approximation
A.2 LDA: local density approximation
A.3 Self-interaction corrected LDA
A.4 BLYP approximation
A.5 Electric dipole, magnetic-dipole and other higher-order interactions
A.6 Code to generate ultrafast pulses
A.7 Code to generate figures in HHG
A.8 Code to compute the cutoff energy in HHG
A.9 Code to compute the C_{60} structure
A.10 Genetic algorithm example
A.11 Special crystal momentum points and lines

A.1 KLI approximation

The Krieger-Li-Iafrate (KLI) approximation is an exchange-only approximation for the exchange-correlation potential, given by [Krieger et al. (1992b)],

$$v_\sigma^{\rm xc}[n_\uparrow, n_\downarrow](\mathbf{r}) \simeq V_\sigma^{\rm KLI}(\mathbf{r}) = V_\sigma^{\rm S}(\mathbf{r}) + \frac{1}{n_\sigma(\mathbf{r})}\sum_{i=1}^{N_\sigma} n_{i\sigma}(\mathbf{r}) C_{i\sigma}, \tag{A.1}$$

where $V_\sigma^{\rm S}$ is the Slater potential [Slater (1951)],

$$V_\sigma^{\rm S}(\mathbf{r}) = \frac{1}{n_\sigma(\mathbf{r})}\sum_{i=1}^{N_\sigma} n_{i\sigma}(\mathbf{r}) v_{i\sigma}^{\rm HF}(\mathbf{r}), \tag{A.2}$$

and $v_{i\sigma}^{\rm HF}$ is the Hartree-Fock exchange potential [Krieger et al. (1992a)]:

$$v_{i\sigma}^{\rm HF}(\mathbf{r}) = \frac{-1}{\psi_{i\sigma}^*(\mathbf{r})}\sum_{j=1}^{N_\sigma}\int d^3\mathbf{r}' \frac{\psi_{i\sigma}(\mathbf{r}')\psi_{j\sigma}^*(\mathbf{r}')\psi_{j\sigma}(\mathbf{r})}{|\mathbf{r}-\mathbf{r}'|}. \tag{A.3}$$

The orbital-dependent constant in Eq. A.1 is defined as

$$C_{i\sigma} \equiv \langle V_{i\sigma}^{\rm KLI}\rangle - \langle v_{i\sigma}^{\rm HF}\rangle, \tag{A.4}$$

which can be calculated non-iteratively as follows [Krieger et al. (1992b)]:

$$C_{i\sigma} = \sum_{j=1}^{N_\sigma}(\mathbf{A}_\sigma^{-1})_{ij}[\langle V_{j\sigma}^{\rm S}\rangle - \langle v_{j\sigma}^{\rm HF}\rangle], \tag{A.5}$$

where

$$(\mathbf{A}_\sigma)_{ij} = \delta_{ij} - \int \frac{n_{i\sigma}(\mathbf{r}) n_{j\sigma}(\mathbf{r})}{n_\sigma(\mathbf{r})} d^3\mathbf{r}, \tag{A.6}$$

and

$$\langle V_{j\sigma}^{\rm S}\rangle = \int V_\sigma^{\rm S}(\mathbf{r}) n_{j\sigma}(\mathbf{r}) d^3\mathbf{r}, \tag{A.7}$$

$$\langle v_{j\sigma}^{\rm HF}\rangle = \int v_{j\sigma}^{\rm HF}(\mathbf{r}) n_{j\sigma}(\mathbf{r}) d^3\mathbf{r}. \tag{A.8}$$

The asymptotic condition: $v_\sigma^{\rm xc} \to 0$ as $r \to \infty$ requires that $C_{i\sigma} = 0$ for the valence electron.

Using the Laplace expansion

$$\frac{1}{|\mathbf{r}-\mathbf{r}'|} = \sum_{\ell,m} \frac{4\pi(-1)^m}{2\ell+1} \frac{r_<^\ell}{r_>^{\ell+1}} Y_\ell^{-m}(\theta,\phi) Y_\ell^m(\theta',\phi'), \qquad (A.9)$$

the Hartree-Fock potential (Eq. A.3) for atomic wavefunction (Eq. 3.25) becomes

$$v_{i\sigma}^{\mathrm{HF}}(r) = \frac{-1}{R_{i\sigma}^{(\ell_i)}(r)} \sum_{j=1}^{N_\sigma} R_{j\sigma}^{(\ell_j)}(r) \sum_{\ell=|\ell_i-\ell_j|}^{\ell_i+\ell_j} \begin{pmatrix} \ell_i & \ell_j & \ell \\ 0 & 0 & 0 \end{pmatrix}^2 W_{ij}^{\sigma,\ell}(r), \qquad (A.10)$$

where $\left(\begin{smallmatrix} \ell_i & \ell_j & \ell \\ 0 & 0 & 0 \end{smallmatrix}\right)$ are Wigner-3j coefficients, and

$$W_{ij}^{\sigma,\ell}(r) \equiv \int \frac{r_<{}^\ell}{r_>{}^{\ell+1}} R_{i\sigma}^{(\ell_i)}(r') R_{j\sigma}^{(\ell_j)}(r') dr', \qquad (A.11)$$

with $r_< \equiv \min(r, r')$ and $r_> \equiv \max(r, r')$. Note that ℓ_i or ℓ_j is not a summation index but specific to each i- or j-th orbital, and hence it is enclosed in parentheses. In practice, $W_{ij}^{\sigma,\ell}(r)$ is obtained by solving the following differential equation:

$$\left[-\frac{d^2}{dr^2} + \frac{\ell(\ell+1)}{r^2}\right] Q_{ij}^{\sigma,\ell}(r) = (2\ell+1) \frac{R_{i\sigma}^{(\ell_i)}(r) R_{j\sigma}^{(\ell_j)}(r)}{r}, \qquad (A.12)$$

where $Q_{ij}^{\sigma,\ell}(r) \equiv r W_{ij}^{\sigma,\ell}(r)$. The boundary conditions are

$$Q_{ij}^{\sigma,\ell}(0) = 0,$$

$$Q_{ij}^{\sigma,\ell}(r_{\max}) = \begin{cases} \delta_{ij} & \text{if } \ell = 0 \\ \frac{1}{(r_{\max})^\ell} \int_0^{r_{\max}} R_{i\sigma}^{(\ell_i)}(r) R_{j\sigma}^{(\ell_j)}(r)\, r^\ell dr & \text{otherwise.} \end{cases} \qquad (A.13)$$

A.2 LDA: local density approximation

The KLI approximation discussed in the previous section is impractical for time-dependent calculations, since the Hartree-Fock potential (Eq. A.3) becomes a complex-valued integral once each wavefunction $\psi_{i\sigma}(\mathbf{r}, t)$ gains a different phase as time progresses. The local density approximation (LDA) is another exchange-only approximation one can use to avoid the complex-valued integral. It was first derived by Dirac in 1930 [Dirac (1930)] based on a free-electron gas as follows. Consider a ground state of N non-interacting electrons in a box of dimension L^3 in a free space, i.e., $\mathcal{V}_{\mathrm{ext}} = 0$ and $\mathcal{V}_{\mathrm{ee}} = 0$. The stationary-state Kohn-Sham equation in this case is simply

$$\frac{-1}{2} \nabla^2 \psi_{i\sigma}(\mathbf{r}) = \varepsilon_{i\sigma} \psi_{i\sigma}(\mathbf{r}). \qquad (A.14)$$

We impose periodic boundary conditions, i.e.,

$$\psi_{i\sigma}(x, y, z) = \psi_{i\sigma}(x + L, y, z), \qquad (A.15)$$

and similarly for the y- and z-directions. Then, the solution of (A.14) is given by

$$\psi_{i\sigma}(\mathbf{r}) = \frac{1}{\sqrt{L^3}} e^{i\mathbf{k}\cdot\mathbf{r}}, \tag{A.16}$$

where $\mathbf{k} = (k_x, k_y, k_z)$ and $\mathbf{r} = (x, y, z)$ must satisfy

$$k_x = \frac{2\pi}{L} n_x, \quad k_y = \frac{2\pi}{L} n_y, \quad k_z = \frac{2\pi}{L} n_z, \tag{A.17}$$

for each $n_x, n_y, n_z = 0, \pm 1, \pm 2, \cdots$, and

$$\varepsilon_{i\sigma} = \frac{k^2}{2} = \frac{(2\pi)^2}{L^2}(n_x^2 + n_y^2 + n_z^2). \tag{A.18}$$

In momentum space, such a free-electron gas occupies the spherical region of minimum energy, called the Fermi sphere, with two electrons of opposite spins in each volume of $(2\pi/L)^3$. The radius k_F of a Fermi sphere is called the Fermi wavevector. The number of electrons in a free-electron gas can therefore be expressed in terms of k_F as

$$N = \left(\frac{4}{3}\pi {k_F}^3\right) \frac{2}{(2\pi/L)^3}. \tag{A.19}$$

It then follows that

$$k_F = (3\pi^2 n)^{1/3}, \tag{A.20}$$

where $n \equiv N/L^3$ is the uniform (or *homogeneous*) electron density of a free-electron gas in configuration space. The Hartree-Fock exchange potential (Eq. A.3) evaluated with the free-electron Kohn-Sham wavefunctions (Eq. A.16) results in [Kohn and Sham (1965)]

$$v_{i\sigma}^{\mathrm{HF}} = -\frac{k_F}{\pi}\left(1 + \frac{{k_F}^2 - k^2}{2kk_F} \ln\left|\frac{k + k_F}{k - k_F}\right|\right), \tag{A.21}$$

which does not depend on the spatial coordinates $\mathbf{r}$. After being averaged over k-vectors, it becomes

$$v_{i\sigma}^{\mathrm{HF}} = -\frac{k_F}{\pi} = -\left(\frac{3}{\pi} n\right)^{1/3}. \tag{A.22}$$

For a closed system in particular, $N_\uparrow = N_\downarrow$ so that $n = 2n_\sigma$, and therefore we obtain the LDA approximation of an exchange-only potential as

$$v_{\mathrm{x},\sigma}^{\mathrm{LDA}}[n_\uparrow, n_\downarrow](\mathbf{r}) = -\left(\frac{6}{\pi} n_\sigma(\mathbf{r})\right)^{1/3}, \tag{A.23}$$

which is orbital independent. In the last expression, the uniform density $n = N/L^3$ of a free electron-gas was generalized into non-uniform electron densities $n_\sigma(\mathbf{r})$ defined by Eq. 3.15. Both KLI and LDA are exchange-only approximation in a sense that they are derived from the Hartree-Fock exchange potential (Eq. A.3).

A.3 Self-interaction corrected LDA

For neutral finite systems, the exact exchange-correlation potential must decay as -1/r, whereas the LDA potential (A.23) falls off exponentially, which leads to underestimated ionization potentials of a system. This erroneous asymptotic behavior is caused by spurious self-interaction and can be removed as follows. The local density approximation with a self-interaction correction (LDA-SIC) is given by [Perdew and Zunger (1981)]

$$v_\sigma^{\mathrm{xc}}[n_\uparrow, n_\downarrow](\mathbf{r}) \simeq v_{\mathrm{x},\sigma}^{\mathrm{LDA}}[n_\uparrow, n_\downarrow](\mathbf{r}) - V_\sigma^{\mathrm{SIC}}(\mathbf{r}), \tag{A.24}$$

where $v_{\mathrm{x},\sigma}^{\mathrm{LDA}}$ is the LDA functional (A.23), and the correction term is given by [Tong and Chu (1997b)]

$$V_\sigma^{\mathrm{SIC}}(\mathbf{r}) = V_\sigma^{\mathrm{SI}}(\mathbf{r}) + \frac{1}{n_\sigma(\mathbf{r})} \sum_{i=1}^{N_\sigma}{}' n_{i\sigma}(\mathbf{r}) C_{i\sigma}, \tag{A.25}$$

where $\sum'$ denotes the summation over all orbitals except for the valence ones, and $V_\sigma^{\mathrm{SI}}(\mathbf{r}, t)$ is the self-interaction potential:

$$V_\sigma^{\mathrm{SI}}(\mathbf{r}) = \frac{1}{n_\sigma(\mathbf{r})} \sum_{i=1}^{N_\sigma} n_{i\sigma}(\mathbf{r}) v_{i\sigma}(\mathbf{r}), \tag{A.26}$$

with

$$v_{i\sigma}(\mathbf{r}) = V_{\mathrm{H}}[n_{i\sigma}](\mathbf{r}) - \left(\frac{6}{\pi} n_{i\sigma}(\mathbf{r})\right)^{1/3}. \tag{A.27}$$

Moreover, $C_{i\sigma} \equiv \langle V_{i\sigma}^{\mathrm{SIC}} \rangle - \langle v_{i\sigma} \rangle$, where

$$\langle V_{i\sigma}^{\mathrm{SIC}} \rangle = \int V_\sigma^{\mathrm{SIC}}(\mathbf{r}) n_{i\sigma}(\mathbf{r}) d^3\mathbf{r}, \tag{A.28}$$

$$\langle v_{i\sigma} \rangle = \int v_{i\sigma}(\mathbf{r}) n_{i\sigma}(\mathbf{r}) d^3\mathbf{r}. \tag{A.29}$$

We can calculate $C_{i\sigma}$ non-iteratively as

$$C_{i\sigma} = \sum_{j=1}^{N_\sigma}{}' (\mathbf{A}_\sigma^{-1})_{ij} [\langle V_{j\sigma}^{\mathrm{SI}} \rangle - \langle v_{j\sigma} \rangle], \tag{A.30}$$

where

$$(\mathbf{A}_\sigma)_{ij} = \delta_{ij} - \int \frac{n_{i\sigma}(\mathbf{r}) n_{j\sigma}(\mathbf{r})}{n_\sigma(\mathbf{r})} d^3\mathbf{r}, \tag{A.31}$$

and

$$\langle V_{j\sigma}^{\mathrm{SI}} \rangle = \int V_\sigma^{\mathrm{SI}}(\mathbf{r}) n_{j\sigma}(\mathbf{r}) d^3\mathbf{r}. \tag{A.32}$$

A.4 BLYP approximation

The BLYP approximation[1] uses the exchange energy functional of Becke [Becke (1988)] and the correlation energy functional of Lee, Yang, and Parr [Lee et al. (1988)]:

$$E_{\rm xc}^{\rm BLYP} = \left(E_{\rm x}^{\rm LDA} - \beta \sum_\sigma \int n_\sigma^{4/3}(\mathbf{r}) \frac{x_\sigma^2(\mathbf{r})}{1 + 6\beta x_\sigma(\mathbf{r}) \sinh^{-1} x_\sigma(\mathbf{r})} \, d^3\mathbf{r} \right) + E_c^{\rm LDA}, \tag{A.33}$$

where $\beta = 0.0042$, and

$$x_\sigma(\mathbf{r}) = \frac{|\nabla n_\sigma(\mathbf{r})|}{n_\sigma^{4/3}(\mathbf{r})}, \tag{A.34}$$

$$E_{\rm x}^{\rm LDA} = -\frac{3}{2} \left(\frac{3}{4\pi} \right)^{1/3} \sum_\sigma \int n_\sigma^{4/3}(\mathbf{r}) \, d^3\mathbf{r}, \tag{A.35}$$

$$\begin{aligned} E_c^{\rm LDA} &= -a \int \frac{\gamma(\mathbf{r})}{1 + dn^{-1/3}(\mathbf{r})} n(\mathbf{r}) \, d^3\mathbf{r} \\ &\quad - 2ab \int \frac{\gamma(\mathbf{r})}{1 + dn^{-1/3}(\mathbf{r})} \left[2^{2/3} C_{\rm F} n^{-5/3} \left(n_\uparrow^{8/3}(\mathbf{r}) + n_\downarrow^{8/3}(\mathbf{r}) \right) e^{-cn^{-1/3}(\mathbf{r})} \right] d^3\mathbf{r}, \end{aligned} \tag{A.36}$$

with $a = 0.04918$, $b = 0.132$, $c = 0.2533$, $d = 0.349$, and $C_{\rm F} = \frac{3}{10}(3\pi^2)^{2/3}$. For simplicity, we adapt the zeroth-order approximation in $E_c^{\rm LDA}$, i.e., the terms that contain ∇n and $\nabla^2 n$ in the Lee-Yang-Parr functional are omitted in Eq. A.36.

The corresponding BLYP functional for the exchange-correlation potential is [Gritsenko et al. (1993)]

$$\begin{aligned} v_\sigma^{\rm xc}[n_\uparrow, n_\downarrow](\mathbf{r}) &\simeq V_\sigma^{\rm BLYP}(\mathbf{r}) = V_{{\rm x},\sigma}^{\rm LDA}[n_\uparrow, n_\downarrow](\mathbf{r}) + V_{\rm c}^{\rm LDA}[n_\uparrow, n_\downarrow](\mathbf{r}) \\ &\quad - \frac{4}{3}\beta n^{1/3}(\mathbf{r}) \left\{ f(y_\sigma) - \frac{3}{8(6\pi^2)^{2/3}} f'(y_\sigma) \left[\frac{\nabla^2 n(\mathbf{r})}{n_\sigma^{5/3}(\mathbf{r})} + \frac{2}{r n^{5/3}(\mathbf{r})} \nabla n(\mathbf{r}) \right] \right. \\ &\quad \left. - \frac{3}{\left(4(6\pi^2)^{2/3}\right)^2 n_\sigma^{10/3}(\mathbf{r})} f''(y_\sigma) \left| \nabla n_\sigma(\mathbf{r}) \right|^2 \left[\frac{\nabla^2 n_\sigma(\mathbf{r})}{n_\sigma(\mathbf{r})} - \frac{4 |\nabla n_\sigma(\mathbf{r})|^2}{3 n_\sigma^2(\mathbf{r})} \right] \right\}, \end{aligned} \tag{A.37}$$

where $y_\sigma = \dfrac{x_\sigma}{4(6\pi^2)^{2/3}}$, and

$$f(y_\sigma) = \frac{y_\sigma^2}{1 + 6\beta y_\sigma \sinh^{-1} y_\sigma}. \tag{A.38}$$

[1]The BLYP functional can be classified as a generalized gradient approximation (GGA), which has a general form of

$$E_{\rm c}^{\rm GGA}[n] = \int f[n, \nabla n, \nabla^2 n](\mathbf{r}) \, d^3 r.$$

for some universal functional f.

The functional $V_{\mathrm{x},\sigma}^{\mathrm{LDA}}$ is given by Eq. A.23, and

$$\begin{aligned} V_{\mathrm{c}}^{\mathrm{LDA}}[n_\uparrow, n_\downarrow](\mathbf{r}) =& -a\left[\frac{\delta F}{\delta n_\uparrow} n(\mathbf{r}) + F[n](\mathbf{r})\right] \\ & - 2^{5/3} abC_{\mathrm{F}}\left[\frac{\delta G}{\delta n_\uparrow}\left(n_\uparrow^{8/3}(\mathbf{r}) + n_\downarrow^{8/3}(\mathbf{r})\right)\right. \\ & \left. + \frac{8}{3} G[n](\mathbf{r}) n_\uparrow^{5/3}\right], \end{aligned} \tag{A.39}$$

where

$$F[n](\mathbf{r}) = \frac{\gamma(\mathbf{r})}{1 + dn^{-1/3}(\mathbf{r})}, \tag{A.40}$$

$$\gamma(\mathbf{r}) = 2\left[1 - \frac{\sum_\sigma n_\sigma^2(\mathbf{r})}{n^2(\mathbf{r})}\right] = \frac{4n_\uparrow(\mathbf{r}) n_\downarrow(\mathbf{r})}{n^2(\mathbf{r})}, \tag{A.41}$$

$$\frac{\delta F}{\delta n_\uparrow} = \frac{4n_\downarrow(\mathbf{r})[n_\downarrow(\mathbf{r}) - n_\uparrow(\mathbf{r})]}{n^3(1 + dn^{-1/3})} + \frac{d}{3} \frac{\gamma(\mathbf{r}) n^{-4/3}}{[1 + dn^{-1/3}(\mathbf{r})]^2} \tag{A.42}$$

$$G[n](\mathbf{r}) = F[n](\mathbf{r}) n^{-5/3} e^{-cn^{-1/3}(\mathbf{r})}, \tag{A.43}$$

$$\frac{\delta G}{\delta n_\uparrow} = G[n](\mathbf{r})\left[\frac{1}{F[n](\mathbf{r})}\left(\frac{\delta F}{\delta n_\uparrow}\right) - \frac{5}{3n(\mathbf{r})} + \frac{c}{3n^{4/3}}\right]. \tag{A.44}$$

The proper asymptotic condition $v_{\mathrm{xc}} \to -1/r$ is not satisfied in the BLYP approximation, and we need to impose instead [van Leeuwen and Baerends (1994)]

$$V_\sigma^{\mathrm{BLYP}}(\mathbf{r}) \to -\frac{1}{\alpha}\frac{1}{r^2} \quad (r \to \infty), \tag{A.45}$$

where $\alpha = 2\sqrt{-2\mu}$ with μ being the chemical potential (negative of the single-ionization energy).

A.5 Electric dipole, magnetic-dipole and other higher-order interactions

$$\begin{aligned} H'_{\alpha\beta} =& \langle\alpha|\hat{H}'|\beta\rangle = -\frac{e}{m_e}\langle\alpha|\mathbf{p}\cdot\mathbf{A}|\beta\rangle + \underbrace{\frac{e^2}{2m_e}\langle\alpha|\mathbf{A}^2|\beta\rangle}_{\approx 0} \\ =& -\frac{e}{m_e}\langle\alpha|\mathbf{p}\cdot\mathbf{A}_0 e^{-i(\frac{\omega}{c}\mathbf{n}\cdot\mathbf{r}-\omega t)}|\beta\rangle \\ =& -\frac{e}{m_e}\mathbf{A}_0\cdot\langle\alpha|\mathbf{p}e^{-i\frac{\omega}{c}\mathbf{n}\cdot\mathbf{r}}|\beta\rangle e^{i\omega t} \\ =& \frac{e}{m_e}\mathbf{A}_0\cdot\langle\alpha|\mathbf{p}(1 - i\frac{\omega}{c}\mathbf{n}\cdot\mathbf{r} + \cdots)|\beta\rangle e^{i\omega t} \\ =& -\frac{e}{m_e}\mathbf{A}_0\cdot\langle\alpha|\mathbf{p}|\beta\rangle e^{i\omega t} + \frac{ie\omega}{m_e c}\mathbf{A}_0\cdot\langle\alpha|\mathbf{p}(\mathbf{n}\cdot\mathbf{r})|\beta\rangle e^{i\omega t} \\ =& \frac{-e}{m_e}\mathbf{A}_0\cdot\langle\alpha|\frac{-im_e}{\hbar}[\mathbf{r}, \hat{H}_0]|\beta\rangle e^{i\omega t} \\ & + \frac{ie\omega A_0}{m_e c}\langle\alpha|(\mathbf{p}\cdot\mathbf{a})(\mathbf{n}\cdot\mathbf{r})|\beta\rangle e^{i\omega t} \end{aligned}$$

$$
\begin{aligned}
&=\frac{ie}{\hbar}\mathbf{A}_0\cdot(E_\beta-E_\alpha)\langle\alpha|\mathbf{r}|\beta\rangle e^{i\omega t}\\
&+\frac{ie\omega A_0}{m_e c}\langle\alpha|p_a r_n|\beta\rangle e^{i\omega t}\\
&=ie\left(\frac{E_\beta-E_\alpha}{\hbar}\right)\mathbf{A}_0\cdot\langle\alpha|\mathbf{r}|\beta\rangle e^{i\omega t}\\
&+\frac{ie\omega A_0}{2m_e c}\langle\alpha|2p_a r_n+p_n r_a-p_n r_a|\beta\rangle e^{i\omega t}\\
&=ie\omega_{\alpha\beta}\mathbf{A}_0\cdot\mathbf{d}_{\alpha\beta}e^{i\omega t}+\frac{ie\omega A_0}{2m_e c}\langle\alpha|p_a r_n+p_n r_a|\beta\rangle e^{i\omega t}\\
&+\frac{ie\omega A_0}{2m_e c}\langle\alpha|p_a r_n-p_n r_a|\beta\rangle e^{i\omega t}\\
&=ie\omega_{\alpha\beta}\mathbf{d}_{\alpha\beta}\cdot\left(\mathbf{A}_0 e^{i\omega t}\right)\\
&+\frac{ie\omega A_0}{2m_e c}\omega_{\alpha\beta}(-im_e)\langle\alpha|2r_a r_n|\beta\rangle e^{i\omega t}\\
&+\frac{ie\omega A_0}{2m_e c}\langle\alpha|\mathbf{L}\cdot(\mathbf{n}\times\mathbf{a})|\beta\rangle e^{i\omega t}\\
&=ie\omega_{\alpha\beta}\mathbf{d}_{\alpha\beta}\cdot\left(\frac{\mathbf{E}}{-i\omega}\right)+\frac{e\omega A_0}{c}\omega_{\alpha\beta}\langle\alpha|\overset{\leftrightarrow}{Q}^{(\mathbf{a},\mathbf{n})}|\beta\rangle e^{i\omega t}\\
&+\frac{ie\omega}{2m_e c}\mathbf{L}_{\alpha\beta}\cdot\left((\mathbf{n}\times\mathbf{a})A_0 e^{i\omega t}\right)\\
&=-\frac{e\omega_{\alpha\beta}}{\omega}\mathbf{d}_{\alpha\beta}\cdot\mathbf{E}+\frac{e\omega\omega_{\alpha\beta}}{c}\left(\frac{-(\nabla\mathbf{E})c}{\omega^2}\right):\overset{\leftrightarrow}{Q}^{(\mathbf{a},\mathbf{n})}_{\alpha\beta}\\
&+\frac{ie\omega}{2m_e c}\mathbf{L}_{\alpha\beta}\cdot\left(\frac{\mathbf{B}c}{-i\omega}\right)\\
&=-\frac{e\omega_{\alpha\beta}}{\omega}\mathbf{d}_{\alpha\beta}\cdot\mathbf{E}-\frac{e\omega_{\alpha\beta}}{\omega}\overset{\leftrightarrow}{Q}_{\alpha\beta}:(\nabla\mathbf{E})-\frac{e}{2m_e}\mathbf{L}_{\alpha\beta}\cdot\mathbf{B},
\end{aligned}
\tag{A.46}
$$

where c is the speed of light, $\mathbf{r}$ is the position operator, $\mathbf{n}$ is the propagation direction of the light, $\mathbf{a}$ is the direction perpendicular to $\mathbf{r}$ and $\mathbf{n}$, E_α and E_β are the energies of states $|\alpha\rangle$ and $|\beta\rangle$, respectively, and $\omega_{\alpha\beta}=\frac{E_\beta-E_\alpha}{\hbar}$. We thus define the vectorial electric-dipole transition matrix element as

$$\mathbf{d}_{\alpha\beta}=\langle\alpha|\mathbf{r}|\beta\rangle, \tag{A.47}$$

the vectorial magnetic-dipole transition matrix element as

$$\mathbf{L}_{\alpha\beta}=\langle\alpha|p_a r_n-p_n r_a|\beta\rangle, \tag{A.48}$$

and the rank-two-tensorial electric-quadrupole transition matrix element as

$$\overset{\leftrightarrow}{Q}_{\alpha\beta}=\langle\alpha|p_a r_n+p_n r_a|\beta\rangle. \tag{A.49}$$

Here p_a and r_n are abbreviations for the projections of $\mathbf{p}$ on $\mathbf{a}$ and of $\mathbf{r}$ on $\mathbf{n}$, respectively. The tensorial product in the perturbation element of the electric dipole is defined as

$$\overset{\leftrightarrow}{Q}_{\alpha\beta}:(\nabla\mathbf{E})=\sum_{i=x,y,z}\sum_{j=x,y,z}\left(\overset{\leftrightarrow}{Q}_{\alpha\beta}\right)_{i,j}\frac{\partial E_i}{\partial r_j}. \tag{A.50}$$

A.6 Code to generate ultrafast pulses

```
1 !this code creates the mode-locking and phase alignment. It also
2 !computes the how the pulse sharpens as the number of modes !increases.
3       implicit double precision (a-h,o-z)
4       integer N
5       parameter(N=20)
6       double precision amplitude(N)
7       pi=dacos(-1d0)
8       phase0=1.34d0
9 !      phase0=0d0
10      w=3d0/4d0*pi
11      wmax=w*N
12 !     write(*,*)rand()
13 !     write(*,*)rand()
14 !     stop
15      aa=10d0
16
17      periodmin=2d0*pi/wmax
18      periodmax=2d0*pi/w
19      step=periodmin/40d0
20      iwidth=2
21 !     open(1,file='twopulse')
22      open(1,file='20pulse')
23 !
24      do t=-iwidth*periodmax,iwidth*periodmax,step
25
26         e=0d0
27
28         do i=1,n
29
30 !          e=e+dexp(-abs(i-n/2)**1/aa)*cos(i*w*t+phase*i+0.0d0*pi*rand(i))
31            phase=phase0*i*i
32            e=e+dexp(-0*abs(i-n/2)**1/aa)*cos(i*w*t+phase+0.0d0*pi
33      $         *rand(i))
34 !            e=e+dexp(-(i-n/2)**2/aa)*cos(i*w*t+phase+0.0d0*pi*rand())
35         enddo
36         c=e
37         e=0d0
38         do i=1,n
39 !          e=e+dexp(-abs(i-n/2)**1/aa)*cos(i*w*t+phase*i+0.0d0*pi*rand(i))
40            phase=phase0*i*i
41            e=e+dexp(-0*abs(i-n/2)**1/aa)*sin(i*w*t+phase+0.0d0*pi
42      $         *rand(i))
43 !            e=e+dexp(-(i-n/2)**2/aa)*cos(i*w*t+phase+0.0d0*pi*rand())
44         enddo
45         s=e
46         write(1,*)t,c**2+s**2
47      enddo
48      close(1)
49      end
```

A.7 Code to generate figures in HHG

```
1 !      Sun Feb  7 11:25:55 EST 2016
2 !      Sun Feb  5 22:02:33 EST 2017
3 !
4 !      We only consider one dimensional hydrogen-like chain.
5        implicit none
6 !nsite is the number of sites in the system
7        integer nm,nsite
8        parameter(nsite=200)
9        double precision lambda   !soc
10 !     t is the kinetic energy operator
11       double precision t ! this is hopping integral
12       double precision e_up,e_dn !this is site energy for spin up and dn
13       double precision efield,damping,pot
14       parameter(nm=1*nsite,lambda=0.00d0)
15       parameter(e_up=0.d0,e_dn=-0.d0,t=0.5d0)
16 !      parameter(efield=0.00d0)
17       double precision h(nm,nm),e(nm),fv1(nm,nm),z(nm,nm),sz(nm)
18       integer ierr
19       integer i,j
20       integer period
21 !     compute the spin moment
22       double precision spin,spint
23       integer nf,ncenter
24       parameter(nf=(nm)/2)
25       damping=1d0
26       open(1,file='hydrogen')
27       do i=1,nm
28          write(1,*)i,-1d0/i**2
29       enddo
30       close(1)
31       period=1
32       do efield=0d0,1d0,0.05d0
33          write(*,*)'Efield',efield
34       do i=1,nm
35          do j=1,nm
36             h(i,j)=0d0                   !this is hamiltonian
37          enddo
38       enddo
39       do i=1,nsite
40          do j=1,nsite
41             if(i.eq.j)then
42                h(i,j)=+2*t
43             else if(abs(i-j).eq.1)then
44                h(i,j)=-t
45             endif
46          enddo
47       enddo
```

```
48        if(period.eq.0)then
49           h(1,nsite)=-t
50           h(nsite,1)=-t
51        endif
52  !     electric field
53           ncenter=(nsite)/2
54        write(*,*)'ncenter',ncenter
55        open(1,file='potential')
56        do i=1,nsite
57  !     potential
58          pot=((i-nsite/2d0))*Efield/nsite-1d0/
59       & (abs(i-ncenter)+damping)
60           h(i,i)=pot+h(i,i)
61           write(1,*)i-nm/2,pot
62        enddo
63        close(1)
64        do i=1,nm
65           do j=1,nm
66              if(h(i,j).ne.h(j,i))stop'wrong'
67              if(h(i,j).ne.0d0)then
68                 write(*,*)i,j,h(i,j)
69              endif
70           enddo
71        enddo
72  !       read(*,*)
73  ! diagonalization routines
74        call tred2(nm,nm,h,e,fv1,z)
75        call tql2(nm,nm,e,fv1,z,ierr)
76        open(1,file='energy')
77        do i=1,nm
78           write(1,*)i,e(i)
79        enddo
80        close(1)
81        open(1,file='wf')
82        do i=1,nm
83           write(1,101)i-ncenter,(z(i,j),j=1,5)
84        enddo
85   101  format(I4,4x,40(f16.10,1x))
86   102  format(40(d16.10,1x))
87        close(1)
88        stop'␣line␣114'
89  !       open(1,file='spin',position='append')
90        open(1,file='spin.up',position='append')
91        open(2,file='spin.dn',position='append')
92        spin=0d0
93        do i=1,nf                        !up to fermi level
94           spint=0d0
95           do j=1,nm
96              spin=spin+z(j,i)*sz(j)*z(j,i)
97              spint=spint+z(j,i)*sz(j)*z(j,i)
98           enddo
99           if(dabs(spint+1d0).le.1d-14)then
100             write(1,101)i,spint,e(i),efield
101          else if(dabs(spint-1d0).le.1d-14)then
102             write(2,101)i,spint,e(i),efield
103          else
104              stop'WRONG'
105          endif
106       enddo
```

```
107      write(2,*)'&'
108      close(2)
109      write(1,*)'&'
110      close(1)
111      open(1,file='energy.versus.field.1',position='append')
112      write(1,102)efield,(e(i),i=1,nf/4,1)
113      close(1)
114      open(1,file='energy.versus.field.2',position='append')
115      write(1,102)efield,(e(i),i=nf/4+1,nf/2,1)
116      close(1)
117      open(1,file='energy.versus.field.3',position='append')
118      write(1,102)efield,(e(i),i=nf/2+1,3*nf/4)
119      close(1)
120      open(1,file='energy.versus.field.4',position='append')
121      write(1,102)efield,(e(i),i=3*nf/4+1,nf,1)
122      close(1)
123      open(1,file='spinchange',position='append')
124      write(1,*)efield,spin/nf
125      close(1)
126      stop
127      enddo
128      end
```

A.8 Code to compute the cutoff energy in HHG

The following is the MatLab code

```
1  f=@(q)cos(q)-cos(pi/180)+(q-pi/180)*sin(pi/180);
2  z=fzero(f,[pi/2, pi*2])
```

where q denotes ωt. In the above code, ωt_i takes $\pi/180$. To find a different root ωt or q, we change ωt_i from $\pi/180$ to $30\pi/180$ by hand. This is tedious.

ωt_i	ωt
$1\pi/180$	5.828569353477918
$2\pi/180$	5.645671396738821
$3\pi/180$	5.506421825449956
$4\pi/180$	5.389426333316624
$5\pi/180$	5.286460173392023
$6\pi/180$	5.193333668414808
$7\pi/180$	5.107573671012561
$8\pi/180$	5.027578453285694
$9\pi/180$	4.952241844522598
$10\pi/180$	4.880762907038256
$11\pi/180$	4.812540234030844
$12\pi/180$	4.747109016916339
$13\pi/180$	4.684101481157128
$14\pi/180$	4.623220898789413
$15\pi/180$	4.564223891094794
$16\pi/180$	4.506908006204815
$17\pi/180$	4.451102771596001
$18\pi/180$	4.396663104586390

ωt_i	ωt
$19\pi/180$	4.343464364505811
$20\pi/180$	4.291398573731649
$21\pi/180$	4.240371487563054
$22\pi/180$	4.190300291447123
$23\pi/180$	4.141111769203121
$24\pi/180$	4.092740829892930
$25\pi/180$	4.045129311301957
$26\pi/180$	3.998224999255151
$27\pi/180$	3.951980817144430
$28\pi/180$	3.906354151001616
$29\pi/180$	3.861306283483076
$30\pi/180$	3.816801916092871

The following is the Fortran code. It is easier but less accurate.

```
1  !this code computes the cutoff energy
2  !      this code finds the solution to the position equation. This is a
3  !      highly nonlinear equation. To get a rough idea where the solution
4  !      locates, we use gnuplot to plot the function first and locate the
5  !      solution. Then when we code this code, we know where the upper
6  !      and lower limits are. Otherwise, the method will converge to a
7  !      wrong solution. This code is very robust, without using any fancy
8  !      method.
9
10 !      p: omega x t_i, where ti is the ionization time when the electron
11 !      merges
12
13 !      q: omega x t
14        implicit double precision (a-h,o-z)
15        integer ip,iq,imin
16        double precision min,last,threshold
17        pi=dacos(-1d0)
18        open(1,file='phase.data')
19 !       do p=0.1d0*pi,pi/2d0,pi/1d2
20        do ip=1,90
21           p=ip/180d0*pi
22           write(*,*)'p␣(degrees)',p,ip
23           min=100d0
24           imin=0
25           do iq=17,3600
26              q=iq/1800d0*pi
27              zero=dcos(q)-dcos(p)+(q-p)*dsin(p)
28              diff=dabs(p-q)       !avoid the trivial p=q point
29              if(dabs(zero).lt.dabs(min).and.diff.gt.0.2d0)then
30                 min=zero
31                 imin=iq
32                 qmin=q
33 !                write(*,*)q,zero,imin,min
34 !                read(*,*)
35              endif
36           enddo
37 2         format(2(f16.12,1x),I5,2f16.12)
38 !      the second variable the kinetic energy in unit of U_p
39 1         write(1,2)p,2d0*(dsin(p)-dsin(qmin))**2,ip,qmin,min
```

```
40        enddo
41        close(1)
42        end
```

Here is the numerical value of our code. The reader can find the cutoff of 3.17 in units of U_p below.

P (omega t_i)	Kin.energy(U_p)	deg	Q(omega t)	residual
.017453292520	.415550223352	1	5.829399701661	.000378802067
.034906585040	.793100308557	2	5.646140130202	.000295260204
.052359877560	1.134756724541	3	5.506513790042	.000069274906
.069813170080	1.441922731580	4	5.389576730158	.000127708208
.087266462600	1.717826052506	5	5.286602304291	.000131728936
.104719755120	1.963099960068	6	5.194099853935	.000759220466
.122173047640	2.184382575626	7	5.106833391335	-.000773545798
.139626340160	2.374843503733	8	5.028293574996	.000779339523
.157079632679	2.544693936230	9	4.951499087908	-.000837751872
.174532925199	2.689549409548	10	4.879940588576	-.000953540371
.191986217719	2.812549534647	11	4.811872747748	-.000791526156
.209439510239	2.915158739150	12	4.747295565425	.000225221090
.226892802759	2.999100115127	13	4.684463712353	.000443572417
.244346095279	3.065104585667	14	4.623377188533	.000193479827
.261799387799	3.114183917565	15	4.564035993965	-.000234467206
.279252680319	3.147564308341	16	4.506440128649	-.000586976980
.296705972839	3.166625353101	17	4.450589592586	-.000645765940
.314159265359	3.172845375744	18	4.396484385774	-.000225107154
.331612557879	3.167753684790	19	4.344124508214	.000830726688
.349065850399	3.149318731650	20	4.291764630654	.000459320746
.366519142919	3.121873151235	21	4.241150082346	.000972618771
.383972435439	3.082661098140	22	4.190535534038	.000292040669
.401425727959	3.036840251663	23	4.141666314983	.000683251459
.418879020479	2.980960000147	24	4.092797095927	.000068691590
.436332312999	2.920824400099	25	4.045673206123	.000657191888
.453785605519	2.852344866714	26	3.998549316319	.000387272983
.471238898038	2.776133098588	27	3.951425426515	-.000654446674
.488692190558	2.698697592834	28	3.906046865963	-.000356980901
.506145483078	2.615180293591	29	3.860668305411	-.000729680432
.523598775598	2.532342071275	30	3.817035074112	.000262338134
.541052068118	2.444896170791	31	3.773401842812	.000655592548
.558505360638	2.353424533580	32	3.729768611512	.000511270306
.575958653158	2.258518243791	33	3.686135380212	-.000111176757
.593411945678	2.167139554720	34	3.644247478164	.000661326824
.610865238198	2.067144795682	35	3.600614246864	-.000788892698
.628318530718	1.971807415500	36	3.558726344816	-.000820453280
.645771823238	1.881565561549	37	3.518583772021	.000489393533
.663225115758	1.784266746740	38	3.476695869973	-.000241568519
.680678408278	1.692883112525	39	3.436553297177	.000368717252
.698131700798	1.601537986105	40	3.396410724381	.000666710235
.715584993318	1.510632823453	41	3.356268151585	.000688874026
.733038285838	1.420552157608	42	3.316125578789	.000470188121
.750491578358	1.331661556927	43	3.275983005993	.000044110666
.767944870878	1.244305762875	44	3.235840433197	-.000557455158
.785398163397	1.164119139706	45	3.197443189654	.000025849224
.802851455917	1.085725511932	46	3.159045946110	.000398408519
.820304748437	1.009368378435	47	3.120648702566	.000586022312
.837758040957	.930510686593	48	3.080506129770	-.000578767730

.855211333477	.863630948354	49	3.043854215478	.000503321760
.872664625997	.794628970437	50	3.005456971934	.000278274774
.890117918517	.728419150079	51	2.967059728390	-.000041204338
.907571211037	.665133386351	52	2.928662484846	-.000435711829
.925024503557	.608605650361	53	2.892010570555	.000074166314
.942477796077	.554774243669	54	2.855358656263	.000453897272
.959931088597	.500385544275	55	2.816961412719	-.000154660049
.977384381117	.452319818001	56	2.780309498427	.000055728453
.994837673637	.407081634652	57	2.743657584135	.000181600968
1.012290966157	.364678979305	58	2.707005669843	.000236299465
1.029744258677	.325103065669	59	2.670353755551	.000232214844
1.047197551197	.288329027680	60	2.633701841259	.000180795372
1.064650843717	.254316704200	61	2.597049926968	.000092557007
1.082104136236	.223011510532	62	2.560398012676	-.000022904414
1.099557428756	.194345389967	63	2.523746098384	-.000156899155
1.117010721276	.169849142248	64	2.488839513344	.000205783528
1.134464013796	.146048037501	65	2.452187599052	.000020301351
1.151917306316	.124619534939	66	2.415535684760	-.000161903753
1.169370598836	.106616052486	67	2.380629099720	.000066338923
1.186823891356	.089456988408	68	2.343977185428	-.000128027324
1.204277183876	.075232686393	69	2.309070600388	.000033045924
1.221730476396	.061879222261	70	2.272418686097	-.000153873647
1.239183768916	.050983705669	71	2.237512101057	-.000038567229
1.256637061436	.041527432102	72	2.202605516017	.000046800944
1.274090353956	.033397778266	73	2.167698930977	.000107049600
1.291543646476	.026055776271	74	2.131047016685	-.000053501942
1.308996938996	.020310961696	75	2.096140431645	-.000007553717
1.326450231516	.015547495839	76	2.061233846605	.000023624640
1.343903524036	.011656351261	77	2.026327261565	.000043037148
1.361356816556	.008531686579	78	1.991420676526	.000053301740
1.378810109076	.005928152119	79	1.954768762234	-.000038918311
1.396263401595	.004070750312	80	1.919862177194	-.000024187316
1.413716694115	.002683781103	81	1.884955592154	-.000014294188
1.431169986635	.001682739326	82	1.850049007114	-.000007938131
1.448623279155	.000990162858	83	1.815142422074	-.000004074198
1.466076571675	.000536235050	84	1.780235837034	-.000001886053
1.483529864195	.000259325036	85	1.745329251994	-.000000758325
1.500983156715	.000106462693	86	1.710422666954	-.000000248585
1.518436449235	.000184845355	87	1.719149313214	.000292426602
1.535889741755	.000343493870	88	1.736602605734	.000643492563
1.553343034275	.000550641200	89	1.754055898254	.000994362509
1.570796326795	.000806034969	90	1.370083462816	-.001344929562

A.9 Code to compute the C_{60} structure

```
1  ! This code computes the C60 molecule structure to any arbitrary
2  ! accuracy. The original idea is from Dr. Allen Broughton of
3  ! Rose-Hulman Institute of Technology who gave GPZ a mathematica
4  ! code.
5        implicit double precision (a-h,o-z)
6        double precision A(3,3),B(3,3),C(3,3) !rotation matrices
7        double precision o(3,3),o1(3,3),o2(3,3),o3(3,3)
8        double precision EA(3),EB(3),EC(3) !eigenvector
9        double precision AEC(3),CEA(3)    !temporary
10       parameter(constant1=0.80901699437494742410d0)
11       parameter(rp=1.45d0,rh=1.40d0,nm=60) ! bond length
12       double precision x(nm),y(nm),z(nm) !coordinates of c60
13       double precision x0(3),y0(3),z0(3)
14       do i=1,nm
15          x(i)=0.d0
16          y(i)=0.d0
17          z(i)=0.d0
18       enddo
19       do i=1,3
20          ea(i)=0.d0
21          eb(i)=0.d0
22          ec(i)=0.d0
23          do j=1,3
24             a(i,j)=0.d0
25             b(i,j)=0.d0
26             c(i,j)=0.d0
27          enddo
28       enddo
29       a(1,1)=1.d0
30       a(2,2)=-1.d0
31       a(3,3)=-1.d0
32       b(1,1)=constant1
33       b(1,2)=0.5d0
34       b(1,3)=constant1-0.5d0
35       b(2,1)=0.5d0
36       b(2,2)=-(constant1-0.5d0)
37       b(2,3)=-constant1
38       b(3,1)=0.5d0-constant1
39       b(3,2)=constant1
40       b(3,3)=-0.5d0
41       c(1,1)=constant1
42       c(1,2)=-0.5d0
43       c(1,3)=constant1-0.5d0
44       c(2,1)=0.5d0
45       c(2,2)=constant1-0.5d0
46       c(2,3)=-constant1
47       c(3,1)=constant1-0.5d0
48       c(3,2)=constant1
49       c(3,3)=0.5d0
50       do i=1,3
51          do l=1,3
52             one=0.d0
53             do j=1,3
```

```
54                do k=1,3
55                   one=one+b(i,j)*b(j,k)*b(k,l)
56                enddo
57             enddo
58             write(*,*)i,l,one,'b^3'
59          enddo
60       enddo
61       do i=1,3
62          do n=1,3
63             one=0.d0
64             do j=1,3
65                do k=1,3
66                   do l=1,3
67                      do m=1,3
68    one=one+c(i,j)*c(j,k)*c(k,l)*c(l,m)*c(m,n)
69                      enddo
70                   enddo
71                enddo
72             enddo
73             write(*,*)i,n,one,'c^5'
74          enddo
75       enddo
76       call operator_mult (b,3,o)
77       do i=1,3
78          do j=1,3
79             write(*,*)i,j,o(i,j),'o'
80          enddo
81       enddo
82       do i=1,3
83          do l=1,3
84             one=0.d0
85             do j=1,3
86                do k=1,3
87                   one=one+a(i,j)*b(j,k)*c(k,l)
88                enddo
89             enddo
90             write(*,*)i,l,one,'abc'
91          enddo
92       enddo
93 c     eigenvector or fixed vector for a,b,c
94       ea(1)=1.d0
95       eb(1)=0.93417235896271569647d0
96       eb(2)=0.35682208977308993193d0
97       write(*,*)eb(1)**2+eb(2)**2,'this␣is␣one'
98       ec(1)=0.850650808352039932l3d0
99       ec(3)=0.525731112119133606O9d0
100      write(*,*)ec(1)**2+ec(3)**2,'this␣is␣one'
101 c    test whether eb,ec are fixed points
102      do i=1,3
103         one=0.d0
104         do j=1,3
105            one=one+b(i,j)*eb(j)
106         enddo
107         write(*,*)i,one,eb(i),'<----last␣2␣columns␣are␣same?'
108      enddo
```

```
109        do i=1,3
110           one=0.d0
111           do j=1,3
112              one=one+c(i,j)*ec(j)
113           enddo
114           write(*,*)i,one,ec(i),'<----last␣2␣columns␣are␣same?'
115        enddo
116 c      find the coordinate for A1 or the first carbon atom
117 !      compute ||AE_c-E_c||
118        do i=1,3
119           aec(i)=0.d0
120           do j=1,3
121              aec(i)=aec(i)+A(i,j)*ec(j)
122           enddo
123           aec(i)=aec(i)-ec(i)
124        enddo
125        aec_=0.d0
126        do i=1,3
127           aec_=aec_+aec(i)**2
128        enddo
129        aec_=dsqrt(aec_)
130        gamma=rh/aec_
131 !      compute ||CE_a-E_a||
132        do i=1,3
133           cea(i)=0.d0
134           do j=1,3
135              cea(i)=cea(i)+C(i,j)*ea(j)
136           enddo
137           cea(i)=cea(i)-ea(i)
138        enddo
139        cea_=0.d0
140        do i=1,3
141           cea_=cea_+cea(i)**2
142        enddo
143        cea_=dsqrt(cea_)
144        write(*,*)1/cea_,1/aec_,'<<<<<<<<<<<<<-'
145        alpha=rp/cea_
146        write(*,*)'----------------------------------------'
147        write(*,*)1.618033988*rp+0.8090169943*rh,0,0.5*rh
148        write(*,*)'----------------------------------------'
149 !      first atom's coordinate
150        x(1)=alpha*ea(1)*1.d0 +gamma*ec(1)
151        y(1)=alpha*ea(2)*1.d0 +gamma*ec(2)
152        z(1)=alpha*ea(3)*1.d0 +gamma*ec(3)
153        r=(dsqrt(x(1)**2+y(1)**2+z(1)**2))
154        do i=1,3
155           write(*,*)
156        enddo
157        write(*,*)'positions␣of␣the␣first␣atom',x(1),y(1),z(1),r
158        do i=1,3
159           write(*,*)
160        enddo
161 !      atoms for layer 1, from 1 to 5
162        do i=1,3
163           do j=1,3
```

```
164            write(*,*)i,j,c(i,j),'::::::::::::'
165         enddo
166      enddo
167      do j=2,5
168         call operator_mult (c,j-1,o)
169         x(j)=o(1,1)*x(1)+o(1,2)*y(1)+o(1,3)*z(1)
170         y(j)=o(2,1)*x(1)+o(2,2)*y(1)+o(2,3)*z(1)
171         z(j)=o(3,1)*x(1)+o(3,2)*y(1)+o(3,3)*z(1)
172      enddo
173 ! write(*,*) dis(x,y,z,1,2),dis(x,y,z,2,3),
174 ! dis(x,y,z,3,4),dis(x,y,z
175 !  ,4,5),dis(x,y,z,5,1)
176 !      atoms for layer 2, from 6 to 30
177      do i=0,4
178         do j=0,4
179            index=5*i+j+6
180            call operator_mult (c,i,o1)
181            call operator_mult (c,j,o2)
182            call matrix_times_matrix(a,o2,o3)
183            call matrix_times_matrix(o1,o3,o2)
184            x(index)=o2(1,1)*x(1)+o2(1,2)*y(1)+o2(1,3)*z(1)
185            y(index)=o2(2,1)*x(1)+o2(2,2)*y(1)+o2(2,3)*z(1)
186            z(index)=o2(3,1)*x(1)+o2(3,2)*y(1)+o2(3,3)*z(1)
187         enddo
188      enddo
189 !      atoms for layer 4, from 31 to 55
190      do i=0,4
191         do j=0,4
192            index=5*i+j+31
193            call operator_mult (c,i,o1)
194            call operator_mult (c,j,o2)
195            call operator_mult (c,2,o3)
196            call matrix_times_matrix(a,o2,o)
197            call matrix_times_matrix(o3,o ,o2)
198            call matrix_times_matrix(a ,o2,o3)
199            call matrix_times_matrix(o1,o3,o )
200            x(index)=o (1,1)*x(1)+o (1,2)*y(1)+o (1,3)*z(1)
201            y(index)=o (2,1)*x(1)+o (2,2)*y(1)+o (2,3)*z(1)
202            z(index)=o (3,1)*x(1)+o (3,2)*y(1)+o (3,3)*z(1)
203         enddo
204      enddo
205 !      atoms  from 56 to 60
206      do j=0,4
207         index=j+56
208         call operator_mult (c,3,o1)
209         call operator_mult (c,2,o2)
210         call operator_mult (c,j,o3)
211         call matrix_times_matrix(a,o3,o)
212         call matrix_times_matrix(o2,o ,o3)
213         call matrix_times_matrix(a ,o3,o2)
214         call matrix_times_matrix(o1,o2,o3)
215         call matrix_times_matrix(a ,o3,o )
216         x(index)=o (1,1)*x(1)+o (1,2)*y(1)+o (1,3)*z(1)
217         y(index)=o (2,1)*x(1)+o (2,2)*y(1)+o (2,3)*z(1)
218         z(index)=o (3,1)*x(1)+o (3,2)*y(1)+o (3,3)*z(1)
219      enddo
220      do i=1,60
221         write(*,125)'C', x(i),y(i),z(i)
222  125    format(a,5x,3(f15.9))
223      enddo
```

```
      open(1,file='c60.allen')
      do i=1,60
         write(1,225)x(i),y(i),z(i)
      enddo
 225  format(3(d24.16,1x))
      close(1)
      x0(1)=x(1)-(x(1)+x(2)+x(3)+x(4)+x(5))/5.d0
      x0(2)=y(1)-(y(1)+y(2)+y(3)+y(4)+y(5))/5.d0
      x0(3)=z(1)-(z(1)+z(2)+z(3)+x(4)+z(5))/5.d0
      y0(1)=x(3)-x(4)
      y0(2)=y(3)-y(4)
      y0(3)=z(3)-z(4)
      zero=0.d0
      do i=1,3
         zero=zero+x0(i)*y0(i)
      enddo
      write(*,*)'zero-xy-->',zero
      read(*,*)
      z0(1)=x0(2)*y0(3)-x0(3)*y0(2)
      z0(2)=x0(3)*y0(1)-x0(1)*y0(3)
      z0(3)=x0(1)*y0(2)-x0(2)*y0(1)
      zero=0.d0
      do i=1,3
         zero=zero+z0(i)*y0(i)
      enddo
      write(*,*)'zero--zy->',zero
      read(*,*)
      zero=0.d0
      do i=1,3
         zero=zero+x0(i)*z0(i)
      enddo
      write(*,*)'zero--xz->',zero
      read(*,*)
      end

      subroutine operator_mult (operator,n,o)
      implicit double precision (a-h,o-z)
!     n is number of order (how many times need compute)
      double precision operator(3,3),o(3,3),ot(3,3)
      if(n.eq.0)then
         do i=1,3
            do j=1,3
               if(i.eq.j)o(i,j)=1.d0
               if(i.ne.j)o(i,j)=0.d0
            enddo
         enddo
         return
      endif
      if(n.eq.1)then
         do i=1,3
            do j=1,3
               o(i,j)=operator(i,j)
            enddo
         enddo
         return
```

```
279         endif
280         do i= 1,3
281            do j=1,3
282               o(i,j)=0.d0
283               ot(i,j)=operator(i,j)
284            enddo
285         enddo
286         iter=n-2
287   1     do i=1,3
288            do j=1,3
289               o(i,j)=0.d0
290               do k=1,3
291                  o(i,j)=o(i,j)+operator(i,k)*ot(k,j)
292               enddo
293            enddo
294         enddo
295         if(iter.ne.0) then
296            iter=iter-1
297            do i=1,3
298               do j=1,3
299                  ot(i,j)=o(i,j)
300               enddo
301            enddo
302            goto 1
303         endif
304         return
305         end
306         double precision function dis(x,y,z,i,j)
307         implicit double precision (a-h,o-z)
308         parameter(nm=60)
309         double precision x(nm),y(nm),z(nm)
310         dis=dsqrt((x(i)-x(j))**2+(y(i)-y(j))**2+(z(i)-z(j))**2)
311         return
312         end
313         subroutine matrix_times_matrix (a,b,o)
314         implicit double precision (a-h,o-z)
315         double precision a(3,3),b(3,3),o(3,3)
316         do i=1,3
317            do j=1,3
318               o(i,j)=0.d0
319               do k=1,3
320                  o(i,j)=o(i,j)+a(i,k)*b(k,j)
321               enddo
322            enddo
323         enddo
324         return
325         end
```

A.10 Genetic algorithm example

The following code needs the Mathematica package. It optimizes the frequency and the duration of a laser pulse for a given few-level system. The most important inputs are the energy levels, and the transition-matrix elements. Note that each run may give different results, since the algorithm is stochastic.

```
(* Below are the properties of the Lambda system,
i.e., transition matrix and unperturbed Hamiltonian
(energies) *)
(* Copyright G. Lefkidis, 2020 *)

nstates=3; (* number of states *)
Dx={{0,1,0},{1,0,1},{0,1,0}};
(* matrix containing the transition matrix elements,
must by hermitian *)
energies={0,1,0};
(* list containing the energies of the states *)
H0=DiagonalMatrix[energies];
(* don't change this line*)

(* Below are the boundary conditions: what we are
looking for, i.e., initial state and target state,
and target fidelity *)
initial=1;target=3;targetfidelity=0.95;
(*fidelity target *)

(* === These are for the genetic algorithm ======= *)
sigmamax=500; (* maximum sigma *)
freqmin=0.5; (*  minimum frequency *)
freqmax=1.5(Max[ energies]-Min[energies]);
(* maximum frequency is set to 1.5 the maximum energy
 difference in the system *)
nindividuals=50;
(* number of individuals in every generation *)
ngenerations=30;
(* maximum number of generations *)
tmax=1000;
(* propagation time is between -tmax and tmax *)
nkept=Round[nindividuals/4];
(* number of kept individuals *)
ncrossed=Round[nindividuals/(2 2)];
(* number of crossed individual couples *)
nmutated = Round[nindividuals/10];
(* number of mutated individuals *)

(* ========= Some initializations =============*)

rho0=Table[0,{i,nstates},{j,nstates}];
rho0[[initial,initial]]=1;
If[HermitianMatrixQ[Dx]False,
  Print["The transition-elements matrix Dx is not
    hermition. Stopping..."];
  Print["Dx = ",MatrixForm[Dx]];Exit[]];
If[Or[initial>nstates ,target>nstates],
 Print["Please check initial and final states.
  Stopping..."];
 Exit[]]
```

```
(* First completely random generation *)
individual=Table[{RandomReal[{freqmin,freqmax}],
    RandomReal[{0,sigmamax}]},{i,nindividuals}];

commutator[x_,y_]:=x.y-y.x
fidelity=Table[0,{j,nindividuals}];
fitness=Table[-1,{j,nindividuals}];
bestfidelity=0;
bestindividual={0,0};

(* ==== propagation of the generation === *)
indsdone=0.;
Print["Current generation ",
 Dynamic[PaddedForm[indsdone,{3,1}]],
 "% done"]
Do[
 Do[
  If[gen>1&&ind<nkept,ind=nkept+1];
   (* after the first generation skip the
  calculation for the kept individuals *)
  omega=individual[[ind,1]];
  sigma=individual[[ind,2]];
  Clear[a,rho,t];
  rho[t_]:=Table[a[i,j][t],{i,nstates},
    {j,nstates}];
  init=rho[-tmax]==rho0;
  sol=Flatten@NDSolve[
     LogicalExpand[
      rho'[t]-I
          commutator[H0+0.01Exp[-(t/sigma)2]
             Cos[omega t]Dx,rho[t]]&&init],
     Flatten[rho[t]],
     {t,-tmax,tmax},MaxSteps?1010];
  fidelity[[ind]]=
   Chop@Re[a[target,target][t]/.sol/.{t?tmax}];
  If[fidelity[[ind]]>bestfidelity,
   bestfidelity=fidelity[[ind]];
   bestindividual={omega,sigma};
   bestgeneration=gen];
  maxpop=
   Max[Table[Re[a[target,target][t]]
       /.sol/.{t?2 tmax/100 i},{i,-20,20}]];
  (* the maximum population of the target
   state during the process *)
  fitness[[ind]]=fidelity[[ind]]2+
    0.7maxpop (1-fidelity[[ind]]);
  (* for the definition see Hartenstein et al.,
  J.Phys.D:Appl.Phys.41,164006 (2008) *)

  indsdone=ind/nindividuals 100.;

  ,{ind,nindividuals}];
 li={fitness,individual,fidelity};
 lo=Transpose[Reverse@SortBy[
     Transpose[li],First]];
```

```
 Print["Best individual in generation ",gen,
  " has fitness = ",lo[[1,1]],
  ", fidelity = ",lo[[3,1]],", omega = ",
  lo[[2,1,1]],", and sigma = ",lo[[2,1,2]]];
 If[lo[[3,1]]>targetfidelity,
  Print["Target fidelity of ",targetfidelity,
    " reached. Exiting."];
  Break[]];
 (* create next generation *)
 Do[individual[[i]]=lo[[2,i]];
  fidelity[[i]]=lo[[3,i]];
  fitness[[i]]=lo[[1,i]],{i,1,nkept}] ;
 (*  keep the 25% best individuals and
 their fidelities *)

 Do[
  individual[[2i-1+nkept]]={ lo[[2,i,1]],
     lo[[2,2ncrossed-i+1,2]]};
  individual[[2i+nkept+1]]=
   { lo[[2,2ncrossed-i+1,1]], lo[[2,i,2]]},
  {i,1,ncrossed}] ;
 (* combine the 50% best individuals *)

 Do[p=RandomReal[{-1,1}];
   If[p>0,
    individual[[i]]={lo[[2,i,1]],
      RandomReal[{0,sigmamax}]},
    individual[[i]]=
     {RandomReal[{freqmin,freqmax}],
      lo[[2,i,2]]}],
   {i,nkept+2ncrossed+1,
    nkept+2ncrossed+nmutated}]

  Do[individual[[i]]=
    {RandomReal[{freqmin,freqmax}],
     RandomReal[{0,sigmamax}]},
   {i,1+nkept+ncrossed+nmutated,
    nindividuals}] ;
 (* create 25% new random individuals *)
 ,{gen,1,ngenerations}]

(* propagate the best individual to plot *)
omega=bestindividual[[1]];
sigma=bestindividual[[2]];
 (* retrieve frequency and sigma from the
best individual of the last generation *)
sol=Flatten@NDSolve[LogicalExpand[
     rho'[t]-I
        commutator[H0+0.01Exp[-(t/sigma)2]
           Cos[omega t]Dx,rho[t]]&&init],
    Flatten[rho[t]],
    {t,-tmax,tmax},MaxSteps?1010];
(* Plot populations *)
Print["Plotting the propagation of the
 best individual (from generation ",
```

```
bestgeneration,") with fidelity ",
bestfidelity]
Plot[Evaluate[Table[Re[rho[t][[i,i]]]/.sol,
   {i,1,nstates}]],{t,-tmax,tmax},
 PlotStyle?Thick]
```

A.11 Special crystal momentum points and lines

The following tables list coordinates of crystal momentum points [Lax (1974)]. They are expressed in conventional reciprocal lattice vectors. Tyler Jenkins entered the tables.

Table A.1
All the coordinates are represented in their conventional reciprocal lattice vectors. Here α changes from 0 to 1.

Point	Triclinic	Simple monoclinic	Center monoclinic	Simple orthorhombic	Base center orthorhombic	Face centered orthorhombic	Body centered orthorhombic
Γ	(0,0,0)	(0,0,0)	(0,0,0)	(0,0,0)	(0,0,0)	(0,0,0)	(0,0,0)
Δ				$(0,\alpha,0)$	$(0,\alpha,0)$	$(0,\alpha,0)$	$(0,\alpha,0)$
Λ		$(0,0,\alpha)$	$(0,0,\alpha)$	$(0,0,\alpha)$	$(0,0,\alpha)$	$(0,0,\alpha)$	$(0,0,\alpha)$
Σ				$(\alpha,0,0)$	$(\alpha,0,0)$	$(\alpha,0,0)$	$(\alpha,0,0)$
A		$(\frac{1}{2},-\frac{1}{2},\frac{1}{2})$	$(\frac{1}{2},0,0)$	$(\alpha,0,\frac{1}{2})$	$(\alpha,0,\frac{1}{2})$	$(\alpha,0,\frac{1}{2})$	
B		$(\frac{1}{2},0,0)$		$(0,\alpha,\frac{1}{2})$	$(0,\alpha,\frac{1}{2})$	$(0,\alpha,\frac{1}{2})$	
C		$(0,\frac{1}{2},\frac{1}{2})$		$(\alpha,\frac{1}{2},0)$	$(\alpha,\frac{1}{2},0)$	$(\alpha,\frac{1}{2},0)$	
D		$(\frac{1}{2},0,\frac{1}{2})$		$(\frac{1}{2},\alpha,0)$	$(\frac{1}{4},\frac{1}{4},\frac{1}{4})$		$(\alpha,\frac{1}{4},\frac{1}{4})$
E		$(\frac{1}{2},-\frac{1}{2},\frac{1}{2})$		$(\alpha,\frac{1}{2},\frac{1}{2})$			
F							
G				$(\frac{1}{2},0,\alpha)$			$(\frac{1}{2},0,\alpha)$
H				$(0,\frac{1}{2},\alpha)$	$(0,\frac{1}{2},\alpha)$	$(0,\frac{1}{2},\alpha)$	
L			$(\frac{1}{2},\frac{1}{4},\frac{1}{4})$			$(\frac{1}{4},\frac{1}{4},\frac{1}{4})$	
M			$(\frac{1}{2},0,\frac{1}{2})$				
N							
P				$(\frac{1}{2},\alpha,\frac{1}{2})$			$(\frac{1}{4},\frac{1}{4},\alpha)$
Q				$(\frac{1}{2},\frac{1}{2},\alpha)$			$(\frac{1}{4},\alpha,\frac{1}{4})$
R	$(\frac{1}{2},\frac{1}{2},\frac{1}{2})$			$(\frac{1}{2},\frac{1}{2},\frac{1}{2})$	$(\frac{1}{4},\frac{1}{4},\frac{1}{2})$		$(\frac{1}{4},0,\frac{1}{4})$
S				$(\frac{1}{2},\frac{1}{2},0)$	$(\frac{1}{4},\frac{1}{4},0)$		$(0,\frac{1}{4},\frac{1}{4})$
T				$(0,\frac{1}{2},\frac{1}{2})$	$(0,\frac{1}{2},\frac{1}{2})$	$(0,\frac{1}{2},\frac{1}{2})$	$(\frac{1}{4},\frac{1}{4},0)$
U	$(\frac{1}{2},0,\frac{1}{2})$	$(\frac{1}{2},-\frac{1}{2},\alpha)$	$(\frac{1}{2},0,\alpha)$	$(\frac{1}{2},0,\frac{1}{2})$		$(\alpha,\frac{1}{2},\frac{1}{2})$	$(\frac{1}{2},\alpha,0)$
V	$(\frac{1}{2},0,\frac{1}{2})$	$(\frac{1}{2},0,\alpha)$	$(0,\frac{1}{4},\frac{1}{4})$				
W		$(0,\frac{1}{2},\alpha)$					$(\frac{1}{4},\frac{1}{4},\frac{1}{4})$
X	$(\frac{1}{2},0,0)$			$(\frac{1}{2},0,0)$			$(\frac{1}{2},0,0)$
Y		$(0,\frac{1}{2},0)$	$(0,0,\frac{1}{2})$	$(0,\frac{1}{2},0)$	$(0,\frac{1}{2},0)$	$(0,\frac{1}{2},0)$	
Z	$(0,0,\frac{1}{2})$	$(0,0,\frac{1}{2})$		$(0,0,\frac{1}{2})$	$(0,0,\frac{1}{2})$	$(0,0,\frac{1}{2})$	

Table A.2
The remaining coordinates for other structures. $\beta = \frac{1}{2} - \alpha$.

Point	Simple tetragonal	BC	SC	FCC	BCC	Rhombohedral	Hexagonal
Γ	$(0,0,0)$	$(0,0,0)$	$(0,0,0)$	$(0,0,0)$	$(0,0,0)$	$(0,0,0)$	$(0,0,0)$
Δ	$(0,\alpha,0)$	$(0,\alpha,0)$	$(0,\alpha,0)$	$(0,\alpha,0)$	$(0,\alpha,0)$		$(0,0,\alpha)$
Λ	$(0,0,\alpha)$	$(0,0,\alpha)$	(α,α,α)	(α,α,α)	(α,α,α)	$(0,0,\alpha)$	$(\alpha,\alpha,0)$
Σ	$(\alpha,\alpha,0)$	$(\alpha,\alpha,0)$	$(\alpha,\alpha,0)$	$(\alpha,\alpha,0)$	$(\alpha,\alpha,0)$	$(\alpha,0,0)$	$(\alpha,0,0)$
A	$(\frac{1}{2},\frac{1}{2},\frac{1}{2})$						$(0,0,\frac{1}{2})$
B							
C							
D					$(\frac{1}{4},\frac{1}{4},\alpha)$		
E							
F					(α,β,α)	$(\frac{1}{2},0,0)$	
G					$(\alpha,\beta,0)$		
H					$(0,\frac{1}{2},0)$		$(\frac{1}{3},\frac{1}{3},\frac{1}{3})$
K							$(\frac{1}{3},\frac{1}{3},0)$
L				$(\frac{1}{4},\frac{1}{4},\frac{1}{4})$		$(\frac{1}{6},\frac{1}{3},\frac{1}{6})$	$(\frac{1}{2},0,\frac{1}{2})$
M	$(\frac{1}{2},\frac{1}{2},0)$	$(\frac{1}{2},\frac{1}{2},0)$	$(\frac{1}{2},\frac{1}{2},0)$				$(\frac{1}{2},0,0)$
N		$(\frac{1}{4},\frac{1}{4},\frac{1}{4})$			$(\frac{1}{4},\frac{1}{4},0)$		
P		$(0,\frac{1}{2},\frac{1}{4})$			$(\frac{1}{4},\frac{1}{4},\frac{1}{4})$	$(\frac{2}{3},\frac{1}{3},\alpha)$	$(\frac{1}{3},\frac{1}{3},\alpha)$
Q		(α,β,α)		$(\frac{1}{4},\beta,\alpha)$			$(\alpha,\alpha,\frac{1}{2})$
R	$(0,\frac{1}{2},\frac{1}{2})$		$(\frac{1}{2},\frac{1}{2},\frac{1}{2})$				$(\alpha,0,\frac{1}{2})$
S	$(\alpha,\alpha,\frac{1}{2})$		(α,α,α)	$(\alpha,\frac{1}{2},\alpha)$			$(\beta,2\alpha,\frac{1}{2})$
T	$(\alpha,\frac{1}{2},\frac{1}{2})$		$(\frac{1}{2},\frac{1}{2},\alpha)$			$(\frac{2}{3},\frac{1}{3},\frac{1}{6})$	$(\beta,2\alpha,0)$
U	$(0,\alpha,\frac{1}{2})$						$(\frac{1}{2},0,\alpha)$
V	$(\frac{1}{2},\frac{1}{2},\alpha)$	$(\frac{1}{2},\frac{1}{2},\alpha)$		$(\alpha,\frac{1}{2},0)$			
W	$(0,\frac{1}{2},\alpha)$	$(0,\frac{1}{2},\alpha)$		$(\frac{1}{4},\frac{1}{2},0)$			
X	$(0,\frac{1}{2},0)$	$(0,\frac{1}{2},0)$	$(0,\frac{1}{2},0)$	$(0,\frac{1}{2},0)$			
Y	$(\alpha,\frac{1}{2},0)$	$(\alpha,\frac{1}{2},0)$				$(\alpha,\frac{1}{3},\frac{1}{6})$	
Z	$(0,0,\frac{1}{2})$		$(\alpha,\frac{1}{2},0)$				

Bibliography

Aeschlimann, M., Bauer, M., Pawlik, S., Weber, W., Burgermeister, R., Oberli, D., and Siegmann, H. C. (1997). Ultrafast spin-dependent electron dynamics in fcc Co. *Phys. Rev. Lett.*, 79:5158.

Al-Naib, I., Sipe, J. E., and Dignam, M. M. (2014). High harmonic generation in undoped graphene: Interplay of inter- and intraband dynamics. *Phys. Rev. B*, 90:245423.

Alebrand, S., Gottwald, M., Hehn, M., Steil, D., Cinchetti, M., Lacour, D., Fullerton, E. E., Aeschlimann, M., and Mangin, S. (2012a). Light-induced magnetization reversal of high-anisotropy TbCo alloy films. *Appl. Phys. Lett.*, 101:162408.

Alebrand, S., Hassdenteufel, A., Steil, D., Cinchetti, M., and Aeschlimann, M. (2012b). Interplay of heating and helicity in all-optical magnetization switching. *Phys. Rev. B*, 85:092401.

Andres, B., Weiss, P., Wietstruk, M., and Weinelt, M. (2015). Spin-dependent lifetime and exchange splitting of surface states on Ni(110). *J. Phys.: Condens. Mat.*, 27:015503.

Angeli, C., Cimiraglia, R., and Malrieu, J. P. (2001). Introduction of n-electron valence states for multireference perturbation theory. *J. Chem. Phys.*, 114:10252–10264.

Ashcroft, N. W. and Mermin, N. D. (1976). *Solid State Physics*. Holt, Rinehart and Winston, 1st edition.

Baierl, S., Mentink, J. H., Hohenleutner, M., Braun, L., Do, T.-M., Lange, C., Sell, A., Fiebig, M., Woltersdorf, G., Kampfrath, T., and Huber, R. (2016). Terahertz-driven nonlinear spin response of antiferromagnetic nickel oxide. *Phys. Rev. Lett.*, 117:197201.

Balistreri, M. L., Gersen, H., Korterik, J. P., Kuipers, L., and van Hulst, N. F. (2001). Tracking femtosecond laser pulses in space and time. *Science*, 294:1080–1082.

Banerjee, C., Teichert, N., Siewierska, K., Gercsi, Z., Atcheson, G., Stamenov, P., Rode, K., Coey, J. M. D., and Besbas, J. (2019). Single pulse all-optical toggle switching of magnetization without Gd: The example of Mn_2Ru_xGa. *arXiv:1909.05809.*

Battiato, M., Carva, K., and Oppeneer, P. M. (2010). Superdiffusive spin transport as a mechanism of ultrafast demagnetization. *Phys. Rev. Lett.*, 105:027203.

Bauer, D., Ceccherini, F., Macchi, A., and Cornolti, F. (2001). C_{60} in intense femtosecond laser pulses: Nonlinear dipole response and ionization. *Phys. Rev. A*, 64:063203.

Baumberg, J. J., Awschalom, D. D., Samarth, N., Luo, H., and Furdyna, J. K. (1994). Spin beats and dynamical magnetization in quantum structures. *Phys. Rev. Lett.*, 72:717.

Beaurepaire, E., Merle, J.-C., Daunois, A., and Bigot, J.-Y. (1996). Ultrafast spin dynamics in ferromagnetic nickel. *Phys. Rev. Lett.*, 76:4250–4253.

Becke, A. D. (1988). Density-functional exchange-energy approximation with correct asymptotic behavior. *Phys. Rev. A*, 38:3098–3100.

Bennett, H. S. and Stern, E. A. (1965). Faraday effect in solids. *Phys. Rev.*, 137:A448–A461.

Berlasso, R., Dallera, C., Borgatti, F., Vozzi, C., Sansone, G., Stagira, S., Nisoli, M., Ghiringhelli, G., Villoresi, P., Poletto, L., Pascolini, M., Nannarone, S., Silvestri, S. D., and Braicovich, L. (2006). High-order laser harmonics and synchrotron study of transition metals $M_{2,3}$ edges. *Phys. Rev. B*, 73:11510.

Bersuker, I. B. (1996). *Electronic Structure and Properties of Transition Metal Compounds*, page 37. Wiley-Interscience.

Bethe, H. (1929). Termaufspaltung in kristallen. *Annalen der Physik*, 395(2):133–208.

Bigot, J.-Y., Vomir, M., and Beaurepaire, E. (2009). Coherent ultrafast magnetism induced by femtosecond laser pulses. *Nat. Phys.*, 5:515–520.

Birss, R. R. (1964). *Symmetry and Magnetism.* North-Holland Publishing Company, Amsterdam.

Bishop, D. M. (1973). *Group Theory and Chemistry.* Clarendon Press, Oxford.

Blaha, P., Schwarz, K., Madsen, G. K. H., Kvasnicka, D., Luitz, J., Laskowski, R., Tran, F., and Marks, L. D. (2018). *WIEN2k, An Augmented Plane Wave + Local Orbitals Program for Calculating Crystal Properties.* Techn. Universität Wien, Austria.

Boeglin, C., Beaurepaire, E., Halte, V., Lopez-Flores, V., Stamm, C., Pontius, N., Dürr, H. A., and Bigot, J.-Y. (2010). Distinguishing the ultrafast dynamics of spin and orbital moments in solids. *Nature*, 465:458.

Borwein, J. and Bailey, D. (2004). *Mathematics by Experiment: Plausible Reasoning in the 21st Century.* Taylor and Francis Group, LLC, Boca Raton, 2nd edition.

Bovensiepen, U. (2006). Ultra-fast dynamics of coherent lattice and spin excitations at the Gd(0001) surface. *Appl. Phys. A*, 82:395.

Bowlan, P., Martinez-Moreno, E., Reimann, K., Elsaesser, T., and Woerner, M. (2014). Ultrafast terahertz response of multilayer graphene in the nonperturbative regime. *Phys. Rev. B*, 89:041408(R).

Boyd, R. W. (1991). *Nonlinear Optics.* Academic Press.

Brorson, S. D., Kazeroonian, A., Moodera, J. S., Face, D. W., Cheng, T. K., Ippen, E. P., Dresselhaus, M. S., and Dresselhaus, G. (1990). Femtosecond room-temperature measurement of the electron-phonon coupling constant γ in metallic superconductors. *Phys. Rev. Lett.*, 64:2172–2175.

Bunge, C. F., Barrientos, J. A., Bunge, A. V., and Cogordan, J. A. (1992). Hartree-Fock and Roothaan-Hartree-Fock energies for the ground states of He through Xe. *Phys. Rev. A*, 46:3691–3696.

Butcher, P. N. and Cotter, D. (1990). *The Elements of Nonlinear Optics.* Cambridge University Press.

Callaway, J. (1976). *Quantum Theory of the Solid State.* Academic Press, Inc.

Carley, R., Döbrich, K., Frietsch, B., Gah, C., Teichmann, M., Schwarzkopf, O., Wernet, P., and Weinelt, M. (2012). Femtosecond laser excitation drives ferromagnetic gadolinium out of magnetic equilibrium. *Phys. Rev. Lett.*, 109:057401.

Carpene, E., Mancini, E., Dallera, C., Brenna, M., Puppin, E., and Silvestri, S. D. (2008). Dynamics of electron-magnon interaction and ultrafast demagnetization in thin iron films. *Phys. Rev. B*, 78:174422.

Chaudhari, P., Cuomo, J. J., and Gambino, R. J. (1973). Amorphous metallic films for magneto-optic applications. *Appl. Phys. Lett.*, 22:337.

Chaudhuri, D., Lefkidis, G., and Hübner, W. (2017). All-spin-based ultrafast nano-logic elements with a Ni_4 cluster. *Phys. Rev. B*, 96:184413.

Chaudhuri, D., Lefkidis, G., Kubas, A., Fink, K., and Hübner, W. (2015). Effect of the variation of the bond length on laser-induced spin-flip scenarios at Ni_2. *Springer Proc. Phys.*, 159:159.

Chaudhuri, D., Xiang, H., Lefkidis, G., and Hübner, W. (2014). Laser-induced ultrafast dynamics in di-, tri- and tetranuclear nickel clusters and the M process. *Phys. Rev. B*, 90:245113.

Chizhova, L. A., Libisch, F., and Burgdörfer, J. (2017). High-harmonic generation in graphene: Interband response and the harmonic cutoff. *Phys. Rev. B*, 95:08436.

Chu, S.-I. and Cooper, J. (1985). Threshold shift and above-threshold multiphoton ionization of atomic hydrogen in intense laser fields. *Phys. Rev. A*, 32:2769–2775.

Ciappina, M. F., Perez-Hernandez, J. A., Landsman, A. S., Okell, W. A., Zherebtsov, S., Förg, B., Schötz, J., Seiffert, L., Fennel, T., Shaaran, T., Zimmermann, T., Chacon, A., Guichard, R., Zair, A., Tisch, J. W. G., Marangos, J. P., Witting, T., Braun, A., Maier, S. A., Roso, L., Krüger, M., Hommelhoff, P., Kling, M. F., Krausz, F., and Lewenstein, M. (2017). Attosecond physics at the nanoscale. *Rep. Prog. Phys.*, 80:054401.

Coey, J. M. D. (2010). *Magnetism and Magnetic Materials.* Cambridge University Press.

Corkum, P. B. (1993). Plasma perspective on strong field multiphoton ionization. *Phys. Rev. Lett.*, 71:1994–1997.

Cox, J. D., Marini, A., and de Abajo, G. J. G. (2017). Plasmon-assisted high-harmonic generation in graphene. *Nat. Comm.*, 8:14380.

Cuberes, M. T., Schlittler, R. R., and Gimzewski, J. K. (1996). Room-temperature repositioning of individual C_{60} molecules at Cu steps: Operation of a molecular counting device. *Appl. Phys. Lett.*, 69(20):3016–3018.

Cywiński, L. and Sham, L. J. (2007). Ultrafast demagnetization in the $sp-d$ model: A theoretical study. *Phys. Rev. B*, 76:045205.

dalla Longa, F., Kohlhepp, J. T., de Jonge, W. J. M., and Koopmans, B. (2007). Influence of photon angular momentum on ultrafast demagnetization in nickel. *Phys. Rev. B*, 75:224431.

Dantus, M., Bowman, R. M., and Zewail, A. H. (1990). Femtosecond laser observations of molecular vibration and rotation. *Nature*, 343:737–739.

David, W. I. F., Ibberson, R. M., Matthewman, J. C., Prassides, K., Dennis, T. J. S., Hare, J. P., Kroto, H. W., Taylor, R., and Walton, D. R. M. (1991). Crystal structure and bonding of ordered C_{60}. *Nature*, 353:147.

Dimitrovski, D., Madsen, L. B., and Pedersen, T. G. (2017). High-order harmonic generation from gapped graphene: Perturbative response and transitions to nonperturbative regime. *Phys. Rev. B*, 95:035405.

Dirac, P. A. M. (1930). Note on exchange phenomena in the Thomas atom. *Proc. Camb. Phil Soc.*, 26:376.

Dong, C. D., Lefkidis, G., and Hübner, W. (2013). Magnetic quantum Diesel engine in Ni_2. *Phys. Rev. B*, 88:214421.

Donnelly, T. D. and Grossman, C. (1998). Ultrafast phenomena: A laboratory experiment for undergraduates. *Am. J. Phys.*, 66(8):677–685.

Dornes, C., Acremann, Y., Kubli, M., Neugebauer, M. J., Abreu, E., Huber, L., Lantz, G., Vaz, C. A. F., Lemke, H., Bothschafter, E. M., Porer, M., Esposito, V., Rettig, L., Buzzi, M., Alberca, A., Windsor, Y. W., Beaud, P., Staub, U., Zhu, D., Song, S., Glownia, J. M., and Johnson, S. L. (2019). The ultrafast einstein–de haas effect. *Nature*, 565:209.

Dörr, M., Potvliege, R. M., and Shakeshaft, R. (1990). Multiphoton processes in an intense laser field: III. Resonant ionization of hydrogen by subpicosecond pulses. *Phys. Rev. A*, 41:558–561.

Dorset, D. L. and McCourt, M. P. (1994). Disorder and the molecular packing of C_{60} buckminsterfullerene: a direct electron-crystallographic analysis. *Acta Crystallogr. A*, 50:344.

Duong, N. P., Satoh, T., and Fiebig, M. (2004). Ultrafast manipulation of antiferromagnetism of NiO. *Phys. Rev. Lett.*, 93:117402.

Dutta, D., Becherer, M., Bellaire, D., Dietrich, F., Gerhards, M., Lefkidis, G., and Hübner, W. (2018). Characterization of the isolated $[Co_3Ni(EtOH)]^+$ cluster by IR spectroscopy and spin-dynamics calculations. *Phys. Rev. B*, 97:224404.

Elliott, R. J. (1954). Theory of the effect of spin-orbit coupling on magnetic resonance in some semiconductors. *Phys. Rev.*, (96):266–299.

El'yashevich, M. A. (1953). *Rare Earth Spectra*, page 456. Gostechteorizdat.

Faisal, F. H. M. and Kamiński, J. Z. (1996). Generation and control of high harmonics by laser interaction with transmission electrons in a thin crystal. *Phys. Rev. A*, 54:R1769–R1772.

Faisal, F. H. M. and Kamiński, J. Z. (1996). Generation and control of high harmonics by laser interaction with transmission electrons in a thin crystal. *Phys. Rev. A*, 54:R1769.

Faisal, F. H. M. and Kamiński, J. Z. (1997). Floquet-Bloch theory of high-harmonic generation in periodic structures. *Phys. Rev. A*, 56:748.

Farkas, G., Tóth, C., Moustaizis, S. D., Papadogiannis, N. A., and Fotakis, C. (1992). Observation of multiple-harmonic radiation induced from a gold surface by picosecond neodymium-doped yttrium aluminum garnet laser pulses. *Phys. Rev. A*, 46:R3605.

Farle, M. (1998). Ferromagnetic resonance of ultrathin metallic layers. *Rep. Prog. Phys.*, 61:755.

Fayer, M. D. (2013). *Ultrafast Infrared Vibrational Spectroscopy*. CRC Press.

Feit, M. D., Fleck, J. A., and Steiger, A. (1982). Solution of the Schrödinger equation by a spectral method. *J. Comput. Phys.*, 47:412–433.

Ferrante, C., Pontecorvo, E., Cerullo, G., Vos, M. H., and Scopigno, T. (2016). Direct observation of subpicosecond vibrational dynamics in photoexcited myoglobin. *Nat. Chem.*, 8:1137–1143.

Ferray, M., L'Huillier, A., Li, X. F., Lompre, L. A., Mainfray, G., and Manus, C. (1988). Multiple-harmonic conversion of 1064 nm radiation in rare gases. *J. Phys. B: At. Mol. Opt. Phys.*, 21(3):L31–32.

Feynman, R. P. (1939). Forces in molecules. *Phys. Rev.*, 56:340–343.

Fiebig, M., Duong, N. P., Satoh, T., Aken, B. B. V., Miyano, K., Tomioka, Y., and Tokura, Y. (2008). Ultrafast magnetization dynamics of antiferromagnetic compounds. *J. Phys. D*, 41(16):164005.

Fiebig, M., Fröhlich, D., Lottermoser, T., Pavlov, V. V., Pisarev, R. V., and Weber, H.-J. (2001). Second harmonic generation in the centrosymmetric antiferromagnet NiO. *Phys. Rev. Lett.*, 87:137202.

Fleischhauer, M., Imamoglu, A., and Marangos, J. P. (2005). Electromagnetically induced transparency: Optics in coherent media. *Rev. Mod. Phys.*, 77:633–673.

Frait, Z. and Gemperle, R. (1971). The g-factor and surface magnetization of pure iron along [100] and [111] directions. *J. Physique*, 32:C1–541.

Furlani, E. P. (1997). The field analysis and simulation of a permanent-magnet bias-field device for magneto-optic recording. *J. Phys. D: Appl. Phys.*, 30:1846.

Gai, F., Hasson, K. C., McDonald, J. C., and Anfinrud, P. A. (1998). Chemical dynamics in proteins: The photoisomerization of retinal in bacteriorhodopsin. *Science*, 279(5358):1886–1891.

Ganeev, R., Bom, L., Abdul-Hadi, J., Wong, M., Brichta, J., Bhardwaj, V., and Ozaki, T. (2009a). Higher-order harmonic generation from fullerene by means of the plasma harmonic method. *Phys. Rev. Lett.*, 102:013903.

Ganeev, R., Bom, L. E., Wong, M. C. H., Brichta, J.-P., Bhardwaj, V., Redkin, P., and Ozaki, T. (2009b). High-order harmonic generation from C_{60}-rich plasma. *Phys. Rev. A*, 80:043808.

Gao, C.-Z., Dinh, P. M., Klüpfel, P., Meier, C., Reinhard, P.-G., and Suraud, E. (2016). Strong-field effects in the photoemission spectrum of the C_{60} fullerene. *Phys. Rev. A*, 93:022506.

Garello, K., Avci, C. O., Miron, I. M., Baumgartner, M., Ghosh, A., Auffret, S., Boulle, O., Gaudin, G., and Gambardella, P. (2014). Ultrafast magnetization switching by spin-orbit torques. *Appl. Phys. Lett.*, 105(21):212402.

Garello, K., Miron, I. M., Avci, C. O., Freimuth, F., Mokrousov, Y., Blügel, S., Auffret, S., Boulle, O., Gaudin, G., and Gambardella, P. (2013). Symmetry and magnitude of spin–orbit torques in ferromagnetic heterostructures. *Nat. Nanotechnol.*, 8:587.

Garg, M., Zhan, M., Luu, T. T., Lakhotia, H., Klostermann, T., Guggenmos, A., and Goulielmakis, E. (2016). Multi-petahertz electronic metrology. *Nature*, 538:359.

Gau, J. S. (1989). Magneto-optical recording materials. *Mater. Sci. Eng. B*, 3:371.

Gerrits, T., van den Berg, H. A. M., Hohlfeld, J., Bär, L., and Rasing, T. (2002). Ultrafast precessional magnetization reversal by picosecond magnetic field pulse shaping. *Nature*, 418(6897):509–512.

Ghimire, S., DiChiara, A. D., Sistrunk, E., Agostini, P., DiMauro, L. F., and Reis, D. A. (2011). Observation of high-order harmonic generation in a bulk crystal. *Nat. Phys.*, 7:138–141.

Gierster, L., Ünal, A. A., Pape, L., Radu, F., and Kronast, F. (2015). Laser induced magnetization switching in a TbFeCo ferrimagnetic thin film: Discerning the impact of dipolar fields, laser heating and laser helicity by XPEEM. *Ultramicroscopy*, 159:508.

Gómez-Abal, R., Ney, O., Satitkovitchai, K., and Hübner, W. (2004). All-optical subpicosecond magnetic switching in NiO(001). *Phys. Rev. Lett.*, 92:227402.

Gong, Y., Kutayiah, A. R., Zhang, X. H., Zhao, J. H., and Ren, Y. H. (2012). Non-thermal excitation and control of magnetization in Fe/GaAs film by ultrafast laser pulses. *J. App. Phys.*, 111:07D505.

Goodenough, J. B. (1958). An interpretation of the magnetic properties of the perovskite-type mixed crystals $La_{1-x}Sr_xCoO_{3-\lambda}$. *J. Phys. and Chem. Solids*, 6(2):287–297.

Gorchon, J., Lambert, C.-H., Yang, Y., Pattabi, A., Wilson, R. B., Salahuddin, S., and Bokor, J. (2017). Single shot ultrafast all optical magnetization switching of ferromagnetic Co/Pt multilayers. *Appl. Phys. Lett.*, 111:042401.

Graf, T., Casper, F., Winterlik, J., Balke, B., Fecher, G. H., and Felser, C. (2009). Crystal structure of new Heusler compounds. *Z. Anorg. Allg. Chem.*, 635:976.

Griffiths, D. J. (1981). *Introduction to Electrodynamics*, chapter 9, page 359. Prentice-Hall Inc.

Gritsenko, O. V., Cordero, N. A., Rubio, A., Balbás, L. C., and Alonso, J. A. (1993). Weighted-density exchange and local-density coulomb correlation energy functionals for finite systems: Application to atoms. *Phys. Rev. A*, 48:4197–4212.

Guidoni, L., Beaurepaire, E., and Bigot, J.-Y. (2002). Magneto-optics in the ultrafast regime: Thermalization of spin populations in ferromagnetic films. *Phys. Rev. Lett.*, 89:017401.

Gutzwiller, M. C. (1963). Effect of correlation on the ferromagnetism of transition metals. *Phys. Rev. Lett.*, 10:159.

Hadri, M. S. E., Hehn, M., Pirro, P., Lambert, C.-H., Malinowski, G., Fullerton, E. E., and Mangin, S. (2016). Domain size criterion for the observation of all-optical helicity-dependent switching in magnetic thin films. *Phys. Rev. B*, 94:064419.

Hafez, H. A., Kovalev, S., Deinert, J.-C., Mics, Z., Green, B., Awari, N., Chen, M., Germanskiy, S., Lehnert, U., Teichert, J., Wang, Z., Tielrooij, K.-J., Liu, Z., Chen, Z., Narita, A., Müllen, K., Bonn, M., Gensch, M., and Turchinovich, D. (2018). Extremely efficient terahertz high-harmonic generation in graphene by hot Dirac fermions. *Nature*, 561:507.

Haken, H. (1984). *Laser Theory*. Springer-Verlag, Berlin/Heidelberg/New York.

Han, Y.-C. and Madsen, L. B. (2010). Comparison between length and velocity gauges in quantum simulations of high-order harmonic generation. *Phys. Rev. A*, 81:063430.

Hartenstein, T., Li, C., Lefkidis, G., and Hübner, W. (2008). Local light-induced spin manipulation in two-magnetic-center metallic chains. *J. Phys. D: Appl. Phys.*, 41:164006.

Hassdenteufel, A., Hebler, B., Schubert, C., Liebig, A., Teich, M., Helm, M., Aeschlimann, M., Albrecht, M., and Bratschitsch, R. (2013). Thermally assisted all-optical helicity dependent magnetic switching in amorphous $Fe_{100-x}Tb_x$ alloy films. *Adv. Mater.*, 25:3122.

Hassdenteufel, A., Schmidt, J., Schubert, C., Hebler, B., Helm, M., Albrecht, M., and Bratschitsch, R. (2015). Low-remanence criterion for helicity-dependent all-optical magnetic switching in ferrimagnets. *Phys. Rev. B*, 91:104431.

Hassdenteufel, A., Schubert, C., Hebler, B., Schultheiss, H., Fassbender, J., Albrecht, M., and Bratschitsch, R. (2014). All-optical helicity dependent magnetic switching in Tb-Fe thin films with a MHz laser oscillator. *Opt. Express*, 22:10017.

Haug, H. and Koch, S. W. (2009). *Quantum Theory of the Optical and Electronic Properties of Semiconductors*. World Scientific.

Hebler, B., Hassdenteufel, A., Reinhardt, P., Karl, H., and Albrecht, M. (2016). Ferrimagnetic Tb-Fe alloy thin films: composition and thickness dependence of magnetic properties and all-optical switching. *Frontiers in Materials*, 3:8.

Hennecke, M., Radu, I., Abrudan, R., T. Kachel, K. H., Mitzner, R., Tsukamoto, A., and Eisebitt, S. (2019). Angular momentum flow during ultrafast demagnetization of a ferrimagnet. *Phys. Rev. Lett.*, 122:157202.

Hentschel, M., Kienberger, R., Spielmann, C., Reider, G. A., Milosevic, N., Brabec, T., Corkum, P., Heinzmann, U., Drescher, M., and Krausz, F. (2001). Attosecond metrology. *Nature*, 414(4):509–513.

Higuchi, T., Kanda, N., Tamaru, H., and Kuwata-Gonokami, M. (2011). Selection rules for light-induced magnetization of a crystal with threefold symmetry: The case of antiferromagnetic NiO. *Phys. Rev. Lett.*, 106:047401.

Hilbert, M. and López, P. (2011). The world's technological capacity to store, communicate, and compute information. *Science*, 332(6025):60–65.

Ho, T.-S. and Chu, S.-I. (1985). Semiclassical many-mode Floquet theory. III. SU(3) dynamical evolution of three-level systems in intense bichromatic fields. *Phys. Rev. A*, 31:659–676.

Hofherr, M., Maldonado, P., Schmitt, O., Berritta, M., Bierbrauer, U., Sadashivaiah, S., Schellekens, A. J., Koopmans, B., Steil, D., Cinchetti, M., Stadtmüller, B., Oppeneer, P. M., Mathias, S., and Aeschlimann, M. (2017). Speed and efficiency of femtosecond spin current injection into a nonmagnetic material. *Phys. Rev. B*, 96:100403.

Hohenberg, P. and Kohn, W. (1964). Inhomogeneous electron gas. *Phys. Rev.*, 136:B864–B871.

Hohenleutner, M., Langer, F., Schubert, O., Knorr, M., Huttner, U., Koch, S. W., Kira, M., and Huber, R. (2015). Real-time observation of interfering crystal electrons in high-harmonic generation. *Nature*, 523:572.

Hohlfeld, J., Matthias, E., Knorren, R., and Bennemann, K. H. (1997). Nonequilibrium magnetization dynamics of nickel. *Phys. Rev. Lett.*, 78:4861.

Hohlfeld, J., Stanciu, C. D., and Rebel, A. (2009). Athermal all-optical femtosecond magnetization reversal in GdFeCo. *Appl. Phys. Lett.*, 94:152504.

Hommelhoff, P. and Higuchi, T. (2015). Harmonic radiation from crystals. *Nature*, 523:541.

Hubbard, J. (1963). Electron correlations in narrow energy bands. *Proc. Roy. Soc. London A*, 276:238.

Hübner, W. and Bennemann, K.-H. (1989). Nonlinear magneto-optical kerr effect on a nickel surface. *Phys. Rev. B*, 40:5973–5979.

Hübner, W. and Bennemann, K. H. (1996). Simple theory for spin-lattice relaxation in metallic rare-earth ferromagnets. *Phys. Rev. B*, 53:3422–3427.

Hübner, W. and Falicov, L. M. (1993). Theory of spin-polarized electron-capture spectroscopy in ferromagnetic nickel. *Phys. Rev. B*, 47:8783–8793.

Hübner, W., Kersten, S., and Lefkidis, G. (2009). Optical spin manipulation for minimal magnetic logic operations in metallic three-center magnetic clusters. *Phys. Rev. B*, 79:184431.

Hübner, W., Kersten, S., and Lefkidis, G. (2010). Using laser-induced spin manipulation to build magnetic nanologic elements. *J. Phys. Conf. Ser.*, 200:042009.

Hübner, W. and Zhang, G. P. (1998). Ultrafast spin dynamics in nickel. *Phys. Rev. B*, 58:R5920–R5923.

Hutchings, M. T. and Samuelsen, E. J. (1972). Measurement of spin-wave dispersion in NiO by inelastic neutron scattering and its relation to magnetic properties. *Phys. Rev. B*, 6:3447.

Iihama, S., Xu, Y., Deb, M., Malinowski, G., Hehn, M., Gorchon, J., Fullerton, E. E., and Mangin, S. (2018). Single-shot multi-level all-optical magnetization switching mediated by spin-polarized hot electron transport. *arXiv1805.02432.*

Jackson, J. D. (1962). *Classical Electrodynamics.* John Wiley & Sons, Inc.

Jana, S. and et al. (2018). Exchange dependent ultrafast magnetization dynamics in $Fe_{1-x}Ni_x$ alloys. *arXiv1810.11001.*

Jin, W., Becherer, M., Belaire, D., Lefkidis, G., Gerhards, M., and Hübner, W. (2014). Infrared and electronic absorption spectra as well as ultrafast spin dynamics in isolated Co_3^+(EtOH) and Co_3^+((EtOH,H_2O) clusters. *Phys. Rev. B*, 89:144409.

Jin, Z., Ma, H., Wang, L., Ma, G., Guo, F., and Chen, J. (2010). Ultrafast all-optical magnetic switching in $NaTb(WO_4)_2$. *Appl. Phys. Lett.*, 96:201108.

John, R., Berritta, M., Hinzke, D., Müller, C., Santos, T., Ulrichs, H., Nieves, P., Walowski, J., Mondal, R., Chubykalo-Fesenko, O., McCord, J., Oppeneer, P. M., Nowak, U., and Münzenberg, M. (2017). Magnetisation switching of FePt nanoparticle recording medium by femtosecond laser pulses. *Sci. Rep.*, 7:4114.

Kachel, T., Pontius, N., Stamm, C., Wietstruk, M., Aziz, E. F., Dürr, H. A., Eberhardt, W., and de Groot, F. M. F. (2009). Transient electronic and magnetic structures of nickel heated by ultrafast laser pulses. *Phys. Rev. B*, 80:092404.

Kahn, O. (1993). *Molecular Magnetism.* VCH Publishers, New York.

Kaindl, R. A., Carnahan, M. A., Hägele, D., Lövenich, R., and Chemla, D. S. (2003). Ultrafast terahertz probes of transient conducting and insulating phases in an electron–hole gas. *Nature*, 423:734.

Kálmán, P. and Brabec, T. (1995). Generation of coherent hard-x-ray radiation in crystalline solids by high-intensity femtosecond laser pulses. *Phys. Rev. A*, 52:R21–R24.

Kampfrath, T., Sell, A., Klatt, G., Pashkin, A., Mährlein, S., Dekorsy, T., Wolf, M., Fiebig, M., Leitenstorfer, A., and Huber, R. (2011). Coherent terahertz control of antiferromagnetic spin waves. *Nat. Photonics*, 5:31.

Kanamori, A. (1963). Electron correlation and ferromagnetism of transition metals. *Progr. Theor. Phys.*, 30:275.

Kanamori, J. (1959). Superexchange interaction and symmetry properties of electron orbitals. *J. Phys. and Chem. Solids*, 10(2):87–98.

Kärtner, F. X. (2004). *Few-Cycle Laser Pulse Generation and Its Applications.* Springer-Verlag, Berlin, Heidelberg, Germany.

Khorsand, A. R., Savoini, M., Kirilyuk, A., Kimel, A. V., Tsukamoto, A., Itoh, A., and Rasing, T. (2012). Role of magnetic circular dichroism in all-optical magnetic recording. *Phys. Rev. Lett.*, 108:127205.

Kim, J.-W., Lee, K.-D., Jeong, J.-W., and Shin, S.-C. (2009). Ultrafast spin demagnetization by nonthermal electrons of TbFe alloy film. *Appl. Phys. Lett.*, 94:192506.

Kimel, A. V., Ivanov, B. A., Pisarev, R. V., Usacgev, P. A., Kirilyuk, A., and Rasing, T. (2009). Inertia-driven spin switching in antiferromagnets. *Nat. Phys.*, 5:727.

Kimel, A. V., Kirilyuk, A., Tsvetkov, A., Pisarevm, R. V., and Rasing, T. (2004). Laser-induced ultrafast spin reorientation in the antiferromagnet $TmFeO_3$. *Nature*, 429:850.

Kimel, A. V., Kirilyuk, A., Usachev, P. A., Pisarev, R. V., Balbashov, A. M., and Rasing, T. (2005). Ultrafast non-thermal control of magnetization by instantaneous photomagnetic pulses. *Nature*, 435:655.

Kirilyuk, A., Kimel, A. V., and Rasing, T. (2010). Ultrafast optical manipulation of magnetic order. *Rev. Mod. Phys.*, 82:2731.

Kittel, C. (1996). *Introduction to Solid State Physics.* John Wiley & Sons, Inc., New York, 7th edition.

Knut, R., Delczeg-Czirjak, E. K., Jana, S., Shaw, J. M., Nembach, H. T., Kvashnin, Y., Stefaniuk, R., Malik, R. S., Grychtol, P., Zusin, D., Gentry, C., Chimata, R., Pereiro, M., Söderström, J., Turgut, E., Ahlberg, M., Akerman, J., Kapteyn, H. C., Murnane, M. M., Arena, D. A., Eriksson, O., Karis, O., and Silva, T. J. (2018). Inhomogeneous magnon scattering during ultrafast demagnetization. *arXiv1810.10994v1.*

Kobayashi, T., Saito, T., and Ohtani, H. (2001). Real-time spectroscopy of transition states in bacteriorhodopsin during retinal isomerization. *Nature*, 414:531–534.

Kohn, W. and Sham, L. J. (1965). Self-consistent equations including exchange and correlation effects. *Phys. Rev.*, 140(4A):1133.

Koopmans, B., Malinowski, G., Longa, F. D., Steiauf, D., Fähnle, M., Roth, T., Cinchetti, M., and Aeschlimann, M. (2010). Explaining the paradoxical diversity of ultrafast laser-induced demagnetization. *Nat. Mater.*, 9:259.

Koopmans, B., Ruigrok, J. J. M., Longa, F. D., and de Jonge, W. J. M. (2005). Unifying ultrafast magnetization dynamics. *Phys. Rev. Lett.*, 95:267207.

Koopmans, B., van Kampen, M., Kohlhepp, J. T., and de Jonge, W. J. M. (2000). Ultrafast magneto-optics in nickel: magnetism or optics? *Phys. Rev. Lett.*, 85:844.

Koseki, S., Schmidt, M. W., and Gordon, M. S. (1998). Effective nuclear charges for the first- through third-row transition metal elements in spin-orbit calculations. *J. Comp. Chem. A*, 102:10430.

Krause, J. L., Schafer, K. J., and Kulander, K. C. (1992). High-order harmonic generation from atoms and ions in the high intensity regime. *Phys. Rev. Lett.*, 68:3535–3538.

Krauß, M., Roth, T., Alebrand, S., Steil, D., Cinchetti, M., Aeschlimann, M., and Schneider, H. C. (2009). Ultrafast demagnetization of ferromagnetic transition metals: The role of the coulomb interaction. *Phys. Rev. B*, 80:180407(R).

Krausz, F. and Stockman, M. I. (2014). Attosecond metrology from electron capture to future signal processing. *Nat. Photon.*, 8:205.

Krieger, J. B., Li, Y., and Iafrate, G. J. (1992a). Construction and application of an accurate local spin-polarized Kohn-Sham potential with integer discontinuity: Exchange-only theory. *Phys. Rev. A*, 45(1):101.

Krieger, J. B., Li, Y., and Iafrate, G. J. (1992b). Systematic approximations to the optimized effective potential: Application to orbital-density-functional theory. *Phys. Rev. A*, 46(9):5453.

Krieger, K., Dewhurst, J. K., Elliott, P., Sharma, S., and Gross, E. K. U. (2015). Laser-induced demagnetization at ultrashort time scales: Predictions of TDDFT. *J. Chem. Theory and Comput.*, 11:4870.

Kruchinin, S. Y., Krausz, F., and Yakovlev, V. S. (2018). Colloquium: Strong-field phenomena in periodic systems. *Rev. Mod. Phys.*, 90:021002.

Kruglyak, V. V., Hicken, R. J., M. Ali, B. J. H., Pym, A. T. G., and Tanner, B. K. (2005). Measurement of hot electron momentum relaxation times in metals by femtosecond ellipsometry. *Phys. Rev. B*, 71:233104.

Kryder, M. H. (1985). Magneto-optic recording technology. *J. Appl. Phys.*, 57:3913.

Kryder, M. H. and Kim, C. S. (2009). After hard drives—what comes next? *IEEE Transactions on Magnetics*, 45(10):3406–3413.

Krylov, A. I. (2008). Equation-of-motion coupled-cluster methods for open-shell and electronically excited species: The hitchhiker's guide to Fock space. *Annu. Rev. Phys. Chem.*, 59:433.

Kulander, K. and Rescigno, T. (1991). Effective potentials for time-dependent calculations of multiphoton processes in atoms. *Comput. Phys. Commun.*, 63(1):523–528.

Kulander, K. C. (1987). Time-dependent Hartree-Fock theory of multiphoton ionization: Helium. *Phys. Rev. A*, 36:2726–2738.

La-O-Vorakiat, C., Turgut, E., Teale, C. A., Kapteyn, H. C., Murnane, M. M., Mathias, S., Aeschlimann, M., Schneider, C. M., Shaw, J. M., Nembach, H. T., and Silva, T. J. (2012). Ultrafast demagnetization measurements using extreme ultraviolet light: Comparison of electronic and magnetic contributions. *Phys. Rev. X*, 2:011005.

Lalieu, M. L. M., Peeters, M. J. G., Haenen, S. R. R., Lavrijsen, R., and Koopmans, B. (2017). Deterministic all-optical switching of synthetic ferrimagnets using single femtosecond laser pulses. *Phys. Rev. B*, 96:220411.

Lambert, C.-H., Mangin, S., Varaprasad, B. S. D. C. S., Takahashi, Y. K., Hehn, M., Cinchetti, M., Malinowski, G., Hono, K., Fainman, Y., Aeschlimann, M., and Fullerton, E. E. (2014). All-optical control of ferromagnetic thin films and nanostructures. *Science*, 345:1337.

Langer, F., Hohenleutner, M., Huttner, U., Koch, S. W., Kira, M., and Huber, R. (2017). Symmetry-controlled temporal structure of high-harmonic carrier fields from a bulk crystal. *Nat. Photonics*, 11:227.

Lax, M. (1974). *Symmetry Principles in Solid State and Molecular Physics*. J. Wiley and Sons, Inc., New York.

Lee, C., Yang, W., and Parr, R. G. (1988). Development of the Colle-Salvetti correlation-energy formula into a functional of the electron density. *Phys. Rev. B*, 37:785–789.

Lefkidis, G. and Hübner, W. (2005). Ab initio treatment of optical second harmonic generation in NiO. *Phys. Rev. Lett.*, 95:077401.

Lefkidis, G. and Hübner, W. (2006). Phononic effects and nonlocality contributions to second harmonic generation in NiO. *Phys. Rev. B*, 74:155106.

Lefkidis, G. and Hübner, W. (2007). First-principles study of ultrafast magneto-optical switching in NiO. *Phys. Rev. B*, 76:014418.

Lefkidis, G. and Hübner, W. (2013). Analytical treatment of ultrafast laser-induced spin-flipping Λ processes on magnetic nanostructures. *Phys. Rev. B*, 87:014404.

Lefkidis, G. and Hübner, W. (2015). Λ-processes induced by chirped lasers. *Springer Proc. Phys.*, 159:128.

Lefkidis, G., Jin, W., Liu, J., Dutta, D., and Hübner, W. (2020). Topological spin-charge gearbox on a real molecular magnet. *J. Phys. Chem. Lett.*, 11:2592.

Lefkidis, G., Zhang, G. P., and Hübner, W. (2009). Angular momentum conservation for coherently manipulated spin polarization in photoexcited NiO: An *ab Initio* calculation. *Phys. Rev. Lett.*, 103:217401.

Lein, M. and Kümmel, S. (2005). Exact time-dependent exchange-correlation potentials for strong-field electron dynamics. *Phys. Rev. Lett.*, 94:143003.

Lejaeghere, K., Bihlmayer, G., Björkman, T., Blaha, P., Blügel, S., Blum, V., Caliste, D., Castelli, I. E., Clark, S. J., Corso, A. D., de Gironcoli, S., Deutsch, T., Dewhurst, J. K., Marco, I. D., Draxl, C., Dułak, M., Eriksson, O., Flores-Livas, J. A., Garrity, K. F., Genovese, L., Giannozzi, P., Giantomassi, M., Goedecker, S., Gonze, X., Granas, O., Gross, E. K. U., Gulans, A., Gygi, F., Hamann, D. R., Hasnip, P. J., Holzwarth, N. A. W., Iusan, D., Jochym, D. B., Jollet, F., Jones, D., Kresse, G., Koepernik, K., Kucukbenli, E., Kvashnin, Y. O., Locht, I. L. M., Lubeck, S., Marsman, M., Marzari, N., Nitzsche, U., Nordström, L., Ozaki, T., Paulatto, L., Pickard, C. J., Poelmans, W., Probert, M. I. J., Refson, K., Richter, M., Rignanese, G.-M., Saha, S., Scheffler, M., Schlipf, M., Schwarz, K., Sharma, S., Tavazza, F., Thunström, P., Tkatchenko, A., Torrent, M., Vanderbilt, D., van Setten, M. J., Speybroeck, V. V., Wills, J. M., Yates, J. R., Zhang, G.-X., and Cottenier, S. (2016). Reproducibility in density functional theory calculations of solids. *Science*, 351(6280):aad3000.

Levine, I. N. (2000). *Quantum Chemistry*. Prentice Hall, Upper Saddle River, New Jersey 07458, 5th edition.

Lewenstein, M., Balcou, P., Ivanov, M. Y., L'Huillier, A., and Corkum, P. B. (1994). Theory of high-harmonic generation by low-frequency laser fields. *Phys. Rev. A*, 49:2117–2132.

L'Huillier, A. (2017). Multiple harmonic conversion of 1064 nm in rare gases. *J. Phys. B*, 50(6):060501.

Li, C., Hartenstein, T., Lefkidis, G., and Hübner, W. (2009). First-principles calculation of the ultrafast spin manipulation of two-center metallic clusters with a CO molecule attached to one center as an infrared marker. *Phys. Rev. B*, 79:180413.

Li, C., Jin, W., Xiang, H., Lefkidis, G., and Hübner, W. (2011). Theory of laser-induced ultrafast magneto-optic spin flip and transfer in charged two-magnetic-center molecular ions: Role of bridging atoms. *Phys. Rev. B*, 84:054415.

Li, C., Lefkidis, G., and Hübner, W. (2014a). Electronic theory of ultrafast spin dynamics in NiO. *J. Nanomater. Mol. Nanotechnol.*, 3(4).

Li, C., Liu, J., Zhang, S., Lefkidis, G., and Hübner, W. (2015). Strain modulation of ultrafast spin switching on $Co_2@C_{60}$ endohedral fullerenes. *Carbon*, 87:153–162.

Li, C., Zhang, S., Jin, W., Lefkidis, G., and Hübner, W. (2014b). Controllable spin-dynamics cycles and erase functionality on quasilinear molecular ions. *Phys. Rev. B*, 89:184404.

Liu, H., Li, Y., You, Y. S., Ghimire, S., Heinz, T. F., and Reis, D. A. (2017). High-harmonic generation from an atomically thin semiconductor. *Nat. Phys.*, 13:262.

Lucchini, M., Sato, S. A., Ludwig, A., Herrmann, J., Volkov, M., Kasmi, L., Shinohara, Y., Yabana, K., Gallmann, L., and Keller, U. (2016). Attosecond dynamical Franz-Keldysh effect in polycrystalline diamond. *Science*, 353(6302):916–919.

Luu, T. T., Garg, M., Kruchinin, S. Y., Moulet, A., Hassan, M. T., and Goulielmakis, E. (2015). Extreme ultraviolet high-harmonic spectroscopy of solids. *Nature*, 521:498.

Mahan, G. D. (2009). Spin shift register from a one-dimensional atomic chain. *Phys. Rev. Lett.*, 102:016801.

Mangin, S., Gottwald, M., Lambert, C.-H., Steil, D., Uhlir, V., Pang, L., Hehn, M., Alebrand, S., Cinchetti, M., Malinowski, G., Fainman, Y., Aeschlimann, M., and Fullerton, E. E. (2014). Engineered materials for all-optical helicity-dependent magnetic switching. *Nat. Mater.*, 13:286.

Marques, M. A. L. and Gross, E. K. U. (2005). Time-dependent density functional theory. *Annu. Rev. Phys. Chem.*, 55:427–55.

Mathias, S., La-O-Vorakiat, C., Grychtol, P., Granitzka, P., Turgut, E., Shaw, J. M., Adam, R., Nembach, H. T., Siemens, M. E., Eich, S., Schneider, C. M., Silva, T. J., Aeschlimann, M., Murnane, M. M., and Kapteyn, H. C. (2012). Probing the timescale of the exchange interaction in a ferromagnetic alloy. *PNAS*, 109:4792.

McPherson, A., Gibson, G., Jara, H., Johann, U., Luk, T. S., McIntyre, I. A., Boyer, K., and Rhodes, C. K. (1987). Studies of multiphoton production of vacuum-ultraviolet radiation in the rare gases. *J. Opt. Soc. Am. B*, 4(4):595–601.

McWeeny, R. (1963). *Symmetry: An Introduction to Group Theory and Its Applications.* Pergamon Press, Oxford, London/New York/Paris.

Meier, D., Maringer, M., Lottermoser, T., Becker, P., Bohatý, L., and Fiebig, M. (2009). Observation and coupling of domains in a spin-spiral multiferroic. *Phys. Rev. Lett.*, 102:107202.

Meijer, P. H. E. and Bauer, E. (1962). *Group Theory: The Application to Quantum Mechanics.* North-Holland Publishing Company, Amsterdam.

Merrick, J. P., Moran, D., and Radom, L. (2007). An evaluation of harmonic vibrational frequency scale factors. *J. Phys. Chem. A*, 111(45):11683–11700.

Moulton, P. F. (1986). Spectroscopic and laser characteristics of Ti:Al_2O_3. *J. Opt. Soc. Am. B*, 3:125–133.

Mücke, O. D. (2011). Isolated high-order harmonics pulse from two-color-driven bloch oscillations in bulk semiconductors. *Phys. Rev. B*, 84:081202(R).

Mukamel, S. (1995). *Principles of Nonlinear Optical Spectroscopy.* Oxford University Press, Inc., New York.

Müller, G. M., Walowski, J., Djordjevic, M., Miao, G.-X., Gupta, A., Ramos, A. V., Gehrke, K., Moshnyaga, V., Samwer, K., Schmalhorst, J., Thomas, A., Hütten, A., Reiss, G., Moodera, J. S., and Münzenberg, M. (2009). Spin polarization in half-metals probed by femtosecond spin excitation. *Nat. Mater.*, 8:56.

Mulliken, R. S. (1933). Electronic structures of polyatomic molecules and valence. IV. Electronic states, quantum theory of the double bond. *Phys. Rev.*, 43:279–302.

Murakami, M. and Chu, S.-I. (2016). Photoelectron momentum distributions of the hydrogen molecular ion driven by multicycle near-infrared laser pulses. *Phys. Rev. A*, 94:043425.

Murakami, M., Korobkin, O., and Horbatsch, M. (2013). High-harmonic generation from hydrogen atoms driven by two-color mutually orthogonal laser fields. *Phys. Rev. A*, 88:063419.

Murakami, M., Zhang, G. P., and Chu, S.-I. (2017). Multielectron effects in the photoelectron momentum distribution of noble-gas atoms driven by visible-to-infrared-frequency laser pulses: A time-dependent density-functional-theory approach. *Phys. Rev. A*, 95:053419.

Murakami, M., Zhang, G. P., Telnov, D. A., and Chu, S.-I. (2018). Photoelectron momentum distribution of ground- and excited-state lithium atoms induced by extreme-ultraviolet photon absorption. *J. Phys. B*, 51(17):175002.

Nakatsuji, H. (1979). Cluster expansion of the wavefunction. Electron correlations in ground and excited states by SAC (symmetry-adapted-cluster) and SAC CI theories. *Chem. Phys. Lett.*, 67:329–333.

Ndabashimiye, G., Ghimire, S., Wu, M., Browne, D. A., Schafer, K. J., Gaarde, M. B., and Reis, D. A. (2016). Solid-state harmonics beyond the atomic limit. *Nature*, 534:520.

Ogata, Y., Chudo, H., Ono, M., Harii, K., Matsuo, M., Maekawa, S., and Saitoh, E. (2017). Gyroscopic g factor of rare earth metals. *Appl. Phys. Lett.*, 110:072409.

Ohkochi, T., Fujiwara, H., Kotsugi, M., Tsukamoto, A., Arai, K., Isogami, S., Sekiyama, A., Yamaguchi, J., Fukushima, K., Adam, R., Schneider, C. M., Nakamura, T., Kodama, K., Tsunoda, M., Kinoshita, T., and Suga, S. (2012). Microscopic and spectroscopic studies of light-induced magnetization switching GdFeCo facilitated by photoemission electron microscopy. *Jpn. J. Appl. Phys.*, 51:073001.

Olsen, J. (2011). The CASSCF method: A perspective and commentary. *Int. J. Quantum Chem.*, 111(13):3267–3272.

Ostler, T. A., Barker, J., Evans, R. F. L., Chantrell, R. W., Atxitia, U., Chubykalo-Fesenko, O., Moussaoui, S. E., Guyader, L. L., Mengotti, E., Heyderman, L. J., Nolting, F., Tsukamoto, A., Itoh, A., Afanasiev, D., Ivanov, B. A., Kalashnikova, A. M., Vahaplar, K., Mentink, J., Kirilyuk, A., Rasing, T., and Kimel, A. V. (2012). Ultrafast heating as a sufficient stimulus for magnetization reversal in a ferrimagnet. *Nat. Commun.*, 3:666.

Ostler, T. A., Evans, R. F. L., Chantrell, R. W., Atxitia, U., Chubykalo-Fesenko, O., Radu, I., Abrudan, R., Radu, F., Tsukamoto, A., Itoh, A., Kirilyuk, A., Rasing, T., and Kimel, A. (2011). Crystallographically amorphous ferrimagnetic alloys: Comparing a localized atomistic spin model with experiments. *Phys. Rev. B*, 84:024407.

Pal, G., Lefkidis, G., and Hübner, W. (2009). Ab initio investigation of Pt dimers on Cu(001) surface. *J. Phys. Chem. A*, 113:12071.

Pal, G., Lefkidis, G., and Hübner, W. (2010). Electronic excitations and optical spectra of Pt_2 and Pt_4 on Cu(001) modeled by a cluster. *Phys. Status Solidi B*, 247:1109.

Parr, R. G. and Yang, W. (1989). *Density-Functional Theory of Atoms and Molecules*. Number 16 in The International Series of Monographs on Chemistry. Oxford Science Publications.

Pearson, W. B. (1958). *A Handbook of Lattice Spacings and Structures of Metals and Alloys*. Pergamon Press, Oxford.

Perdew, J. P. and Zunger, A. (1981). Self-interaction correction to density-functional approximations for many-electron systems. *Phys. Rev. B*, 23(10):5048.

Pershan, P. S., van der Ziel, J. P., and Malmstrom, L. D. (1966). Theoretical discussion of the inverse faraday effect, raman scattering, and related phenomena. *Phys. Rev.*, 143(2):574.

Petrovykh, D. Y., Altmann, K. N., Höchst, H., Laubscher, M., Maat, S., Mankey, G. J., and Himpsel, F. J. (1998). Spin-dependent band structure, fermi surface, and carrier lifetime of permalloy. *Appl. Phys. Lett.*, 73:3459.

Pickel, M., Schmidt, A. B., Giesen, F., Braun, J., Minar, J., Ebert, H., Donath, M., and Weinelt, M. (2008). Spin-orbit hybridization points in the face-centered-cubic cobalt band structure. *Phys. Rev. Lett.*, 101:066402.

Pitaevskii, L. P. (1961). Electric forces in a transparent dispersive medium. *Sov. Phys. JETP*, 125:1008.

Plaja, L. and Roso-Franco, L. (1992). High-order harmonic generation in a crystalline solid. *Phys. Rev. B*, 45:8334.

Pronin, K. A., Bandrauk, A. D., and Ovchinnikov, A. A. (1994). Harmonic generation by a one-dimensional conductor: Exact results. *Phys. Rev. B*, 50:3473.

Qiu, Z. Q. and Bader, S. (1998). Kerr effect and surface. In Bennemann, K. H., editor, *Nonlinear Optics in Metals*. Oxford University Press.

Radu, I., Stamm, C., Eschenlohr, A., Radu, F., Abrudan, R., Vahaplar, K., Kachel, T., Pontius, N., Mitzner, R., Holldack, K., Föhlisch, A., Ostler, T. A., Mentink, J. H., Evans, R. F. L., Chantrell, R. W., Tsukamoto, A., Itoh, A., Kirilyuk, A., Kimel, A. V., and Rasing, T. (2015). Ultrafast and distinct spin dynamics in magnetic alloys. *SPIN*, 5:1550004.

Radu, I., Woltersdorf, G., Kiessling, M., Melnikov, A., Bovensiepen, U., Thiele, J.-U., and Back, C. H. (2009). Laser-induced magnetization dynamics of lanthanide-doped permalloy thin films. *Phys. Rev. Lett.*, 102:117201.

Razdolski, I., Alekhin, A., Ilin, N., Meyburg, J. P., Roddatis, V., Diesing, D., Bovensiepen, U., and Melnikov, A. (2017). Nanoscale interface confinement of ultrafast spin transfer torque driving non-uniform spin dynamics. *Nat. Commun.*, 8:15007.

Rhie, H.-S., Dürr, H. A., and Eberhardt, W. (2003). Femtosecond electron and spin dynamics in Ni/W(110) films. *Phys. Rev. Lett.*, 110:247201.

Rosker, M. J., Wise, F. W., and Tang, C. L. (1986). Femtosecond relaxation dynamics of large molecules. *Phys. Rev. Lett.*, 57:321–324.

Roth, L. M. (1964). Theory of the faraday effect in solids. *Phys. Rev.*, 133:A542–A553.

Runge, E. and Gross, E. K. U. (1984). Density-functional theory for time-dependent systems. *Phys. Rev. Lett.*, 52(12):997.

Saito, S., Miura, M., and Kurosawa, K. (1980). Optical observations of antiferromagnetic s domains in NiO(111) platelets. *J. Phys. C: Solid State Phys.*, 13(8):1513–1520.

Sakurai, J. J. (1994). *Modern Quantum Mechanics*, chapter 5, page 305. Addison Wesley.

Satoh, T., Cho, S.-J., Iida, R., Shimura, T., Kuroda, K., Ueda, H., Ueda, Y., Ivanov, B. A., Nori, F., and Fiebig, M. (2010). Spin oscillations in antiferromagnetic NiO triggered by circularly polarized light. *Phys. Rev. Lett.*, 105:077402.

Satoh, T., Duong, N. P., and Fiebig, M. (2006). Coherent control of antiferromagnetism in NiO. *Phys. Rev. B*, 74:012404.

Schafer, K. J. (2008). Numerical methods in strong field physics. In Brabec, T., editor, *Strong Field Laser Physics*, chapter 6, page 111. Springer Science+Business Media, LLC.

Scholl, A., Baumgarten, L., Jacquemin, R., and Eberhardt, W. (1997). Ultrafast spin dynamics of ferromagnetic thin films observed by fs spin-resolved two-photon photoemission. *Phys. Rev. Lett.*, 79:5146.

Schubert, C., Hassdenteufel, A., Matthes, P., Schmidt, J., Helm, M., Bratschitsch, R., and Albrecht, M. (2014a). All-optical helicity dependent magnetic switching in an artificial zero moment magnet. *Appl. Phys. Lett.*, 104:082406.

Schubert, O., Hohenleutner, M., Langer, F., Urbanek, B., Lange, C., Huttner, U., Golde, D., Meier, T., Kira, M., Koch, S. W., and Huber, R. (2014b). Sub-cycle control of terahertz high-harmonic generation by dynamical bloch oscillations. *Nat. Photonics*, 8:119.

Schwarz, K.-H. (1986). CrO_2 predicted as a half-metallic ferromagnet. *J. Phys. F: Met. Phys.*, 16:L211.

Shank, C. V., Bigot, J. Y., Portella, M., Fragnito, H. L., Shoenlein, R., and Becker, P. C. (1990). Excitation of nonstationary states in molecules with 6-fs optical pulses. In *International Quantum Electronics Conference, Anaheim, California*, page QTHP1.

Shen, Y. R. (1984). *The Principles of Nonlinear Optics.* John Wiley and Sons, Hoboken, New Jersey.

Shieh, H. P. D. and Kryder, M. H. (1986). Magneto-optic recording materials with direct overwrite capability. *Appl. Phys. Lett.*, 49:473.

Si, M. S. and Zhang, G. P. (2010). Resolving photon-shortage mystery in femtosecond magnetism. *J. Phys.: Condens. Mat.*, 22:076005.

Sibbett, W., Lagatsky, A. A., and Brown, C. T. A. (2012). The development and application of femtosecond laser systems. *Opt. Express*, 20:6989.

Silfvast, W. T. (2003). Lasers. In Guenther, A., Pedrotti, L. S., and Roychoudhuri, C., editors, *Fundamentals of Photonics.* University of Connecticut.

Simanek, E. and Heinrich, B. (2003). Gilbert damping in magnetic multilayers. *Phys. Rev. B*, 67:144418.

Slater, J. C. (1951). A simplification of the Hartree-Fock method. *Phys. Rev.*, 81(3):385.

Stamm, C., Pontius, N., Kachel, T., Wietstruk, M., and Dürr, H. A. (2010). Femtosecond x-ray absorption spectroscopy of spin and orbital angular momentum in photoexcited Ni films during ultrafast demagnetization. *Phys. Rev. B*, 81:104425.

Stanciu, C. D., Hansteen, F., Kimel, A. V., Kirilyuk, A., Tsukamoto, A., Itoh, A., and Rasing, T. (2007). All-optical magnetic recording with circularly polarized light. *Phys. Rev. Lett.*, 99:047601.

Steil, D., Alebrand, S., Hassdenteufel, A., Cinchetti, M., and Aeschlimann, M. (2011). All-optical magnetization recording by tailoring optical excitation parameters. *Phys. Rev. B*, 84:224408.

Stodolna, A. S., Rouzée, A., Lépine, F., Cohen, S., Robicheaux, F., Gijsbertsen, A., Jungmann, J. H., Bordas, C., and Vrakking, M. J. J. (2013). Hydrogen atoms under magnification: Direct observation of the nodal structure of stark states. *Phys. Rev. Lett.*, 110:213001.

Stöhr, J. and Siegmann, H. C. (2006). *Magnetism: From Fundamentals to Nanoscale Dynamics.* Springer-Verlag, Berlin/Heidelberg/New York.

Strelkov, V. V., Khokhlova, M. A., Gonoskov, A. A., Gonoskov, I. A., and Ryabikin, M. Y. (2012). High-order harmonic generation by atoms in an elliptically polarized laser field: Harmonic polarization properties and laser threshold ellipticity. *Phys. Rev. A*, 86:013404.

Sultan, M., Atxitia, U., Melnikov, A., Chubykalo-Fesenko, O., and Bovensiepen, U. (2012). Electron- and phonon-mediated ultrafast magnetization dynamics of Gd(0001). *Phys. Rev. B*, 85:184407.

Sultan, M., Melnikov, A., and Bovensiepen, U. (2011). Ultrafast magnetization dynamics of Gd(0001). *Phys. Status Solidi B*, 246:2323.

Szabo, A. and Ostlund, N. S. (1996). *Modern Quantum Chemistry: Introduction to Advanced Electronic Structure Theory.* Dover Publications, Mineola, New York.

Teichmann, M., Frietsch, B., K. Döbrich, R. C., and Weinelt, M. (2015). Transient band structures in the ultrafast demagnetization of ferromagnetic gadolinium and terbium. *Phys. Rev. B*, 91:014425.

Telnov, D. A., Sosnova, K. E., Rozenbaum, E., and Chu, S.-I. (2013). Exterior complex scaling method in time-dependent density-functional theory: Multiphoton ionization and high-order-harmonic generation of ar atoms. *Phys. Rev. A*, 87:053406.

Tengdin, P., You, W., Chen, C., Shi, X., Zusin, D., Zhang, Y., Gentry, C., Blonsky, A., Keller, M., Oppeneer, P. M., Kapteyn, H. C., Tao, Z., and Murnane, M. M. (2018). Critical behavior within 20 fs drives the out-of-equilibrium laser-induced magnetic phase transition in nickel. *Sci. Adv.*, 4:eaap9744.

Thomas, L. H. (1926). The motion of the spinning electron. *Nature*, 117(2945):514–514.

Tong, X.-M. and Chu, S.-I. (1997a). Density-functional theory with optimized effective potential and self-interaction correction for ground states and autoionizing resonances. *Phys. Rev. A*, 55:3406–3416.

Tong, X.-M. and Chu, S.-I. (1997b). Theoretical study of multiple high-order harmonic generation by intense ultrashort pulsed laser fields: A new generalized pseudospectral time-dependent method. *Chem. Phys.*, 217:119–130.

Tong, X.-M. and Chu, S.-I. (1998). Time-dependent density-functional theory for strong-field multiphoton processes: Application to the study of the role of dynamical electron correlation in multiple high-order harmonic generation. *Phys. Rev. A*, 57(1):452.

Tong, X.-M. and Chu, S.-I. (2001). Multiphoton ionization and high-order harmonic generation of he, ne, and ar atoms in intense pulsed laser fields: Self-interaction-free time-dependent density functional theoretical approach. *Phys. Rev. A*, 64:013417.

Torlina, L., Morales, F., Kaushal, J., Ivanov, I., Kheifets, A., Zielinski, A., Scrinzi, A., Muller, H. G., Sukiasyan, S., Ivanov, M., and Smirnova, O. (2015). Interpreting attolock measurements of tunnelling times. *Nat. Phys.*, 11:503.

Tzschaschel, C., Otani, K., Iida, R., Shimura, T., Ueda, H., Günther, S., Fiebig, M., and Satoh, T. (2017). Ultrafast optical excitation of coherent magnons in antiferromagnetic NiO. *Phys. Rev. B*, 95:174407.

Vahaplar, K., Kalashnikova, A. M., Kimel, A. V., Gerlach, S., Hinzke, D., Nowak, U., Chantrell, R., Tsukamoto, A., Itoh, A., Kirilyuk, A., and Rasing, T. (2012). All-optical magnetization reversal by circularly polarized laser pulses: Experiment and multiscale modeling. *Phys. Rev. B*, 85:104402.

Vahaplar, K., Kalashnikova, A. M., Kimel, A. V., Hinzke, D., Nowak, U., Chantrell, R., Tsukamoto, A., Itoh, A., Kirilyuk, A., and Rasing, T. (2009). Ultrafast path for optical magnetization reversal via a strongly nonequilibrium state. *Phys. Rev. Lett.*, 103:117201.

Vampa, G., Hammond, T. J., Thire, N., Schmidt, B. E., Legare, F., McDonald, C. R., Brabec, T., Klug, D. D., and Corkum, P. B. (2015). All-optical reconstruction of crystal band structure. *Phys. Rev. Lett.*, 115:193603.

van Leeuwen, R. and Baerends, E. J. (1994). Exchange-correlation potential with correct asymptotic behavior. *Phys. Rev. A*, 49:2421–2431.

Varjú, K., Johnsson, P., Mauritsson, J., L'Huillier, A., and López-Martens, R. (2009). Physics of attosecond pulses produced via high harmonic generation. *Am. J. Phys.*, 77:389.

Varro, S. and Ehotzky, F. (1994). Higher-harmonic generation from a metal surface in a powerful laser field. *Phys. Rev. A*, 49:3106.

Vignale, G. (1995). Center of mass and relative motion in time dependent density functional theory. *Phys. Rev. Lett.*, 74:3233–3236.

Villeneuve, D. M., Hockett, P., Vrakking, M. J. J., and Niikura, H. (2017). Coherent imaging of an attosecond electron wave packet. *Science*, 356:1150.

Vomir, M., Albrecht, M., and Bigot, J.-Y. (2017). Single shot all optical switching of intrinsic micron size magnetic domains of a Pt/Co/Pt ferromagnetic stack. *Appl. Phys. Lett.*, 111:242404.

Vomir, M., Andrade, L. H. F., Guidoni, L., Beaurepaire, E., and Bigot, J.-Y. (2005). Real space trajectory of the ultrafast magnetization dynamics in ferromagnetic metals. *Phys. Rev. Lett.*, 94:237601.

von der Linde, D., Engers, T., Jenke, G., Agostini, P., Grillon, G., Nibbering, E., Mysyrowicz, A., and Antonetti, A. (1995). Generation of high-order harmonics from solid surfaces by intense femtosecond laser pulses. *Phys. Rev. A*, 52:R25.

Vonesch, H. and Bigot, J.-Y. (2012). Ultrafast spin-photon interaction investigated with coherent magneto-optics. *Phys. Rev. B*, 85:180407.

Wadley, P., Howells, B., Železný, J., Andrews, C., Hills, V., Campion, R. P., Novák, V., Olejník, K., Maccherozzi, F., Dhesi, S. S., Martin, S. Y., Wagner, T., Wunderlich, J., Freimuth, F., Mokrousov, Y., Kuneš, J., Chauhan, J. S., Grzybowski, M. J., Rushforth, A. W., Edmonds, K. W., Gallagher, B. L., and Jungwirth, T. (2016). Electrical switching of an antiferromagnet. *Science*, 351(6273):587–590.

Weber, A., Pressacco, F., Günther, S., Mancin, E., Oppeneer, P. M., and Back, C. H. (2011). Ultrafast demagnetization dynamics of thin Fe/W(110) films: Comparison of time- and spin-resolved photoemission with time-resolved magneto-optic experiments. *Phys. Rev. B*, 84:132412.

Wein, H. P. J. (1988). *Landolt-Börnstein, New Series*, volume III/19a. Springer, Berlin.

Wienholdt, S., Hinzke, D., and Nowak, U. (2012). THz switching of antiferromagnets and ferrimagnets. *Phys. Rev. Lett.*, 108:247207.

Wietstruk, M., Melnikov, A., Stamm, C., Kachel, T., Pontius, N., Sultan, M., Gahl, C., Weinelt, M., Dürr, H. A., and Bovensiepen, U. (2011). Hot-electron-driven enhancement of spin-lattice coupling in Gd and Tb $4f$ ferromagnets observed by femtosecond x-ray magnetic circular dichroism. *Phys. Rev. Lett.*, 106:127401.

Wijn, H. P. J. (1991). *Magnetic Properties of Metals: d-Elements, Alloys and Compounds.* Springer, Berlin.

Wildenauer, J. (1987). Generation of the ninth, eleventh, and fifteenth harmonics of iodine laser radiation. *J. Appl. Phys.*, 62(1):41–48.

Wohlfarth, E. P. (1980). *Ferromagnetic Materials: a Handbook on the Properties of Magnetically Ordered Substances*, volume 1. North Holland.

Wüstenberg, J.-P., Steil, D., Alebrand, S., Roth, T., Aeschlimann, M., and Cinchetti, M. (2011). Ultrafast magnetization dynamics in the half-metallic Heusler alloy $Co_2Cr_{0.6}Fe_{0.4}Al$. *Phys. Status Solidi B*, 248:2330.

Xiang, H., Lefkidis, G., and Hübner, W. (2012). Unified theory of ultrafast femtosecond and conventional picosecond magnetic dynamics in the distorted three-center magnetic cluster Ni_3Na_2. *Phys. Rev. B*, 86:134402.

Xiang, H., Lefkidis, G., and Hübner, W. (2013). First-principles study of the ultra-fast, laser-induced spin transferability in the multi-center magnetic cluster Ni_3Na_2. *J. Supercond. Nov. Magn.*, 26:2001.

Xie, X. D., Jiang, P., and Lu, F. (1984). *Group Theory and Its Application in Physics.* China Science Publishing & Media Ltd. Beijing.

Yafet, Y. (1963). *g* factors and spin-lattice relaxation of conduction electrons. *Solid State Phys.*, 14:1.

Yamamoto, S., Taguchi, M., Fujisawa, M., Hobara, R., Yamamoto, S., Yaji, K., Nakamura, T., Fujikawa, K., Yukawa, R., Togashi, T., Yabashi, M., Tsunoda, M., Shin, S., and Matsuda, I. (2014). Observation of a giant kerr rotation in a ferromagnetic transition metal by M-edge resonant magneto-optic kerr effect. *Phys. Rev. B*, 89:064423.

Ye, J. and Cundiff, S. T. (2005). *Femtosecond Optical Frequency Comb: Principle, Operation and Applications.* Springer, Boston.

Yoshikawa, N., Tamaya, T., and Tanaka, K. (2017). High-harmonic generation in graphene enhanced by elliptically polarized light excitation. *Science*, 356:736.

Zewail, A. (2000). Femtochemistry. past, present, and future. *Pure Appl. Chem.*, 72:2219.

Zhang, G. P. (2005). Optical high harmonic generations in C_{60}. *Phys. Rev. Lett.*, 95:047401.

Zhang, G. P. (2011). Microscopic theory of ultrafast spin linear reversal. *J. Phys.: Condens. Mat.*, 23:206005.

Zhang, G. P. and Bai, Y. H. (2019). Magic high-order harmonics from a quasi-one-dimensional hexagonal solid. *Phys. Rev. B*, 99:094313.

Zhang, G. P. and Bai, Y. H. (2020). High-order harmonic generation in solid C_{60}. *Phys. Rev. B*, 101:081412(R).

Zhang, G. P., Bai, Y. H., and George, T. F. (2015a). A new and simple model for magneto-optics uncovers an unexpected spin switching. *EPL*, 112:27001.

Zhang, G. P., Bai, Y. H., and George, T. F. (2016a). Switching ferromagnetic spins by an ultrafast laser pulse: Emergence of giant optical spin-orbit torque. *EPL*, 115:57003.

Zhang, G. P., Bai, Y. H., Jenkins, T., and George, T. F. (2018a). Laser-induced ultrafast transport and demagnetization at the earliest time: First-principles and real-time investigation. *J. Phys.: Condens. Matter*, 30:465801.

Zhang, G. P., Callcott, T. A., Woods, G. T., Lin, L., Sales, B., D. Mandrus, D., and He, J. (2002a). Electron correlation effects in resonant inelastic x-ray scattering of NaV_2O_5. *Phys. Rev. Lett.*, 88:077401.

Zhang, G. P. and George, T. F. (2006). Ellipticity dependence of optical harmonic generation in C_{60}. *Phys. Rev. A*, 74:023811.

Zhang, G. P. and George, T. F. (2007a). Manifestation of electron-electron interactions in time-resolved ultrafast pump-probe spectroscopy in C_{60}. *Phys. Rev. B*, 76:085410.

Zhang, G. P. and George, T. F. (2007b). Origin of ellipticity anomaly in harmonic generation in C_{60}. *J. Optical Society of America B*, 24:1150.

Zhang, G. P. and George, T. F. (2013). Thermal or nonthermal? That is the question for ultrafast spin switching in GdFeCo. *J. Phys.: Condens. Mat.*, 25:366002.

Zhang, G. P., Gu, M., and Wu, X. S. (2014). Ultrafast reduction in exchange interaction by a laser pulse: alternative path to femtomagnetism. *J. Phys.: Condens. Matter*, 26:376001.

Zhang, G. P. and Hübner, W. (2000). Laser-induced ultrafast demagnetization in ferromagnetic metals. *Phys. Rev. Lett.*, 85:3025.

Zhang, G. P., Hübner, W., Beaurepaire, E., and Bigot, J.-Y. (2002b). Laser-induced ultrafast demagnetization: Femtomagnetism, a new frontier? *Topics Appl. Phys.*, 83:245.

Zhang, G. P., Hübner, W., Lefkidis, G., Bai, Y., and George, T. F. (2009). Paradigm of the time-resolved magneto-optical Kerr effect for femtosecond magnetism. *Nat. Phys.*, 5:499.

Zhang, G. P., Jenkins, T., Bennett, M., and Bai, Y. H. (2017). Manifestation of intra-atomic 5d6s-4f exchange coupling in photoexcited gadolinium. *J. Phys.: Condensed Mat.*, 29:495807.

Zhang, G. P., Latta, T., Babyak, Z., Bai, Y. H., and George, T. F. (2016b). All-optical spin switching: A new frontier in femtomagnetism – A short review, and a simple theory. *Mod. Pys. Lett. B*, 30:1630005.

Zhang, G. P., Lefkidis, G., Hübner, W., and Bai, Y. (2011). Ultrafast demagnetization in ferromagnets and magnetic switching in nanoclusters when the number of photons is kept fixed. *J. Appl. Phys.*, 109:07D303.

Zhang, G. P., Lefkidis, G., Hübner, W., and Bai, Y. (2012). Manipulating femtosecond magnetization in ferromagnets and molecular magnets through laser chirp. *J. Appl. Phys.*, 111:07C508.

Zhang, G. P. and Murakami, M. (2019). All-optical spin switching under different spin configurations. *J. Phys.:Condens. Mat.*, 31:345802.

Zhang, G. P., Murakami, M., Bai, Y. H., George, T. F., and Wu, X. S. (2019). Spin-orbit torque-mediated spin-wave excitation as an alternative paradigm for femtomagnetism. *J. Appl. Phys.*, 126:103906.

Zhang, G. P., Murakami, M., Si, M. S., Bai, Y. H., and George, T. F. (2018b). Understanding all-optical spin switching: Comparison between experiment and theory. *Mod. Phys. Lett. B*, 32:1830003.

Zhang, G. P., Si, M. S., Bai, Y. H., and George, T. F. (2015b). Magnetic spin moment reduction in photoexcited ferromagnets through exchange interaction quenching: Beyond the rigid band approximation. *J. Phys.: Condens. Mat.*, 27:206003.

Zhang, G. P., Si, M. S., and George, T. F. (2015c). Laser-induced ultrafast demagnetization time and spin moment in ferromagnets: First-principles calculation. *J. App. Phys.*, 117:17D706.

Zhang, G. P., Si, M. S., Murakami, M., Bai, Y. H., and George, T. F. (2018c). Generating high-order optical and spin harmonics from ferromagnetic monolayers. *Nat. Commun.*, 9:3031.

Zhang, G. P., Zhu, H. P., Bai, Y. H., Bonacum, J., Wu, X. S., and George, T. F. (2015d). Imaging superatomic molecular orbitals in a C_{60} molecule through four 800-nm photons. *Int. J. Mod. Phys. B*, 29:1550115.

Zhang, Q., Nurmikko, A. V., Miao, G. X., Xiao, G., and Gupta, A. (2006). Ultrafast spin-dynamics in half-metallic CrO_2 thin films. *Phys. Rev. B*, 74:064414.

Zhang, W., He, W., Peng, L.-C., Zhang, Y., J.-W. Cai, R. F. L. E., Zhang, X.-Q., and Cheng, Z.-H. (2018d). The indispensable role of the transversal spin fluctuations mechanism in laser-induced demagnetization of Co/Pt multilayers with nanoscale magnetic domains. *Nanotechnology*, 29:275703.

Zvezdin, A. K. and Kotov, V. A. (1997). *Modern Magnetooptics and Magnetooptical Materials.* Institute of Physics Publishing, London, U. K.

Index

$3.17U_p$, 94
$I_p + 3.17U_p$, 85, 92
Λ process, 204
γ matrices, 75
λ, 33

absorbing function, 54, 97
adiabatic approximation, 50
all-optical switching, 37
angular frequency, 12
antiferromagnetic, 189
AO, 58, 59
AO-HDS, 161, 162, 164
AO-HIDS, 162, 164
AOS, 157
associated Legendre
functions, 46
atomic orbitals, 58
attosecond, 3
attosecond physics, 37
aufbau principle, 61
augmented planewave
method, 44
Autler-Townes effect, 197
axial vector, 78

bacteriorhodopsin, 28
band index, 73
band structure, 122
basis function, 58
Bloch theorem, 73
Bloch wavefunction, 70
BLYP approximation, 45
Bohr radius, 7
bond angles, 102
Boolean logic, 224
Born-Oppenheimer
approximation, 103,
217
Born-von Karman boundary
condition, 71
Bose-Einstein distribution,
145
broad bandwidth, 21

carrier frequency, 16
CAS-SCF, 198
cavity, 12, 20
centripetal force, 7
chirped laser pulses, 240
CI, 60
CIP, 180
CIS, 62
cis, 8
CISD, 62
configuration interaction, 60
contraction, 58
conversion factor, 122
Corkum's theory, 92
Coulomb potential, 89
coupled-cluster, 63
CPP, 180
Crank-Nicolson, 98
CW, 150
cw, 15

degenerate four-wave mixing,
18
demagnetization barrier, 147
demagnetization time, 126
density matrix, 28
density of states, 122
difference frequency
generation, 16
dipole approximation, 95
Dirac equation, 75
Dirac notation, 58
domain size, 118
dye laser, 33
Dzyaloshinskii-Moriya
interaction, 200

eigenstates, 44
eigenvalues, 44
electron binding energy, 127
electron-phonon coupling, 33,
141
electronic correlations, 211
electronic temperature, 33
element specificity, 127
ellipticity, 96
endofullerenes, 221
energy conversion efficiency,
37
envelope function, 96
EOM-CCSD, 64
ERASE functionality, 238
Euler angles, 80
exchange interaction, 117
exchange-correlation hole, 62
exchange-correlation
potential, 44
excitation operator, 63

Faraday effect, 19, 167
femtochemistry, 5
femtomagnetism, 35, 37, 127
femtosecond, 3
Fermi energy, 122
Fermi surface, 124
Fermi velocity, 124
ferrimagnet, 157
fluence, 158
four-wave mixing, 18
functional, 44
functional derivative, 45
fundamental frequency, 13

gas jet, 38
gauge transformation, 96
Gaussian primitives, 58
GMR, 180
gold coated grating, 37

Hübner time, 126
Hückel model, 139
Hartree potential, 44
Hartree-Fock approximation,
59
HDD, 181
Heisenberg exchange, 200
Heisenberg model, 117
Hellmann-Feynman theorem,
215
Hessian matrix, 216
Heusler, 164
hexagons, 102
HHG, 37, 85
high harmonic generation, 85
Hohenberg-Kohn theorem, 42
HOMO, 105
Hubbard, 141

index of refraction, 157

intraband transition, 108
inversion, 80
ionization, 90
IR marker, 218

Kerr effect, 19, 35
KLI approximation, 45
Kohn-Sham wavefunctions, 44
KTP, 85

Landau-Lifshitz-Gilbert equation, 200
LAPW, 111
laser intensity, 88
lattice constant, 67
lattice temperature, 33
LCAO, 59
LDA approximation, 45
LDA-SIC approximation, 45
left circularly polarized light, 167
Legendre polynomials, 97
Legendre transform, 52
length gauge, 95
light cone, 6
linear reversal, 172
Liouville equation, 103, 111
LLG, 200
LUMO, 105

M processes, 235
MAE, 119
magnetic anisotropy, 187
magnetic domains, 118
magnetic field, 78
magnetic moment, 122
magnetic phases, 230
magnetization, 118, 122
magneto-optical recording, 157, 182
magnon, 145
majority spins, 122, 141
many-electron Schrödinger equation, 44
Maxwell equations, 78
MCD, 164
mechanical strain, 221
metals, 33
minority spins, 122, 141
mirror plane, 80
Mn_2RuGa, 164
MO, 59
mode locking, 20
mode-locked, 33
MOKE, 149
MOSFET, 181
MRAM, 181
muffin-tin sphere, 44

NiO, 189, 195
nonlinear optics, 15, 87
nonperturbative, 37
nonvolatile, 181
number of photons, 158

optical rectification, 16
orbital angular momentum, 167

Pauli exclusion principle, 103
Pauli matrices, 80
penetration depth, 157
pentagons, 102
permalloy, 127
perturbative, 37
phenomenological spin relaxation, 233
phonons, 33
photoisomerization, 8, 28
photon shortage, 157
picosecond, 3
planewave basis, 44
plateau, 85
PMA, 163
polar vector, 78
polarization, 15
ponderomotive energy, 91
ponderomotive energy U_p, 85
position space, 104
Poynting vector, 158
pseudopotential, 44
pulse duration, 157
pump-probe, 27

quantum chemistry, 58
quantum mechanics, 3
quintets, 234

Rabi oscillation, 197
rebirth, 91
relative detuning, 241
relativity, 6
remanence, 118
retinal, 8
rhodopsin, 8, 28
right circularly polarized light, 167
RIXS, 105
rotating-wave approximation, 196, 206
rotation, 80
rotation matrix, 78
Rowland circle, 37
Runge-Gross theorem, 50
RWA, 208

SAC-CI, 64
second harmonic generation, 16
selection rules, 59, 196
self-interaction errors, 62
semiconductors, 33
shutter, 21
signal, 27
single active electron approximation, 95
singlet, 204
size consistency, 63
size consistent, 63
Slater determinants, 60
SOC, 75
space group, 67
special **k** points, 75
spectrometer, 37
spherical harmonics, 46, 97
spin polarized, 122
spin switching, 157
spin wave, 141, 145
spin wavelength λ_s, 147
spin-orbit coupling, 168
spintronics, 236
split-operator, 97
spontaneous emission, 14
standing wave, 12
state space, 104
stimulated emission, 14
stimulated Raman scattering, 17
Stokes vector, 206
Stoner, 141
SU(2), 78
sum frequency generation, 16, 85
superconductors, 33
susceptibility, 15

TDDFT, 50, 148
TDLDFT, 148
TDSE, 95
temperature, 145
tensors, 15
third-order harmonic generation, 16
time-dependent Schrödinger equation, 103
time-independent

Kohn-Sham
equations, 44
trans, 8
triplet, 204
TRMOKE, 126, 149
two-photon absorption, 17

unit cell, 149

vector potential, 78
velocity gauge, 96
velocity map imaging, 5
vibration, 33, 103
vibrons, 217
virtual orbitals, 61
VMI, 5
Voigt effect, 19
volatile, 181

Wannier functions, 139
wavelength, 12
wavenumber, 4
wavevector, 27
window functions, 114
Wyckoff positions, 67